MW00895711

About the Author

John Madden was born in Vancouver, BC, in 1939 and, after an absence of about twenty years, has lived in Vancouver since 1980. He studied physics at the University of British Columbia, and completed a D. Phil. in nuclear physics at Oxford on a Rhodes Scholarship.

His working life has included technical research and development and research and development management at a senior level in industry and in the Canadian government primarily in the fields of telecommunications and computing. He has served on a number of boards, including the National Research Council of Canada, Simon Fraser University, Science World, International Submarine Engineering Research Ltd. and Microtel Pacific Research Ltd. (of which he was the founding President).

Madden was a co-author (with Alphonse Ouimet, Dave Godfrey and Doug Parkhill) of *Gutenberg Two: The New Electronics and Social Change* (Press Porcepic, first edition, 1979, through to a fourth edition in 1985).

The Davey Dialogues . . .

is an easy-to-understand history and background to key discoveries which have altered our perception of our universe and of ourselves, couched as a dialogue between the writer and a mysterious extra-terrestrial being with an agenda that may surprise you.

For more information please go to: www.thedaveydialogues.com.

A note about the cover photographs.

Front: This famous photo of Earthrise over the lunar horizon was taken from the orbiting Apollo 8 spacecraft on December 28, 1968. By visiting http://www.nasa.gov/mission_pages/LRO/news/apollo8-retrace.html it is possible to view an ingenious reconstructed video of the earthrise complete with the spontaneous comments made by the astronauts as they marvelled at the earth rise. Once this picture became widely distributed, membership in the Flat Earth Society went into serious decline. (Website accessed on October 30, 2012)

Back: Stephan's Quintet is in the constellation Pegasus. The photo was taken from the upgraded Hubble spacecraft (photo courtesy of NASA, ESA and the Hubble SM4 ERO Team). Four of the five galaxies form the first compact galaxy group ever discovered (in 1877 by Édouard Stephan at the Marseilles Observatory). The group is about 300 million light years distant. The fifth galaxy is only 40 million light years away, and does not interact with the other four. This photo, even though only partly visible on the back cover, is a beautiful example of the stunning views of the heavens possible with telescopes of very high magnification.

THE DAVEY DIALOGUES

AN EXPLORATION OF THE SCIENTIFIC FOUNDATIONS OF HUMAN CULTURE

John C. Madden

STC Enterprises Vancouver

Published by STC Enterprises Inc.
Vancouver, BC, Canada

© John C. Madden 2012

First Edition November, 2012
Print on Demand Edition February, 2013

All rights reserved. No part of this publication may be reproduced, stored in a retrieval system or transmitted in any form or by any means – electronic, mechanical, photocopying, and recording or otherwise –without the prior written permission of the author, except for brief passages quoted by a reviewer in a newspaper or magazine. To perform any of the above is an infringement of copyright law.

Note for Librarians: A cataloguing record for this book is available from Library and Archives Canada at www.collectionscanada.ca/amicus/index-e.html

ISBN – 978-0-9917675-1-9

To my grandchildren

Alex

Jamie

Andrew

Lilyana

Simon

Contents

Tables

iii

Figures

PREFACE

*The aim of science is not to open the door to everlasting wisdom
but to set a limit on everlasting error.*
BERTOLT BRECHT, *THE LIFE OF GALILEO*

This book is an exploration of some of the science that has altered our world view and that may reasonably influence our assessment of ourselves and the place of humanity in the universe we inhabit.

At least since the time that the fabled Adam bit into the apple, knowledge has proven to be both a blessing and a curse. Life is, on the whole, much longer and pleasanter for most of us than it was for our hunter-gatherer forebears. But our huge advances in knowledge bring with them new challenges, some of which will put in question the existence of our species in ways much more complex and difficult than the invention of the hydrogen bomb has ever been able to do.

Our survival as a species, not to mention the happiness of our grandchildren and succeeding generations, will likely depend on enough of us understanding enough about ourselves and our environment to sort out which, if any, of the inevitable opposing views on vital issues is most likely to be correct.

In this endeavour, quite appropriately, most of us are influenced by the thinking of ancient philosophers, literary greats and religious thinkers. It is often challenging to sort out which parts of this cornucopia of imparted wisdom are truly timeless and which should be modified or even ignored in the light of our current understanding of the universe we live in. This book is intended to provide the reader with a tool set to help in this endeavour.

In some ways, therefore, the book deals as much in the history of science as in science itself, for, by and large, the savants of old were as influenced by the then current state of knowledge as we are. Some of their conclusions would clearly have been different had they known what we know today.

We humans are notoriously bad at choosing options that will make us happier, a fact discussed entertainingly and at some length by Dan Gilbert in his book, *Stumbling on Happiness*. This fault is at least partially attributable to the rather quixotic way in which our brains have evolved over the past 500 million years and the rather different lifestyles led by ourselves as opposed to our hunter-gatherer ancestors of only a few thousand years ago from whom our inherited set of instincts have come, with little time for species evolution to bring about material alteration adequate to match our very different circumstances.

Our conscious decisions are clearly influenced by our instincts, as well as by unconsciously acquired behaviour modification brought about by our experiences in life, especially experiences from early childhood when our brains were still actively developing. In the circumstances, the concept of "free will" seems ill-fitted to the triage process that apparently takes place in an ancient part of our brain called the *amygdala*, where instinctual and adapted behavioural biases mingle (unbeknownst to us) with conscious thought to determine action.

As unconscious as these biases may be to us, we can be sure that advertisers, politicians, producers of electronic games, movies and TV programs, casino owners, drug dealers, not to mention preachers and novelists, are all, to a degree, cognisant of what drives us and shape their pitches to resonate with the instinctual desires and needs of their audience. Because the merry band of persuaders listed above is much more likely to be aware of the instincts that drive us than are we, I have attempted to redress some of the imbalance in this book. Nevertheless most of us are at least vaguely aware that our conscious inputs are operating in prevailing winds and currents that are capable of overpowering our conscious desires and sending our frail vessel off in unintended directions. It is my hope that this book will help the reader to better understand this process amongst others. The discourse is intended to be intelligible to my grandchildren when they reach their late teens or early twenties. It does not require a background in science.

Some of the questions addressed by philosophy and religion over the ages, and which are also addressed in the dialogues include:

Are humans different in essence from animals or merely one of many species?
How old is Earth? How big and how old is the universe?
How did life come about?
How do our brains work? Is this very different from the way other animal brains work?
What is happiness? What makes us happy?
Do we have a free will?

What is the scientific basis for morality?

Science certainly does not yet have all the answers to such questions, but especially in the last fifty years or so hundreds of thousands of motivated and intelligent people have together furnished us with a mutually re-enforcing array of facts that together provide us with a framework out of which new answers to old questions are emerging. This dialogue draws primarily on the scientific developments I believe are most relevant to those who want to develop their own answers to such questions.

In order to keep the book to a reasonable length, it was necessary to select topics with care and to summarize unmercifully, while still maintaining interest by occasional excursions into greater detail. The general layout is as follows:

Part 1: Introduction.
Part 2: How our understanding of space and time evolved, along with our theories of the universe.
Part 3: Darwin, evolution and early lessons from the study of animals and human societies.
Part 4: The workings of cellular life and the coding of it in DNA. How might life have emerged?
Part 5: How does the brain remember, learn and associate seemingly disparate thoughts? Is there a hidden agenda?
Part 6: How do the real drivers of human happiness differ from what we might think they are? Why might this be so?
Part 7: Some concluding advice from Davey – the cool outside participant in the dialogues, who is pessimistic about the future of humanity but who has a challenging problem of his own to address.

Early readers of this book varied substantially in what they wanted to know about the way in which important new scientific discoveries came about. One or two said, "Just give me the facts, and spare me the details." For others, the inclusion of some details of the route to new discoveries was more than just interesting. For them, these details represented the essential background needed for the dialogue to be credible, especially where the discussion challenged a long-held belief. It is difficult, if not impossible, to please everyone in this respect. But those readers who find uninteresting the details of how particular new discoveries were made are encouraged to make a guilt-free skip of the more detailed descriptions and look for the factual summary that occurs near the end of most of the dialogues.

The urge to skip may be particularly strong in the sections dealing with the biosciences. All life forms turn out to be extraordinarily intricate and complex chemical factories. Sadly, but inevitably, rather lengthy and confusing vocabularies have blossomed to describe the essential components of these factories, resulting in some heavy reading for the novice. A brief glossary to help the reader over rough spots is included at the back of the book.

The introduction of Davey as an interlocutor in the proceedings was strongly welcomed by some early readers but rejected by others who nonetheless enjoyed reading "the scientific parts". If you find yourself agreeing with the latter readers, please do as they did, and move on quickly, though you may find that toward the end of the book, Davey lays out some rather interesting ideas.

There is a wealth of good science writing out there that treats in greater detail than I could cover all of the topics raised in this book. Many of my favourite books and writers appear in the Bibliography and can serve as useful routes to a better understanding of topics that have been rather summarily treated herein.

If there is a message in this book it is that we are most fortunate to live at a time when, perhaps for the first time ever, we can delight in a coherent story about us and our universe. What the physicists and the philosophers, the economists and the ecologists, the anthropologists and the animal behaviouralists, the religious historians, the neuroscientists and the biochemists are telling us by and large falls into place to provide a consistent story of the development of our universe and of humanity. Of course the story is not yet complete. There is much more to discover and understand, but the outlines are there in ample detail.

You, dear reader, will want to draw your own conclusions about what, if anything, this story is saying to you about how you should live your life. Whatever you conclude, it is my earnest hope that what you read herein will not only pique your interest and feed your curiosity, but will also help you to lead a happier and more rewarding life.

Acknowledgements

Many people have contributed to making this book much better than it would otherwise have been.

Special recognition goes to the late J. Fraser Mustard, who died in November 2011 after a stellar career in medical research, research management more generally and the development of public policy, particularly in the realms of population health and early childhood education and care. He was a brilliant and innovative medical educator and the founder and first president of the Canadian Institute for Advanced Research (CIFAR), an international collaborative research centre without walls. Fraser's energy and determination, coupled with his intellect and curiosity, sparked a career that earned him the rare distinction of being named a Companion of the Order of Canada. He is much missed by his many friends and supporters around the world.

Fraser was kind enough to read an early draft of this book in March 2010, but he didn't just read the book and make a few comments! He summarized it in his own words, expressed strong and warm support, and added in a lot of judicious and helpful criticism – by far the most comprehensive and helpful commentary I received. With Fraser's permission, a few of his words about the book are reproduced on the back cover.

Many others have provided less comprehensive but nonetheless extremely helpful commentary and analysis. I would particularly like to thank Valerie Dunsterville, David Gallop, John Helliwell, Richard Hooper, Bill McKerlich, J.B. Molson, Ian Robertson, Basil Rolfe, Arthur Scace, Mary Shakespeare, Lorne Whitehead, Jan Whitford and, of course, my wonderful family, with my wife Sidney at the head of the list.

Finding appropriate illustrations and gaining permissions to print them in a book can be a daunting task. I count myself particularly fortunate to have been able to track down an outstanding illustrator (and scholar) who encouraged me to draw on his vast store of bioscientific illustrations, hassle free. If you find that you appreciate his illustrations as much as I do, please join me in thanking Keith Roberts, whose artistic works are to be found in Dialogues 11, 13, 16 and 18.

Finally, I would like to thank my editor, Meaghan Craven, for her intelligent and diligent approach to editing the book. I can assure you that her attention to detail and thoughtful comments have done much to enhance the enjoyment of the reader.

In some circles, the offering of an apology in advance is viewed as a sign of weakness. I am not in that circle, at least in this instance. The writing of this book has meant that I had the opportunity to study many interesting fields of research in which I am not expert. Inevitably there will be some mistakes, some errors of omission and some misplaced emphasis.

I therefore apologise in advance for these transgressions. Should you detect any, please know that I would welcome any comments that you might care to make. Who knows, but there may be another edition which will afford an opportunity to make appropriate corrections.

Comments should be sent to: info@thedaveydialogues.com.

Thank you.

x

PART I

THE DIALOGUE BEGINS

This section is introductory in nature. It describes the circumstances of Peter's first encounter with Davey, as well as the challenges that each faced in achieving a mutual understanding.

The Arrival of the Voice

Ignorance more frequently begets confidence than does knowledge: it is those who know little, and not those who know much, who so positively assert this or that problem will never be solved by science.
CHARLES DARWIN, Introduction, *The Descent of Man*, 1871

– Do you really want me to start out with a discussion of creation myths?

It was the beginning of our first session, only a week after my first encounter with the alien voice. I was feeling edgy and insecure, and was still questioning myself on the wisdom of taking on the project that I then thought of as its education. But I was fairly certain that a discussion of creation myths was not a promising place to start our dialogue.

– Please yourself. How you choose to begin our conversation is of great interest to me.

This, I came to understand, was a typical remark. It was at once accommodating and very judgmental.

But I am getting ahead of myself already. I must first tell you how I encountered this strange disembodied voice.

My name is Peter Alexander. I am a retired consultant, latterly employed by companies and governments to provide advice about scientific and technical research. I live in Vancouver, BC, with my wife Margaret. Our two boys have long since left home and are helping to bring up families of their own.

Every once in a while, I like to go hiking by myself. I proceed at my own pace, take any detours I want, and use the combination of exercise and quietude to carry out some uninterrupted thinking.

About two years ago, on just such a hike, I reached a prime viewpoint near the summit of Hollyburn Mountain, one of several mountains that crowd Vancouver from the north. It was time for lunch, so I sat down on a rocky knoll and extracted a special old cheddar cheese sandwich from my backpack. This particular cheddar is a favourite of mine.

It was what tourist guides call "a glorious autumn day". The sun's warmth was perfectly moderated by a light breeze from the west as I looked down on the city below me. There was a clear view of the Strait of Georgia to the west and south, its mixture of sky-blue water infused with gyres of muddy-brown water from the Fraser River. Below me, a large blue and white ship stacked high with multicoloured containers was slipping under the Lions Gate Bridge into Vancouver Harbour, likely loaded with industrial goods from China destined for transport by rail to the US Midwest.

I forget what I was thinking. Perhaps I was simply drinking in the view, the fresh air and the sunshine.

I was startled from my reverie by a voice out of nowhere. It was singing an old (but once very popular) song called "Zip-A-Dee Doo Dah". It was a cultured and musical baritone voice that I heard.

"Zip-A-Dee-Doo-Dah", the voice intoned and then added another round of Zip-A-Dee nonsense before going on to observe just what I was thinking, namely that it was a wonderful day and that there was "plenty of sunshine", all (or almost all) of which seemed to be "comin' my way".

Not very inspiring words, but the tune is lively and pleasant, and it certainly got my attention. However something was not quite right with the singing voice. It was masculine, clear and tuneful, but it had a hint of a hollow ring to it.

I turned to look for the singer, but there was no one to be seen. What is more, the voice sounded as though it came from someone sitting beside me. I don't believe in ghosts or goblins, but I do clearly remember feeling disconcerted.

Soon after the song came to an end, the voice said:

– Hello there. I'm visiting from another universe, and I'm trying to understand yours a bit better. I am particularly interested in you humans, why you congregate the way you do, and why you do some of the peculiar things you do. Would you be able to help me? If so, perhaps we could start with creation myths.

Obviously there were just a few things wrong. As I write these words, cosmologists are by no means agreed that there is another universe completely separate and different from the one they explore so assiduously with telescopes, satellites and ingenious instruments. And those who do think there are other universes believe it extremely unlikely, in fact virtually impossible, that there could be any communication between one universe and another.* I heard myself speaking.

– You must be kidding! No one in his or her right mind would ask such questions, unannounced, of a total stranger. Anyway, as you likely know, they are complicated questions. If you were to ask a hundred people, you would get a hundred different answers. You're wasting your time with me and probably with anyone else you may decide to ask. In any case, you cannot be from another universe. So, who or what are you?

– Sorry! Sorry! Sorry! I was a little sudden. Abrupt, you might even say. You are correct that I cannot sit beside you in person, but do your theoretical physicists really say that my voice cannot be heard in your universe?

– Well, I'm not actually sure.

– I thought not.

The voice sat silent for what seemed to be quite a while. Eventually, it said:

– By the way, I enjoy listening to your grandchildren.

– You just keep my grandchildren out of this!

I spoke as emphatically as I could. Now it really had my attention. This was unnerving to say the least. My grandchildren ranged in age from one to nine. I certainly did not want to have them haunted by a personless voice whose intentions were not at all clear.

Up to that point, I had been idly wondering if I was hallucinating. Perhaps the combination of cheese sandwich and cranberry juice had affected my brain. Even if the voice was real, I wondered whether I really cared enough to tolerate this nagging intrusion on my solitude any longer. The voice soothed:

– Sorry again! I won't bother your grandchildren. I was just thinking though that what I want to learn from you is likely just the kind of information you have been preparing for your grandchildren. Please don't be too surprised. While I cannot actually see or read anything, my hearing is close to being perfect. I learnt about you by listening

---

* See Dialogue 6 for an elaboration on this subject.

to a couple of your friends talking about you. One of them thought you would never finish the book you are writing. The way one of them described the proposed book, I thought it sounded interesting, so here I am. I could perhaps help you by providing just the incentive you need to finish it. I might even help you to draw sensible conclusions!

For my part, I just want to understand how you came into being, and just what motivates human behaviour. There is a lot of history to it, I know, but there does not seem to be a whole heap of agreement on the facts amongst the lot of you. I could use your help to figure it all out!

– Why me? I will grant you that I am writing a book for my grandchildren, but there must be thousands of others like me.

– Of course you're right. Most of the more than seven billion people currently on Earth share both your curiosity and your love of grandchildren – provided they live long enough to get to know them. Indeed I am already in touch with several thousand others. Since your society has not yet reached any consensus on some important parts of your history, I am left with little option but to talk to quite a few of you humans in order to understand your origins and motivations. Even then I may fail, since none of you may be right, but there doesn't seem to be much choice! Especially in your case, I see some advantages to both of us from such consultations.

Why "especially in my case", I wondered.

To add to my bewilderment, I didn't really believe yet that the voice came from another universe – so what was it, and why was it lying? I realized I needed time to think, so I told it to contact me in a few days' time if it was still interested.

Silence was restored, but it was not the same silence as before. I got up and paced about. At some point, I set off back down the mountain, though I have no recollection of the return journey, as my brain and I tried to come to grips with this strange experience.

Somehow I knew that ultimately I would accede to the voice's request. At the time I was still struggling to connect the pieces that illuminate the puzzle of human existence into a semblance of order, both for myself and for my grandchildren. Wherever the voice came from, and whatever it was, it looked as though it could be the trigger to force me to get serious about laying out my version of the salient facts in as coherent and compelling a manner as I could muster. And after all, just maybe it could help me. It certainly seemed to think it could.

Four days later, as I was at my desk answering e-mail and sipping a mid-morning hot chocolate, the voice returned.

– So glad you agreed to help me. When can we start?

– Just you wait a darned minute! There are a few things I need to know before we begin.

I was nettled that it had read my mind so easily. I sensed, but of course did not actually see, a condescending smile on the faceless voice.

– Fire away, but there is not much I can tell you about our universe just yet. Nor would it help you much if I could tell you. You humans don't seem to understand your own universe all that well. Mine is very different.

– First of all, I still don't believe you are from another universe. Just exactly what are you?

– Actually, your experts are simply mistaken. I do inhabit another universe. Believe me, this is not the only miscalculation your cosmologists have made.

– Whether or not what you say is true, I need to know more about you. How much do you already know about me? Can you read my mind? Do you know the history of humankind as well as my history?

– All good questions. Let me put your mind at rest. There is not much more to tell about me. Just as you cannot see me, I cannot see you. By a strange coincidence of nature, some aspects of our physics are the same as yours. This means that I can actually hear all the sounds made on your Earth, and, after careful study, I can attribute a source to each sound. I have made a study of all the noise emanating from human beings for what to you will seem like a long time, but for me is a very reasonable space of time. This has allowed me to decode and understand all languages currently spoken on Earth. Having heard your friends talk about you, and then tracked you down, I have been listening in on your conversations.

I could feel the blood rush to my face, as I tried to recall what I might have said that was embarrassing.

– Of course, as I have already told you, I am also holding conversations with others who share your interests.

– So, what would happen to your project if we all got together and gave you the same story?

I heard a hollow laugh.

– Not much chance of that. First of all, you have to find the others, none of whom so far has admitted to anyone that they are talking with me – and almost no one would believe them if they did. Secondly, you would all have to agree on what to tell me. Of course, I would be listening in on your conversation, and, in any case, I can assure you that there would be precious little agreement in the group of individualists I am talking to! You would be as intransigent as any of them.

I bristled at this.

– Nonsense! I don't read a lot just so I can confirm what I now believe!

– True to a degree, but you are not really very tolerant of views you consider to be unscientific nonsense, are you?

Like most people, I rather flatter myself that I am very tolerant of the views of others, perhaps especially their religious views. I tried to think back on past conversations that might have led the disembodied voice to draw such a conclusion but gave up when the voice said it would like to get started, and could I please tell it how I planned to proceed.

I was ready for the request.

– Before we begin, Mr. Voice From Afar, what can you say or do to prove to me that you are real and not just a sign that I am going mad?

– Yet another doubting Thomas, as some of you would say! About half the people I talk to ask that question. The rest would probably like to ask but are afraid to. Well, the short answer is that there is likely not much I can do to convince you I am real. I have no power to influence events in your universe, except insofar as I might tell you something that would cause you to do something you might not otherwise have done. That being the case, I have to be pretty sure I understand what is going on before I give any advice!

But if you think about it a little, and especially if you decide to continue our discussion, you will likely come to the conclusion that your own mind would be incapable of inventing our discussion. If you think I may be the voice of another human sending messages to you by extra-sensory perception (as some of my other human contacts were inclined to do), you will then have to decide if any other human might reasonably come up with a story anything like what I am going to tell you, and, if you thought they might do so, you should then wonder why they would go to all that trouble.

– Okay, then you won't mind if I ask you a few questions, like: "Who is your best friend? And what does he or she look like?" "Do you have a sun and moon where you live?" "Do you have literature and art and music?" "Do you fall in love?" "What do you eat? And where does your energy come from?" "Do your children go to sch . . ."

– Wait! Wait! Wait! I can answer most of these questions for you eventually, but not now. Any answer I give you now will be just as strange and incredible to you as what you know of me already.

Isn't it enough for you to know that what you tell me may have an important influence in helping me to understand my universe? When we know each other better, you will learn quite a bit about me, I promise you.

This was the sort of reply I had expected to receive. In thinking about it after my first encounter with the voice, I had concluded that if I were in its position, visiting another unfamiliar universe, I, too, would not want to disclose very much at first, for fear that my universe could be harmed as a result. I had also concluded that there was not much to be lost from proceeding to the next stage. But the next stage, I had vowed to myself, would be a small step.

– All right, let's proceed. You'll have to give me a week or ten days to prepare, and then I will tell you all I know about creation myths. It will take me about an hour. Is that too long and drawn out for you?

There was silence. At length it spoke.

– This time it is you who must be kidding! Make me a break! Your proposal does not begin to provide enough information to explain your wars, your partisanship, your duplicity, your nobility, your avarice and cruelty, your thirst for knowledge or your literature. Yes, your literature, as well. Why do humans write and read books? Did Darwin provide an explanation for that?

Please don't imagine that I am in a hurry. Time is a feature of your universe, and for you humans I know it can seem to pass quickly. I have no reason to be in a hurry. How often do you think I have heard you expound on Shakespeare and on your revered scientists and philosophers, on the properties of the brain and of DNA, on the nature of your universe and on exciting research on the real sources of human happiness. Are you going to try to tell me that these subjects are not relevant to an understanding of mankind? I think not!

Remember that I cannot read your books and journals; I can only listen in on conversations. Most of these don't lead anywhere. As often as not, at least one of the speakers is enjoying herself or himself, while the others are too busy thinking about what to say next to listen to what the first speaker says. As a result I am taking in a jumble of information, most of it outdated and repetitive. I need a framework to fit it all into – but useful frameworks are not just hastily assembled beams leaning against each other at odd angles, they are carefully built segments of a larger structure whose overall form is determined by the assemblage of its parts so as to provide the backbone of a living functional structure.

– Hold on! I don't need a lecture on a framework for understanding ourselves. What you propose is a major task. Even if I were to agree to proceed as you wish, what good would it do you? If you really are from another universe, it will almost certainly be useless – unless you have some evil designs on our universe. I have better ways of spending my time than talking to you. Please just go away and talk to some of the other thousands you claim to be conversing with.

An unworldly strangled sound filled the room, a sound straight from Hollywood. I had an almost uncontrollable urge to laugh at the incongruity of it all. But somehow the mid-ranges of the sound conveyed feelings of anguish and pathos, so I sat motionless waiting for the sound to cease.

– I suppose I should not be surprised to be so totally misunderstood.

The voice paused and then continued, quiet but firm.

– It is likely impossible for you to understand, but our universe has suffered an unimaginable tragedy. A whole species has vanished. Quite gone. This was a species I loved. Its disappearance has caused me great sorrow and great difficulty. I cannot describe to you just how serious this situation is. It never ever occurred to me that such a catastrophe could happen.
Although you humans are not really at all like my vanished, once-noble friends, there are some resemblances. I am hoping that by understanding you better I may come upon some strategy for bringing my friends back from nothingness.

Incredible, I thought, a part of me profoundly moved, and another part just as profoundly suspicious.

– I already know how difficult it is for you humans to believe what I say. That is why I really did not want to mention my friends. Even if you don't entirely believe me, I hope you will find it worthwhile from your own point of view to engage in a prolonged conversation with me. I flatter myself that you will be rewarded with some important insights into your own condition.

I beg you therefore to tell me about the way your species has gained the understanding it now has of itself and its surroundings. I need to understand how the gradually increasing knowledge about your surroundings and yourselves has affected your behaviour and your beliefs. Frankly, I fear that you humans will suffer a fate similar to that of my friends. I don't know why I should care, but for some reason that I don't understand, I do. My loss in my universe is beyond measure. I despair more than a little that you might also disappear from your universe.

So, please agree to continue our dialogue in a meaningful way. You will not regret it! I will come back to learn your decision next week.

Silence again . . . this time I somehow knew I was alone with my thoughts.

The voice had gone even before I could correct one of its very rare mistakes in the use of English. "Make me a break" touched my funny bone. Could the disembodied voice have been mistranslating from another language? It is after all only human to make such an error. But the voice was decidedly not human. Perhaps, I thought, it made the error deliberately to help spare any embarrassment I might feel over its dismissal of my proposal. I shall likely never know for certain.

After a lot of soul searching, I eventually decided to continue with my dialogue with the disembodied voice. In hindsight I shudder to think that I might easily have chosen not to proceed.

So it was that our ten month series of dialogues began.

My immediate problem was to decide how best to begin the real task. The voice wanted to know about creation myths! How could I sensibly convey the long march of humanity through a bewildering variety of gods, goddesses, demons, devils and prophets?

After several, long, solitary walks, I finally put together a starting strategy.

A Speculative Tale

> Glendower: *I can call spirits from the vasty deep.*
> Hotspur: *Why, so can I, or so can any man;*
> *But will they come when you do call for them?*
> WILLIAM SHAKESPEARE, Henry IV, Part I, Act III, Scene 1

I had already decided in principle that I would like to continue my conversations with the voice. Even though it would involve a lot of work on my part, the opportunity was just too fascinating to miss, and, being retired, I certainly had the time to do the work.

What, I wondered, would I want to know about human beings if I were in the voice's position? Why was the voice so interested in understanding our beliefs about our origins? Was there a clue there? Perhaps the voice had something to do with our origins? But that seemed unlikely. Perhaps it was in search of its own origins? I would have to try and find out.

It seemed to me that it would want to understand a bit about the history of our accretion of knowledge, especially since progress in one field can strongly influence developments in another. So, I listed what I thought were the key discoveries that had affected the ways in which mankind viewed itself, and I sorted them into a rough order. What was not very clear to me at the time was the length of time I would need to prepare myself in each of these subject areas, but I decided that if I needed more time along the way, I would simply ask for it. The voice did not seem particularly hurried, though it was clearly worried.

A week after our last encounter the voice returned as it had promised. I was at work in my study when it announced its arrival with a sound remarkably like the clearing of a human throat.

– Well, I hope I am correct in concluding that you have agreed to continue our discussions.

– Almost. This idea of yours will take up a lot of my time. I want your assurance that you will be open with me, and tell me about your universe.

– You have my word. But I do need to listen to you first. That way I can better judge how what I say can help you. As you know, what I have learned from other humans does not leave me optimistic about your species. My vanished friends were a much better species than you are, and look what happened to them!

I was doubtful about the voice's promise, but I decided to proceed.

– I suppose I can take your word of honour, assuming that your universe has such a thing as honour. And somewhere along the way, you will have to convince me that your vanished friends really were superior to us.

By the way, you may have noticed that it is our custom to adopt names by which we can refer to each other. I don't know if you have a name that you use?

There was silence. So, I prompted.

– Surely many, if not most, of the other humans you are talking to have given you a name? Why don't you just pick one you like?

– If you don't mind, I would rather you chose a name for me.

I was more than a little taken aback by this, thinking to myself that this was a rather unhuman thing to request. Nonetheless, I searched my mind for a suitable name.

– In that case, would you mind if I called you Davey? You see, from my point of view, you are a Disembodied Alien Voice. I thought of Dav for short, but to my ears at least, Davey sounds better, especially since your voice is quite low, and definitely not feminine.

– That is satisfactory with me, though I can change my voice to sound like a woman if you would prefer it. Then you could call me Davina.

I balked at Davina. Perhaps he was just joking! It was hard to tell. Anyway, I was a long way from thinking of Davey as even close to divine. So, Davey it was for the duration of our dialogues.

Davey jumped in straight away.

– Let's proceed immediately. All week I have been trying to guess how you will start.

– Do you really want me to start out with a discussion of creation myths?

– Please yourself. How you choose to start our conversation is of great interest to me.

I decided to interpret his response as a "Yes" and started in. At least I knew I would have his attention at the start of our long discourse.

– Imagine yourself to have been born as a human (*Homo sapiens*) about 150,000 years ago, which is to say very soon after *Homo sapiens* became a distinct species. Current scientific thinking is that you almost certainly were born in Africa, perhaps somewhere near the Olduvai Gorge in the eastern Serengeti Plains, just south of the equator. It is probable that you lived and died as part of a group of perhaps twenty people, all or almost all of whom were family members. You had intermittent contact with other similar groups. Sometimes these contacts were friendly (perhaps friendly enough for new spouses to be introduced to the group), and some decidedly hostile (which could result in reproduction through rape and slavery).

Your life expectancy likely was about twenty-five years, but a few members of your group may have lived a lot longer. You were a hunter-gatherer. Agriculture had not yet been developed. You had access to some crude tools, such as spears and stone knives. Writing had not yet been invented. History was all orally transmitted. Your spoken language was likely quite well developed, though your vocabulary was limited by your fairly restricted range of knowledge and experience, as judged by today's standards.

The fight of your ancestors for survival amongst the competing mammals, reptiles, disease and hostile tribes, and while faced with variations in climate and environment over months, years and millennia – including cataclysmic events such as floods, fires and volcanic eruptions – meant that you had a number of innate behavioural penchants (almost all of which are subject to variations and modifications across the species) that helped you adapt to changing circumstances.

Insofar as mammalian competitors are concerned, your physical attributes were unexceptional. You were not especially strong or big, not particularly fleet of foot, nor could you climb trees rapidly, burrow in the ground, or swim with much finesse. In such circumstances, survival depended on co-operative behaviour practiced by groups of humans. These groups had to be large enough to yield survivable odds against predators and to improve chances of success when hunting, but not so large as to exhaust available sources of food within easy walking range.

You regarded most members of your small tribe with respect and affection. Others you disdained, though not so much that, for the most part, you wished them gone. You were part of a hierarchy, or pecking order. Your chief and his wife decidedly had the upper hand. The chief directed the hunt and took for himself, and his immediate family, the best part of the available food and supplies. His chief wife supervised the domestic chores (and may well, by dint of superior intelligence and drive, have been the *de facto* head of the tribe). No one was left without food or shelter unless he or she had unsuccessfully challenged the chief, or otherwise transgressed tribal rules in an unforgivable manner.

While your chief may have had a favourite spouse, he especially – and other males to a lesser extent – copulated with a number of females both in the group and outside it. Nonetheless most group members formed marriage-like relationships of a semi-permanent or permanent nature. Children born into longer-term relationships generally had an enhanced likelihood of survival into adulthood.

The pecking order in the group was clear. Those at the bottom were given the menial jobs and got the worst accommodation and the least tasty food. There was a constant jockeying within the group, as younger people especially attempted to improve their social position.

Now, think of yourself as having been a young woman or a young man, with, let us suppose, lots of drive and intelligence, but a pretty junior ranking in the group, a ranking that you badly wished to improve. The chief, let us again suppose, looked pretty unassailable. He was physically powerful and a good hunter. He could be vengeful and cruel, but most of the tribe respected his decisions. Those at the bottom of the pecking order feared him. His first wife was very cunning. She kept her ear to the ground for any signs of discontent, and she saw to it that the wishes of her husband were carried out. Both their children were especially favoured. The boy was likely to become the next leader. Not much hope for you there, unless (if you were a girl), you could become the first wife to the son – but age difference alone made this unlikely in your case.

It had been a disastrous last two years for the group. The rainy season hardly happened in either year, and the local water hole had dried up. Game, which had been plentiful, had moved on. Thankfully, most of the lions and leopards had moved on, as well, but some had stayed and developed a liking for human flesh. Two young children had been lost to lions in the last two months. Guards had to be posted all night, placing an additional strain on the small community. The group was surviving largely on plants still growing around the dried-up water hole. Small quantities of water were still available thanks to a second hole the women dug in the dried mud of the hole. Recently, signs of other humans in the area had been detected, and the tribe was worried about an attack from these humans.

Your mother had told you the story of the great Earth Goddess. It was generally believed that the tribe was suffering from her displeasure over some imagined slight unwittingly perpetrated by the tribe. You were dubious of this story and had seen no evidence of any earth goddess. But what other dark and mysterious force could be the cause of all your misery, and what might be done about it? You wished you knew!

It turns out that from early childhood you had been fascinated by the sun and moon. Somehow, you thought, it is the sun and moon together that decide how much rain will fall, and it is rain and sunshine together that seem to decide the health of plants and animals. Unlike the Earth Goddess – an invisible lurking shadow, imagined but not seen – the sun and moon are visible presences. If only you could communicate with them, you might learn how to arrange for more rain, so that the game animals would come back and the grass would once again be green and tender.

In order to better understand these distant but visible gods, as you thought of them, you embarked on a period of study. You found a quiet spot on a small rise of land about twenty minutes' walk from your camp, and you spent many hours there, day and night, observing these aloof gods. You observed carefully the patterns on the moon, and thought you could discern the face of a goddess. You found a small clearing where you could drive a stake into the ground and observe the shadows cast by both the moon and the sun, and you devised a way to record the changes in shadow length and direction, and tried to correlate these with the weather. You tried saying prayers to both sun and moon. Sometimes you complimented them on their wisdom and strength, and thanked them for their gifts to you and your tribe. Other times you begged them for their help and offered them such gifts as you had that you thought might interest them, an animal hide, perhaps, or some food. All of this work took its toll on you, since you were still expected to contribute your share of labour to the daily life of your tribe, and your study of the moon in particular, being primarily a nocturnal pursuit, was both dangerous and tiring.

However, your mother encouraged you to keep going and helped you by carrying out some of your daily chores for you. Your interest in the sun and moon came to the attention of the chief quite early on – there was little in tribal life that did not – and after thinking about it, he decided to encourage you. Invocations to the Earth Goddess after all, had not been very effective, and it looked as though the tribe would have to move off south in search of better land, with all the perils that that involved. Just maybe, he thought, you were onto something that could help the tribe. Soon, all the tribe took an interest in what you were doing and came to you with suggestions. Have you tried eating some particular healing herbs before communing with the sun and moon? Have you tried going without food? What about sacrifices? You listened to all of these suggestions, especially to those coming from the chief, who was over twice your age and had some knowledge of the customs of neighbouring tribes. The chief's first wife, unfortunately, was a problem for you. She thought your ideas would offend the Earth Goddess, and the goddess would curse the tribe even more. Fortunately, at least for a while, the chief was on your side, but several other tribe members agreed with his wife.

One night, after two days of fasting, you fell into an exhausted sleep after hours of prayer and invocation to your new gods. Times were especially harsh. The last drop of water would soon be gone from the remaining water hole, and the grass was tinder dry. A brush fire had almost demolished the campsite and its inhabitants a few days earlier. The group was becoming desperate. While you slept the Sun God came to you in the form of a handsome youth. He thanked you for your prayers to him, and told you he would answer your prayers, that he would bring rain, provided the tribe began to observe certain practices to please him. He advised you that he was the most powerful of the gods, and that the moon was his wife, and that she ruled the night. You are pleased to be told that the Earth Goddess did indeed exist and was a daughter of the sun and moon. It was therefore appropriate to worship her, too, but first obeisance must be to him, the Sun God. He commanded you to convince the tribe to worship him each morning at sunrise and to leave him some berries to eat here where you were sleeping. In the evening, you were to bring a gourd of water for him to drink. You alone were authorized to do this. He promised that rain would come on the third day of such observances, and that the animals would return and the tribe would prosper.

You awoke with a start and, looking up, you saw a male kudu, its noble horns spiralling upward in the moonlight, standing stock still about ten feet from you. After a pause that seemed to be an eternity, it turned and walked slowly away. You rushed back to camp and woke the chief. Soon everyone was roused, and you told them about your vision, for you were convinced that what you saw and heard was as real as the dry grass under your feet.

There was much discussion after you recounted your dream. The chief's first wife was alarmed. She was convinced that it was suicidal to downgrade the all-powerful Earth Goddess to inferior status, and several others chimed in in strong agreement. There was even a suggestion that you were the incarnation of an evil spirit and should be banished or worse. But the chief quieted the discussion. He pointed out that obeisance to the Earth Goddess had been of little help of late, and that the tribe was in desperate circumstances. He suggested that the group heed the instructions and conduct a three-day trial to see if rain came. He even went so far as to praise you for your efforts on the tribe's behalf. After three days, he suggested, the group would reconvene to judge the success of the experiment. If it failed to rain as promised by the Sun God, it would be obvious that the tribe should look elsewhere for solutions to its desperate problems.

Your pleasure at this success, and the huge increase in your status in the tribe, was tinged with a nagging fear. If rain did not result, you realized that you would be in some danger.

In the event, it was a near-run thing. Rain did finally come, but late at night on the fourth day. Late enough for you to have experienced the righteous resentment of the chief's wife, and to hear her suggestions for the painful manner in which you should be disposed of to satisfy the angry Earth Goddess. But, after all, the rain arrived early enough to save your skin. Indeed, the rains continued for long enough to restore the water supply and to bring back the animals. Your reputation was made, and it spread to neighbouring tribes. The

Figure 2.1 – *The kudu is a large member of the antelope family. It lives in southern and eastern Africa. The long horns, found only on the male, average 1.2 metres long, make two or three complete twists of a spiral and diverge slightly. The kudu's striking appearance has made it an animal of special affection and*

chief was able to negotiate treaties of friendship and co-operation with some of these, which made it possible to devote more resources to the worship of the sun and moon, a religious exercise of which you were now the chief practitioner, assisted by medicine men and women from the other tribes, and, later, by one of your children.

You were lucky. During your lifetime, the drought was never again as bad as it had been on that fateful day when you convinced the tribe to shift to sun and moon worship, while still paying obeisance to the Earth Goddess. Whenever the drought threatened, if prayers and sacrifices seemed to have no effect, you grew adept at finding earthly reasons for the failure. Perhaps someone in the tribe had jinxed the prayers, or perhaps the Sun God was otherwise occupied or the Moon Goddess had run off with someone else.

Through all the years, you never lost your interest in better understanding the world about you, for, in an all too real sense, your survival depended on knowing more than others. You kept looking for new ways to convince people to take you seriously, for the task of invoking rain or of predicting events of any kind beyond the sunrise and the changing of the seasons was fraught with error.

Somewhere along the line, you came up with the idea of promising immortality. Everyone was aware that they were going to die, and no one really wanted to die. Indeed, most would go to extreme lengths to stay alive. Why not promise those who adhered strictly to your religion the chance of an afterlife? At first, the chance for immortality was offered only to the priest class and the chiefs. You noted how effective it was in convincing the chiefs to do some things you wanted them to do, and it virtually guaranteed ready access to consultations with your own chief.

Some years later, when times were bad, and the people were restive, you had another revelatory dream, in which it became clear to you that every member of the tribe could enjoy everlasting life in the hereafter provided he or she met certain basic criteria. The criteria, which you reviewed with the chief and your council of priests, included obedience to religious authority and respect for one's father and mother. The concepts of sin and morality were developed some time later by other members of your priestly class.

– Peter, all this is very well, but you haven't yet told me how you or any of your ancestors thought you had come about. I thought I asked for your creation myth, not how religions got started!

– Quite true. You did ask for my creation myth. But you have to realize that for most religions the story of creation is not the central focus. The main goal of religion was to provide adherents, particularly group leaders, with a safer and a better life. For the early religions, this was a particularly difficult task as humankind knew very little about the world around them. Even the source of rain was a mystery.

As time went on, and scientific understanding increased, new religions better tailored to existing knowledge and the changing needs of increasingly organized societies came into being.

I plan to provide you with enough background on our universe and ourselves that you will be able to discern for yourself the most probable roots of our creation. It would spoil the fun to tell you now!

– Okay for now, but please keep in mind that I am not here for fun. I simply want to understand humans as best I can.

In that regard, I note that your invented religious leader had very simplistic religious rules for members of her tribe. The list doesn't begin to compare with the rules of contemporary religions that I have been hearing about in some of my conversations with others. Are you implying that robbery, adultery, brutality, fraud and lying (to mention but a few categories of current thinking on sinful behaviour that I hear about) were all tolerated?

– Very likely. Remember that the groups were small, so the chief would generally set the rules without the need or desire for religious sanctions to back up his orders. Murder could seriously affect the success of the hunt and the requirement for the critical numbers needed for self-defence, so a religious sanction against it would likely have been introduced early in the evolution of religion. The sanction of course would probably have only held for other members of the same group. The murder of humans who were members of other groups was much more likely to have been permitted, and doubtless was actively encouraged in some cases.

Anthropologists have documented many quite recent primitive societies where deception, lying and cheating are or were the norm. They may not have been pleasant societies to live in, but they worked well enough to permit the tribe to survive. For example, in *Patterns of Culture*, Ruth Benedict describes the lifestyle of the Dobu Islanders of eastern New Guinea. She writes that "the social forms which obtain in Dobu put a premium upon ill-will and treachery and make of them the recognised virtues of their society."[1] The Dobuan society was actually matriarchal, perhaps an exception to a widely held belief that if only women were running affairs, things would be a lot better!

The life of our early forebears was almost certainly nasty, brutish and short, just as the seventeenth-century philosopher Thomas Hobbes famously suggested it was.

I prepared this imaginary tale because I think it illustrates well how the pursuit of science likely got its start. Our forebears simply observed that there were a lot of survival-affecting things going on that they did not understand. These were largely understood to be acts of God or of gods, so they quite sensibly tried to find her or them and address them in the hopes of getting some help. Since no gods were observed walking on Earth, a study of the heavens seemed to be the next best place to look for them.

This was a very sensible approach to take. Over time, it led to some fascinating discoveries, as you will see. After all, it does not take a genius to realize that many of the phenomena that matter to survival are regulated by or come from the sky. Daylight and warmth come from the sun, night light almost entirely from the moon, while thunder and rain and wind all seem to be "heaven sent". It was also natural that a tool as powerful as religion would be used for other purposes. Over history, our religions have served four primary purposes: explaining otherwise inexplicable phenomena; serving as an instrument of power; serving as a means to establish and enforce social harmony (particularly useful as tribes grew in size so that it became increasingly difficult for a chief to maintain control); and finally as a source of comfort and solace when things did not go well, and, increasingly, even when things did go well. Thus religion very quickly became much more than just a means of understanding otherwise inexplicable events.

Davey seemed pleased. I thought I detected faint notes of gratitude and excitement in his voice when he said:

– Already you have helped me! I have found this whole business of religion amongst you humans rather confusing. It is obviously very important to most of you, yet I was not aware of any religious practices amongst my superhuman friends before their disappearance. They had ancestors just as you do. I wonder if they were ever afflicted with religion?

– You might want to be careful about assuming that religion is an affliction. As I just mentioned, for many people religion is a great joy and comfort. It suffuses their lives.

– So I have noticed. But it doesn't make sense to me. Can you explain why different people profess faith in different gods? You would have expected that over time one god would have won out over the others. After all, they cannot possibly all be omnipotent at once!

– Such behaviour is completely rational in the context of human development, but hold on, I would prefer to answer your question later on. For now, I hope you will be as amazed as I am at what the ancients were able to learn about the heavens, driven in large part by curiosity about what the gods were doing up there.

As far as we know, Galileo fashioned the world's first astronomical telescope in 1609. Prior to 1609 astronomy was practiced exclusively with the unaided eye assisted by some instruments that permitted the user to measure direction and the angle of inclination to a particular object in the sky.

How might you and I have fared in sorting out the riddles of the heavens in such circumstances?

Might we have been like the Sumerians, deemed by some to be the earliest astronomers of recorded history, who, as early as 4000 BC considered the sun, the moon and Venus to be the homes of gods? If we were both lucky and crafty, we might have become priests in a temple system with hundreds of "staff." Perhaps we, like them, would have designed a calendar; identified the basic cycles of the sun, moon, planets and stars; and divided the year into twelve months based on the moon's twelve cycles during a year, though I find it hard to imagine that I, at least, would have been smart enough to do that!

Perhaps, like some ancient Britons at Stonehenge in about 2450 BC, we would have added some very large stones to an existing structure to indicate the alignment of the sun at the time of summer solstice, and thus marked the start of a decline toward a winter whose dampness and cold were uncomfortable and life-threatening.

Mankind's first known record of an eclipse of the sun was made in China in 2136 BC. Might we, like the Chinese, have attached particular importance to the constellation Ursa Major, and looked on the North Star as the keystone of the heavens?

Would we, like Aryabhatta of India (said to have been born in 476 AD) have concluded that Earth orbited the sun (and not vice versa as most of the "civilized" world believed at the time), and then gone on to predict eclipses as well?

Maybe we would have been more like the Mayans in Central America. They were able to figure out that the planet Venus has a 584-day cycle of appearance as seen from Earth. They also learned how to predict lunar eclipses and developed a detailed calendar based on a twenty-day month that accounted for seasonal changes as well as lunar cycles.

And I could go on with many more examples from other parts of the globe. It turns out that many if not most peoples of the ancient world had an interest in and a fascination for astronomy, and some learned a surprising amount while pursuing that interest.

– Are you just being rhetorical, or are you really asking me these questions? Frankly speaking, if it were up to me, I certainly would have made all those discoveries you speak of, and many more besides. I would have thought that having an understanding of one's surroundings was pretty basic, yet you humans seem to have been quite slow to catch on.

I felt frustrated by Davey's response at this point, after all my work at finding good examples of ancient discoveries, and I felt it was important that I express it.

– You have me at a disadvantage. You tell me nothing about yourself and how you came about, while you seem to imagine that human intelligence was such that as soon as it had evolved, it would necessarily learn all about its surroundings almost immediately. That is not the way things happen in our universe. I shall be interested to learn how things got going in yours.

– You are right to complain. I apologise. I sometimes forget how different your development has been from mine. Please continue.

– Well, that pretty well wraps up what I wanted to say about early synergies between religion and science. Like other human institutions, religions have undergone a continuing evolution as the human situation has changed. For example, as Richard Dawkins has succinctly pointed out, Christian doctrine today is worlds apart from that of the early Christians.[2]

While there is no shortage today of writers promoting atheism, Dawkins notably included, available polling evidence strongly suggests that religions will continue to have an important influence on human activity for years to come. After all, the founders of the major religions were all very wise men, whose insight into the foibles of human nature rightly won them many adherents. Furthermore, the pastoral work of the clergy and the social networks provided by attendance at religious ceremonies are, for the most part, important contributors to social cohesion and happiness.

Nonetheless, religious tenets and practices will need to continue to adapt if religion is to survive.

You can count on it that many if not most major religions will therefore adapt.

– Well, I am not at all sure you are right. I continue to find your religions very puzzling, though I think I now understand much better how they probably got started. My friends in my universe did not seem to have any need for a religion. Nor do I. But you humans do seem to have such a need, though I wonder if that need is permanent, or merely a phase in your evolution.

Very puzzling, indeed.

– I hope that as our discussions progress, the mainsprings of human religion will become clearer to you. Next week, I plan to talk a bit about some other non-scientific sources of knowledge concerning humans and the environment we live in. That should set us up for the scientific pilgrimage I have in mind.

– I, too, hope you are correct in predicting that it will all become clearer for me. At this time it seems to me that religion is a potentially damaging and unnecessary luxury.

I look forward to next week.

With those few sceptical words, Davey apparently took his leave. I quickly discovered that I always knew when he left, but I never did quite understand how I knew.

I clearly recall that I felt both relieved and exhausted after this first real session ended. It was evidently more stressful than I had anticipated or even realized as the dialogue progressed. It was surely a privilege to be able to have a dialogue with Davey, but it was a very odd privilege – and it was quite some time before I developed any feelings of kinship or affection for Davey.

By the next morning I felt rested and imbued with a renewed sense that Davey's education was a challenge I could and should meet. At the very least it seemed important to understand him better.

DIALOGUE 3

Setting the Stage

All the world's a stage,
And all the men and women merely players;
They have their exits and their entrances . . .
WILLIAM SHAKESPEARE, *As You Like It*, Act II, Scene 7

– Let's get going. I was a bit disappointed last week when you informed me that humans are likely to continue to indulge in a variety of disparate religions for some time to come. I hope you have some better news for me this week!

Davey's words broke in on my thoughts about the possibility of a winter holiday to find respite from Vancouver's gloomy rain storms.

It was my turn to be enigmatic.

– Maybe so, maybe not.

It is a reasonable assumption that human interest in creation myths and religion more generally was motivated by a belief that knowledge of our origins will assist us and our family groups to lead secure and happy lives. In ancient times, this largely seems to have meant knowing who the powerful gods were, whom they liked and disliked, and what one had to do to keep them happy and constructively engaged in your welfare, while bestowing famine and pestilence on your enemies.

Although many of us no longer believe in deities who control our destiny, we modern humans are no less anxious to acquire knowledge that will help us to lead happier lives. We want to know what new theories our scientists have deduced and how those theories have withstood challenges from others proposing alternative theories. We also look to our literature to shed light on our own inner natures, as well as those of others in our society. Some humans still look to their religions for the answers to the most challenging and difficult questions. Others choose to search more widely for the answers they seek.

For example, let me read you what the American anthropologist and poet Ruth Benedict once wrote:

> The trouble with life isn't that there is no answer, it's that there are so many answers. There's the answer of Christ and of Buddha, of Thomas à Kempis and of Elbert Hubbard, of Browning, Keats and of Spinoza, of Thoreau and of Walt Whitman, of Kant and of Theodore Roosevelt. By turn their answers fit my needs. And yet, because I am not any one of them, they can none of them be completely mine.[3]

This quotation helps to illuminate her feelings, as well as our state of knowledge about ourselves in her lifetime, which spanned 1887–1948. Today, I am inclined to rephrase Benedict as follows:

> The wonder of life is that all that we can observe in many fields of science, in literature and philosophy as well as in the study of religion, points to a singular and fascinating path of development of the human body and the human mind. By turns the path is lit by Newton, Einstein and Shakespeare, by Darwin and by religious scripture, by Benedict, Tinbergen and Lorenz, by Watson and Crick, Woese and Ramachandran, as well as by those leading today's studies of subjective well-being. And yet, we still do not have all the answers but rather an emerging, still somewhat blurry picture of how we came about and what we need to do to be happy.

The creation story I believe in is not a simple narrative. It has been written by many bright and thoughtful people working independently in many different fields of study. In a way, it is a surprise that such disparate work in such large quantities, should all fit together so nicely, but then again, if it is true that there is but one course of history leading to our existence, it is not so surprising. Everything should fit together.

At this point I told Davey that I planned to introduce him to my personal choice of the key ideas and the key progenitors of those ideas that I believed would give him the foundation he needed to understand us. I went on to warn him that some of the selections I made would have been on almost everyone's list of key ideas and would likely therefore not be new to him. Others would be unapologetically idiosyncratic.

I waited for a comment. Eventually he said:

– What you propose is interesting, but I can't help noticing that you, as well as all the others I am conversing with, seem to claim that many of the most revered humans lived in a time I have been led to believe was well before people understood much about their environment or even about themselves. For instance, you just mentioned Newton, Shakespeare and Darwin. Is this what some seem to call "ancestor worship?" How long has this been going on, and will it continue?

Davey went on to comment that amongst all the animals only humans seemed to take the trouble to develop a philosophy of life at all. The rest simply lived out their lives. Why was it useful for humans when other life forms seemed to find it unnecessary?

My reply was that it is not necessary to develop a personal philosophy but that I believe my chances of leading a satisfactory and happy life are enhanced because I have devoted the effort to develop such a framework for living. I illustrated my point with an analogy that I was not sure he would be able to grasp.

– Many people drive cars without much, if any, understanding of how they work. But if you are one of those people, and you hear a strange noise near the back axle, how do you know if it is something trivial like a stone stuck in the tread of the tire, or something more serious like a brake problem or a worn wheel bearing? And if the car won't start, how might you know if it is a problem of engine flooding, a malfunctioning solenoid associated with the starter, a faulty battery or an intermittent electrical connection?

In a human lifetime, a person may suddenly find that he or she has fallen in love, or someone he or she doesn't particularly care for may have fallen in love with her. A person may feel a strong urge to buy a house, to divulge a secret, to change jobs, to show off, to start an argument or to procrastinate. One may wonder if satisfying those urges will or will not contribute to long-term happiness. Understanding oneself better may not resolve a particular conundrum, but it certainly won't hurt. Furthermore, by understanding more about oneself, it is easier to understand others, and thus, hopefully, to be in a position to enjoy happier and more constructive relationships with them.

As you suggested, Davey, the strongest counter-argument to this position is that we are, after all, just animals like all the rest of them, and most of us strongly doubt that any other animals attempt to define a philosophical framework to guide their lives. As far as we know an ant performs most, if not all, its actions instinctively, with narrowly defined options, such as "fight or flee" determined by inherited genes. Even chimpanzees, which are generally thought to be the closest animal to humans both genetically and in terms of intelligence, appear to be driven primarily by innate behaviour, though with much greater room for behaviour modification through assessment of real-world circumstances than ants have. So, why should we be different? If chimps have not studied philosophy, why do we need to?

I looked earnestly to where Davey's voice seemed to originate (a habit I found hard to shake), sensed no response and continued.

– You will discover that I am amongst those who believe that an important part of the reason that we need to better understand ourselves lies in the fact that although our lives bear little resemblance to those of our prehistoric ancestors, we are nonetheless still bound by essentially the same set of instincts that they evolved to survive while living a tribal "prehistoric" existence.[*] Many of our inherited behaviour patterns are no longer appropriate to our long-term happiness or even to our survival, and get in the way of the very large, well-ordered civil societies that have done so much to enhance our standard of living and our happiness. In a very real sense we are intensely tribal animals for whom the rewards of living in meta-tribal civilizations are so attractive that we suppress a lot of instinctual behaviour, though each of us, in a very real sense, still lives and reacts primarily with his or her own "tribe" of relatives, friends and acquaintances.

To take a simple example, it is rare these days that we have to make a fight-or-flee decision, yet such decisions are central, largely instinctive reactions in most animals. Indeed, a substantial fraction of military training is devoted to shaping our fight-or-flee instincts in favour of fighting, even when our instinct is crying, "Flee!"

More important, democracy has a very different practical meaning in a society of less than several hundred people than in a society with millions of individuals. In a tribal society, everybody knows everyone else quite well and therefore knows a lot about the personal qualities and viewpoints of other members of the tribe. In large societies we rely on second-hand (or worse) information that notoriously can be "spun" to mislead us about personalities and to favour particular viewpoints. Our instinctual baggage leaves us badly adapted to life in the large agglomerations of people needed for the efficient operation of today's societies.

– If what you say is correct, then it seems that you humans are very poorly adapted to your life on Earth. Perhaps you should seriously consider returning to a hunter-gatherer existence.

– An interesting idea and not an original one. It has been tried, but generally speaking the enhancements to survival and happiness enjoyed by

---

[*] I consciously decided not to confuse him by adding that those same prehistoric ancestors themselves were likely endowed with some, perhaps even many, instincts that predated the evolution of mankind from earlier ancestors.

mankind in a civil society make the return to hunter-gatherer or even isolated rural agricultural subsistence very unattractive, as evidenced, for example by a flurry of independently conceived communes established by the "hippie" generation in the 1960s, which usually failed, and which have certainly not been widely adopted.

The enduring challenge of designing a social structure that provides the many advantages offered by large societies with resources to spend on a multitude of highly specialized activities tailored to human comfort, security and happiness while still allowing us to live amidst a much smaller tribe, almost certainly numbering less than five hundred people, all of whom we know personally, without dependence on outside assistance or intervention, appears to be a long way from achievement.

What about you, Davey? Do you live in a utopia full to the brim with peace, order and good government, not to mention eternal happiness?

– I think we probably did at one time, but then something happened, and our happy society vanished. We have not recovered from this disaster so far.

– I want to know more. Perhaps there are lessons for us in what you endured. How was your society structured? Were you, too, adapted to live in small tribes, and did you find yourselves forming into much larger social groups in order to enjoy a higher standard of living? And what . . . ?

– Peter, stop asking me all those questions! I have already told you that I am unable to answer such questions until I know a lot more about you humans than I now know. It is just too dangerous!

– Well, I certainly hope you change your mind soon. We badly need to find a lasting solution that will permit us to better adapt our society to our instincts, or perhaps, someday, vice versa!

– I think we had better change the subject.

As I mentioned two weeks ago, I have not yet heard a convincing explanation of the function of literature. What is your explanation for what seems to be a widespread addiction to fictional tales?

– Its role seems very obvious to most of us. We simply enjoy a good read. The author may lead us to greater insights into the variability and plasticity of human nature, inform us about the nature of the world around us, make us laugh or cry or feel indignant or disgusted, all without having to experience directly the crises and cares of the fictional characters. It is a wonderfully economical and pleasurable way to learn about life. I understand that you cannot read. Would it help if I acquired some talking books for you and played them on my cassette recorder?

– Perhaps later. For now I would prefer that you give me some examples of good literature.

– Okay. Perhaps we should start with Shakespeare. As you likely know already he is the best-known writer in the English-speaking world, and perhaps the best-known writer anywhere. He provides an excellent example of the sometimes almost magical insight of our great writers into the human condition.

– I think I am disappointed already. As you mentioned last week, it is easy to find people to talk to who, because of their participation in one of the major religions, believe that they understand the meaning of their life. When I approached you, I did so because I thought you would come with a different perspective – not unique, but different from most of the others I am talking to. Now you start by invoking Shakespeare, where they start by invoking Allah or Buddha or Jesus Christ. Is Shakespeare then just another god? You humans seem to have no shortage of gods!

I laughed out loud. I had never before thought of Shakespeare in such a light. I have to admit that there is a sense in which he and his fellow literary greats share with the great religious figures a sparkling insight into human nature, often accompanied by a wrapping of the insights in colourful and commanding language. But Shakespeare as a deity required a bit of more imagination than I could command at such short notice. After laughing, I paused and continued.

– Well, no. Shakespeare is no god, though some would claim that his poetry is divine. His insights are not the product of meticulous science but rather of a keen assessment of everyday human conduct. Indeed, as you likely have already observed, not everyone even admires or likes Shakespeare. Their choices to illustrate the influence and importance of literature would be different from mine.

But let me return to Shakespeare.

Take for example this morsel from Act 1, Scene 1 of *Much Ado About Nothing*.

Beatrice: *I had rather hear my dog bark at a crow than a man swear he loves me.*

Benedick: *God keep your ladyship still in that mind; so some gentleman or other shall 'scape a predestinate scratched face.*

Beatrice: *Scratching could not make it worse, an 'tween such a face as yours.*

Why is it that both Benedick and Beatrice, each very much in love with the other, strenuously deny that love? Most of us laugh at this point in the play, because we recognize their actions as typical human behaviour. The problem of declaring love is as full of anguish today as it was in Shakespeare's time.

But why do you suppose this is? Is there some evolutionary advantage to be discerned from such behaviour? Do other animals have similar experiences? Or is the anguish and folderol a necessary by-product of civilization?

And what of the tragedies? Of the *"carnal, bloody and unnatural acts, of accidental judgements, casual slaughters; Of deaths put on by cunning and forc'd cause"*, as Shakespeare wrote in *Hamlet*.[4] Crime and wars reported in my time have many of the same elements as those reported by Horatio to Fortinbras in Shakespeare's play. Are these innate to humanity or simply products of mismanaged civility or malfunctioning minds?

And then there is Polonius's advice to his son, Laertes, also taken from Shakespeare's *Hamlet*.[5] It was part of Shakespeare's genius that he assigned such wonderful perceptive lines to a garrulous old man. His advice is given while his son is impatiently champing at the bit to leave his father and get to Paris with all its famous and infamous attractions.

As a young boy I was required to memorize Polonius's speech, and quite a few others, as well. When no one was listening, I would practice reciting the lines in what I imagined was an appropriate manner – chock full of gravitas and sober reflection. You may imagine my chagrin when I went to see the play and discovered that the actor playing Polonius rattled off the lines in a river of careless and swiftly flowing words, stunning the play-goer by the sheer wantonness of inattention to such good lines. Of course, Shakespeare knew well that had those same lines been treated as seriously as I had thought to be appropriate, he would likely have put his audience to sleep!

So, I hope you get the idea that good literature does many things for us humans. It usually amuses and entertains us, but it also informs us about the limits of human experience. It allows us to see and experience vicariously situations that we could not or would not otherwise see or understand.

– That is interesting, I suppose. But I would have thought that sticking to the facts was much more useful. There is usually only one version of the facts, but even I could invent millions of versions of pure fiction.

– Did your superhumans have no literature then? Did they not dream? Did they have no emotions?

– Oh, yes! They had emotions all right. Their emotions were the root cause of their tragic disappearance!

– Surely that can't be true! Here on Earth our emotions are our guide to survival. Without them, we humans would surely have been extinct long ago.

– That may be true for you humans, but it is not true for my superhumans. Perhaps you will come to understand why later.

– I certainly hope so.

I was torn by a desire to follow up this lead and a wish to complete my thoughts about the key role of literature. I opted to return to Shakespeare.

– By the way, Polonius's advice ends with the following three lines:

This above all: to thine own self be true,
And it must follow, as the night the day,
Thou canst not then be false to any man.

If this is good advice (as I believe it is), it begs the question of how we humans know what is involved in being true to ourselves. A crucial first step in getting to the answer was carved in stone at the temple of the oracle at Delphi in ancient Greece. The carved words were very simple: "*Know thyself*".

But that invocation, too, leaves unanswered questions. How do you get to "know yourself," or, from your viewpoint, Davey, come to know and understand us?

Literature and philosophy primarily illuminate the *what* of human nature and often suggest to us (directly or indirectly) good strategies for successful behaviour. But ultimately, insofar as we are able, we need to understand *why* we do the things we do. A complete answer to the *why* question will often lead us back to the need for an understanding of *how* the universe, including us, came about and *how* it works. It is thus primarily in the realm of science that we seek and, especially in recent times, sometimes find, the answers to the *why* and sometimes also the *how* questions.

Despite the fact that we keep telling ourselves that each human being is unique, we all know that it is only true up to a point. We are defined as a species not only by similarities in our outward appearance but also by similarities in behaviour. Hence, without in any way denigrating our individual differences – a mixed source of joy and frustration for most of us – it is beneficial to our long-term happiness, as well as long-term species survival, if we understand what science has learned about shared innate human characteristics and behaviour.

After some thought on the subject, it seems to me that you, Davey, probably seek the same knowledge in order to better understand us and perhaps as a result, to better understand the fate of your vanished friends.

That is all I want to say today. At our next meeting you will see that as humans became better organized, and job specialization made a steadily improving standard of living possible, science became better organized, as well.

But despite your apparent surprise at our ignorance, for us humans the story of our accelerating understanding of the world about us is dramatic and full of surprises along the way. Inevitably, it has shaped the way most of us view ourselves and our role on Earth.

Davey replied with a drawn-out "Mmmmm", and then his voice faded away.

PART 2

ORIGINS

Part 2 is a record of my discussions with Davey that relate to humankind's attempts to determine the age and extent of the universe and of our planet Earth using what came to be called scientific methods.

Once human civilizations reached a stage where some individuals had both the means and the time to pursue science on a near full-time basis, (the early scientists were almost all amateurs), discoveries started to come apace. It is hardly surprising that some of the discoveries shattered previously accepted conceptions of our universe, and of how and when it grew to be what we observe today.

In the last five hundred years or so, humanity has moved from a simple and unsophisticated view of the age and extent of our universe to a gradual realization that it is very much older than human memory, and very much larger than Aristotle's concept of the spherical Earth plus a smallish outer layer holding all the celestial objects. This transformative expansion in humankind's horizons was not accomplished quickly or easily.

By definition, we have no records of life in prehistoric times, other than some bones and artifacts left or lost by early humans. Like us, our forebears would almost certainly have possessed an active curiosity, and must have been continually looking for ways to improve the odds of survival. I had already told Davey that it is not unreasonable to speculate that religion and

33

scientific curiosity initially developed together, as humankind sought explanations for rain, sunshine, the moon and stars, as well as drought and pestilence, disease and a thousand other afflictions suffered by *Homo sapiens* in those dangerous times. During the Renaissance in Europe (roughly the period between the fourteenth and seventeenth centuries), religion and science drew apart, as scientific investigators began to make discoveries that were antithetical to established religious dogma, which quite naturally was largely based on earlier scientific investigations.

The first dialogue in this part (Dialogue 4) skips lightly over some of the most important findings in astronomy, many of which are standard topics in high-school science classes today but all of which were portentous and exciting at the time, and proved to be of great interest to Davey.

But there is still a lot we don't know about our universe. For example, we don't know if the universe explored by our astronomers and astrophysicists is the only universe, or if there are other universes, perhaps in countless numbers, which are invisible to us. Even in our own universe, our scientists are challenged by the mysteries of "dark matter" and "dark energy", phenomena that the observed behaviour of the stars and galaxies seem to require to explain their behaviour, but that, so far at least, are invisible to us.

Some of the ideas and the evidence in their favour are difficult to comprehend, but I was sure that I had to make the effort, since humanity's view of itself and of its own importance has been, and continues to be, strongly influenced by its beliefs about the role of humans in the universe as we see it. The succeeding dialogues in this part were attempts to equip Davey with the tools to better understand these ongoing mysteries.

DIALOGUE 4

How Big Is My Universe?

The highest endowments do not create – they only discover. All transcendent genius has the power to make us know this as utter truth. Shakespeare, Beethoven – it is inconceivable that they have fashioned the works of their lives; they only saw and heard the universe that is opaque and dumb to us. When we are most profoundly moved by them, we say, not "O superb creator" – but "O how did you know! Yes it is so."
RUTH BENEDICT, In Margaret Mead's *Ruth Benedict*, 1974

– Well, I hope you have a good story to tell me today. I have just been listening to an astronomer give me his version of what I hoped you were going to say. He was very nice, but he fed me a lot of equations, and showed me a lot of pictures, which as you know I cannot see. All astronomers seem star-struck by the beauty of the heavens. I gather from him that what they see through their telescopes far surpasses the beauty that often causes wonder to non-astronomers when viewing the sky at night while far removed from city lights.

Davey and I had not spoken for a week, during which I had a wonderful time reminding myself about the historical peaks of astronomical research.

– I have to agree that for most of us humans, the night sky can be staggeringly beautiful, and for astronomers, who commonly see relative close-ups of a great variety of stellar events, the sights are even more moving and beautiful. Personally, I find it hard to improve on the photographs of our own planet taken from outer space or from the moon.

35

I have worked hard in the past week to assemble an overview of advances in human understanding of the heavens since those first discoveries of a thousand or more years ago. However, it appears that you may have heard about this already, and I don't want to bore you.

Figure 4.1 – *Earthrise over the lunar surface as photographed from the Apollo 8 spacecraft in 1968.*

I was feeling just a little bit piqued that despite my hard work over the past week, he was evidently not relying overly much on what I might tell him!

– On the contrary, Peter. As I told you, the astronomers I spoke with did not quite give me the perspective on their work that I was seeking. I am expecting that you will do better. Perhaps we should get started.

I was partially mollified but once again nervous that I, too, would fail to provide Davey with information he could understand.

– Very well, but be sure to stop me if I am not telling you what you want to know.

At the dawn of civilization there was a lot about the world around us – as well as about ourselves – that seemed unknowable. Gradually, and with accelerating success, we have pushed back the barriers to our understanding. But barriers still remain. One classic barrier has been our understanding of the origins and evolution of the universe. In my lifetime an amazing amount has been learned, and we have reached some totally unexpected conclusions. Not surprisingly, the knowledge we have gained about the origins of our universe has only led to speculation as to whether or not there may be other universes! Today I shall summarize what I think were the key milestones in expanding our knowledge of our universe, and close with some very speculative conclusions that bear on how we might put our own species in an appropriate perspective.

With the possible exception of the Greeks, the major driver of our forebears' interest in astronomy was their belief that by their studies they could learn more about the gods whose capricious will could bring disaster upon them. Nonetheless, their accomplishments were very impressive, and a good lesson in how much enquiring minds can learn without the benefit of much in the way of special equipment.

However, there is another lesson to be learnt from examining the more recent history of astronomy: with a combination of enquiring minds and ingenious equipment one can learn a lot more. What we have learnt about the heavens in the past fifty to one hundred years would have knocked the socks off the ancients!

– I'm surprised you think the ancients wore socks. When I last asked what the ancients wore, I was told either that they ran around barefoot wearing hairy skins, or were clad in simple cloth and wore sandals on their bare feet.

Was Davey attempting to be funny, or was he serious? I wondered. Perhaps I should avoid phrases that required a deeper knowledge of the English language than Davey might have.

– Well , um, yes.

– Don't be so serious. Of course I know what you mean.

He still sounded strangely serious. I couldn't think of a good comeback. He must have been joking, but if so, he had not yet mastered a joking tone of voice or even a muffled giggle.

After a brief pause to collect my thoughts, I carried on.

– Aristarchus of Samos, who was born in 310 BC, twelve years after the death of Aristotle, is generally credited as having been the first to conclude that Earth orbits the sun, rather than vice versa (thus beating out Indian mathematician-astronomer Aryabhata by about eight hundred years). Aristotle (384–322 BC), who was for twenty years a student of Plato and who later tutored Alexander the Great, concluded, as did most others at the time, that Earth was at the centre of the universe, with the sun and the moon orbiting us in circular paths, and the observable planets also orbiting Earth, but in orbits that were peculiarly hard to fathom. Aristotle believed that Earth was spherical, based in part on observing that the shadow of Earth on the moon, as seen during a lunar eclipse, is curved. He also believed that a sphere is the most perfect shape.[6] The stars, he concluded, were a static display stuck to an outer sphere that also rotated about Earth. Many of Aristotle's writings, including some related to astronomy, were conserved in Athens up to the time of the Roman conquest of Athens by Sulla in 86 BC, when they were taken to Rome, where they attracted considerable scholarly attention.

In the late medieval period, Roman Catholic Church scholars "rediscovered" Aristotle, who became the bedrock philosopher for the re-awakening Church. Unfortunately, in the process, Aristotle's views on astronomy and the universe were adopted as a part of the package, and became integrated into Church dogma. It is more than a little ironic that the Church should have adopted as its paramount philosopher a man who believed that "the true spirit of science and philosophy is born when problems are studied for their intrinsic interest, *detached* from practical interests."[7] The Church at that time was still firmly of the view that it was the ultimate arbiter of the "creation story," at least as far as the Western world was concerned. The task of scientists was to confirm the words of God, as interpreted by senior Church officials.

For almost no one was this attitude more unfortunate than Galileo Galilei (1564–1642), who in 1609 was the first to use a telescope for astronomical purposes. Galileo's cardinal sin was that his observations seemed to confirm that Earth orbited the sun, consistent with Copernicus's rediscovery of what Aristarchus had observed about seventeen centuries before!

By the early 1900s, with lots of assistance from other early pioneers such as Tycho Brahe, Johannes Kepler, Isaac Newton and Giordano Bruno, a lot was known about our solar system and rather less about the stars and galaxies beyond it. The time was ripe for some breathtaking changes in mankind's perception of time and distance.[§]

Starting in 1863, some very powerful new tools and new theories fuelled remarkable advances in astronomy. The new tools, and the accompanying observations and theories, have enabled us to gain both a qualitative and a quantitative understanding of our universe.

In 1863 Sir William Huggins, an amateur English astronomer, was the first to examine the spectra of visible stars. Only four years previously, two Germans, Gustav Kirchoff and Robert Bunsen, had discovered that each gas has its own spectrum of lines of different colours, which are emitted when the gas is heated (in their case, in a Bunsen burner). Huggins found the spectral lines characteristic of hydrogen and helium in many stars, as

---

[§] Giordano Bruno was undoubtedly less fortunate than Galileo in tangling with Church authorities. He was burnt at the stake in 1600 for his belief that not only was Earth a satellite of our sun but also that there were many other suns with planets. So even the sun was not at the centre of his universe! Religious authorities were doubly scandalized by the heresy. Galileo, who was a friend of Pope Urban VIII, suffered the much milder penalty of house arrest for the rest of his life starting in 1633, after having first been forced to recant his Copernican views.

well as in nebulas and comets. For the first time, humans knew that the stars were predominantly composed of these two gasses. Thenceforth astronomers focussed a lot of attention on examinations of the spectral lines present in the light coming from the various sources of light in the sky. This was the start of an accelerating series of discoveries, not the least of which was a vital helping hand that spectral analysis provided in calculating the distance to many visible stars.

There is an old joke about a physicist, an engineer and a businessman who, just after the First World War, somehow got involved in a bet about who could first determine the height of the steeple of a distant church. They all three raced over to the church. The engineer quickly got his transit out of the trunk of his Model T Ford car, measured off an appropriate distance over level ground from the steeple (d), and then measured the angle (θ) from where he stood up to the steeple top. He then used high-school trigonometry to determine the steeple height (h), i.e. $h=d.\tan\theta$. The physicist, on the other hand, stopped only to pick up a few stones, before scrambling to the steeple top. He threw the stones off the top and timed their descent. Since he knew the force of gravity (g), and he now knew how long the rocks took to fall (t), he was able to calculate the height (h) from Newton's famous equation $h=gt^2/2$. The businessman, too, went straight to work. He went into the church and asked the verger how high the steeple was. Of course, he won the bet hands down!

If the same threesome had instead bet on who would first discover the distance from Earth to a certain star, they would have been at a loss. The businessman would have found no one to give him an authoritative answer. The physicist, too, would obviously have been out of luck. If the engineer was really lucky, and one of the sun's closest stars had been chosen for measurement, he would have found that with great care it was possible to measure the distance to such a star by observing the apparent change in position of the star as Earth moves in orbit around the sun. The radius of Earth's orbit is fairly constant, varying between 149 and 152 million kilometres (or about eight light minutes*). The principle would have been the same as the one he used to measure the height of the church steeple, but even if he took his measurements at the most appropriate six-month interval, the difference in angle that he would measure for the

---

* I had thought that I might have to explain to Davey that a light minute is the distance travelled by light in one minute, just as a light year is the distance travelled by light in one year. It turned out he already knew. The speed of light is almost 300,000 km/sec. A light minute is thus 18,000,000 km. A light year is about 9,500,000,000,000 km.

closest star (*Proxima Centauri*) would be only 1.534 seconds of arc, or about four hundreths of a degree! No wonder most observers through history have assumed that stars hold their same relative position in the sky indefinitely.

At the current time, the limit of accuracy for the angular measurement of celestial objects is about 0.05 seconds of arc, a situation that allows the distance to less than two thousand of our closest stars to be measured using geometry.

I knew I could not show Davey any sort of image that would help him understand the method of measurement I described, so I had to hope that I'd explained it adequately. However I have shown a very rough diagram of the geometry involved in Figure 4.2 below. The representation is clearly not to scale. The distance from the sun to Earth is shown as "r". If the distance to *Proxima Centauri* (d) were to scale, it would be about 2.7 kilometres off to the left of the page! The angle "a" shown on the diagram is half the difference in apparent position of the star from the two different viewing points (technically called the parallax angle) of 1.534 arc seconds referred to above.

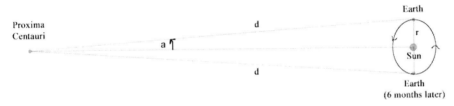

Figure 4.2 – *Measurement of the distance to* Proxima Centauri *by the parallax method.*

– Davey, I wonder if you ever studied trigonometry in your equivalent of our high schools? If so, you may recall that the sine of the angle a is r/d. Hence d, the distance to *Proxima Centauri* is r/sin a, which works out to 4.2 light years.

– Peter, you continue to underestimate my intelligence. I suppose that humankind has been the most intelligent species on Earth for so long that it cannot admit to itself that there may be other beings that are more intelligent than humans. I'll let you know if you ever tell me something I cannot understand!

– Thanks a lot! Perhaps I should call you Mr. Smartypants! I suppose I should have known that you would think you are a lot cleverer than I am, especially after all you have told me about yourself!

Davey noticed my sarcasm. He at once apologised and asked me to continue. His apology did not at all address his claim to be much smarter than me, a claim I was not yet ready to concede.

– Distance measurements are obviously fundamental to gaining an understanding of our universe. Some method of measuring our separation from more distant objects was badly needed. Solutions to the problem came from unexpected directions.

In 1895 Henrietta Leavitt was given a job at the Harvard University observatory. In accordance with the practices of the day, because she was a woman, she was not given full status as an astronomer but was instead given leadership of the department that studied the photographic images of stars to determine their magnitude. During her career there, she discovered over 2400 stars whose brightness varied periodically in magnitude, usually with a period ranging from several hours to several tens of days. These variable stars are called *cepheids* after the name of the first variable star discovered (in 1784), *Delta cephei*. Leavitt was able to determine that there is a direct correlation between the brightness of these stars and the period of magnitude variations. This was a very significant discovery, since it permitted observers to determine both the actual brightness of a star and to measure its apparent brightness when seen from Earth. Since brightness is known to decrease as the square of the distance from the light source, it became possible to determine the distance to the star, having first calibrated cepheids, which were close enough to measure using the previously described geometric (or parallax) method to provide the baseline distances.[8]

However, it is the third link in the chain of astronomical distance measuring techniques that is the most spectacular.

In the 1920s, the vast majority of astronomers believed that the universe and the Milky Way galaxy were one and the same thing. Nebulas were thought to be stars (in the Milky Way) in the process of formation. The universe was thought to be a static entity, though the theory included the idea that the birth and death of stars was part of a continual regeneration process.

The first real clue that this might not be so came from an American astronomer, Vesto Melvin Slipher, who started in 1909 to study nebulas in the hopes of learning more about how our sun came about.[9] Slipher looked first at the Andromeda nebula and observed that the spectral lines of light from the nebula, including the lines from excited hydrogen gas, were displaced from their proper position on the frequency scale.

As you may already be aware, Davey, both light and sound from a source that is moving relative to the observer will undergo an apparent frequency shift audible or visible to the observer, a shift that is referred to as the Doppler shift (after the effect's discoverer, Christian Doppler [1803–53], an Austrian mathematician and physicist). As it happens, the Andromeda nebula is moving toward us, so the frequency of the spectral lines were shifted toward the blue, or high-frequency, end of the visible spectrum. However, by 1914 Slipher had discovered that some other nebulas were apparently receding from Earth at very high velocities, since the spectral lines were shifted down in frequency, i.e. toward the colour red. Calculations showed that one nebula was apparently receding from us at an astounding 1100 km/sec, or about 4 million km/hr! How could something moving this fast be in our galaxy?

It was Edwin P. Hubble who provided the evidence to resolve this dilemma. Hubble was an American Rhodes Scholar who studied law at Oxford (to please his father), was briefly a school teacher, served in the US Army during the First World War, and, finally, in 1919 returned to his first and enduring passion, astronomy, when he accepted an offer of employment at Mount Wilson Observatory in California. In 1923 Hubble launched into a study of very bright stars (*novae*) in spiral nebulas. To his surprise, he came across a cepheid in the Andromeda nebula, and, by applying the by then standard calculations, discovered that this nebula was a million light years from us, putting it well outside the Milky Way. In rapid order, he found other cepheids, and found that many of them, too, were well removed from our own galaxy. The implication was clear. The Milky Way was not the only galaxy in town! There were many others. In fact the current estimate is that there are over one billion galaxies in our universe, each with, on average, about 100 billion stars! There are thus about 10^{20} (i.e. 100,000,000,000,000,000,000) stars in the universe. [*]

But there was yet another major surprise still in store.

Slipher's data had shown that some nebulas were moving away from us at high speeds. There were two physicists who dared to wonder aloud whether this phenomenon might be due to the expansion of the universe. The first was Alexander Friedmann, a Russian World War One fighter pilot and professor in St. Petersburg, and the second was Georges Lemaître, a young Belgian abbé and professor. To begin with, the evidence was spotty. Hubble changed all that. By 1929 he had collected enough evidence to show that the rate at which a galaxy was apparently receding from Earth was directly proportional to its distance from us. The constant of

---

[*] Only a few years earlier, Harlow Shapley had made his reputation by determining that the Milky Way had a diameter of 300,000 light years, about ten times larger than previously thought.

proportionality came to be called "the Hubble constant."* Alexander Friedmann and Georges Lemaître were right.

Humanity thus took another great leap toward insignificance, or at least to lack of centrality. Not only was our planet not at the centre of the universe, neither was the sun, nor even our galaxy. Our sun is only one of a hundred billion billion stars, and who knows how many habitable planets there are in the universe? Furthermore, like all the other planets and stars, we resulted from a cosmic Big Bang since calculated to have taken place about 13.7 billion years ago!

The idea of the Big Bang rose directly from Hubble's observation that the universe seemed to be exploding outward. By running the events in reverse, everything seemed to come back together to a time of origin of one, huge, "universal" explosion 13.75 billion years ago.

All this took some time to digest, even amongst the scientists. Arthur Eddington, the leading astronomer of the day, wrote about these findings.

I paused to find my quote from Eddington.

– "... it seems to require a sudden and peculiar beginning of things.... As a scientist I simply do not believe that the present order of things started out with a bang; unscientifically I feel equally unwilling to accept the implied discontinuity of the divine nature." [10] Others have since found the concept that everything started with one Big Bang equally uncomfortable. Was there really nothing at all prior to the Big Bang? We all have some freedom to speculate here, since, as of now, there is really not much hope of making any observations outside of our universe.

At this point I waited deferentially for a comment from Davey, but none was forthcoming. Perhaps he was sleeping? In the circumstances, I would bet money that while the outer boundaries of our powers of observation and comprehension seem to be the boundaries of the universe we inhabit, it is likely that there are now, have been and will be other universes. But let's be clear that this is not a scientific observation. In my case, it is based in part at least, on my unprovable, but for me very stimulating and real, conversations with Davey.

– In spite of Arthur Eddington's unease, there was at least one very significant member of the cheering section for Hubble and his findings. Albert Einstein had discovered that his General Theory of Relativity predicted that the universe was not in a stable state but was either

---

* The actual value of this constant is still debated. It appears that the constant is about 74.3 km/sec/megaparsec. One megaparsec is 3.262 million light years.

expanding or contracting. This was directly counter to the generally held belief in a stable universe prior to Hubble, a fact that had persuaded Einstein that his theory must be wrong. He therefore introduced a *cosmological constant* whose magnitude was calculated to ensure stability of the universe. Six years later when he first heard of Hubble's discoveries, he revised his equations to remove the constant and referred to his earlier introduction of the constant as the biggest blunder of his life. He even made a special trip up Mount Wilson to thank Hubble for his work. By that time he had already taken pains to give special recognition and thanks to Friedmann and Lemaître.[11]

It is perhaps ironic that the discovery in 1998 that there is a small but continuous acceleration in the rate at which the universe is expanding has caused a revival of Einstein's cosmological constant. The leading explanation for this surprising discovery is that the universe is suffused with a constant (but so far undetected) energy filling all space homogeneously. This energy has been labelled "dark energy", and can be conveniently thought of as a cosmological constant. If you find this confusing, don't be discouraged. There will likely be more changes coming soon. These are exciting times in the field of cosmology.

Apparently Davey, too, had noticed the anomaly. Finally, a response!

– I find it very interesting that even very distinguished scientists such as Einstein seem to incorporate new variables into their theories just to make them fit the facts, without, it seems, any real notion of what in the physical world that particular variable represents. There seems to be a lot your physicists don't know about your surroundings.

– On the contrary, I think you should be amazed at how much they do know. It has taken us centuries to accumulate the knowledge we have of our environment, but in most cases we are now able to understand and to forecast what is going on in our universe to an astounding degree of accuracy. For instance, the Standard Model theory of particle physics was able to forecast not just the existence of new particles as yet undiscovered, but also predicted their mass correctly to better than four decimal places, as measured after experimental physicists had searched for and found the particles. When a theory is able to make predictions such as these, one has real confidence that the theory is on the right track.

– I suppose you may be right. But what exactly is the Standard Model of which you speak so highly. I don't recall that you have mentioned it before.

– It is an intellectual triumph - a masterpiece of co-operative scientific endeavour stretching over sixty years or so. The model encompasses a theory of almost everything in that it describes all known forces except the force of gravity, and successfully predicts the properties of all known sub-atomic particles.

– What you seem to be telling me is that this great model of yours describes and predicts a lot of things, but it is of little, if any, help if we want to understand your universe, where the force of gravity reigns supreme.

– There is an element of truth to what you say, I conceded, but you should know that the nuclear reactions taking place in the billions and billions of stars in the universe are very accurately predicted and described by the Standard Model. That said, the Standard Model, while useful in providing an understanding of the life and death of stars, is currently of little help when it comes to understanding the overall structure and history of our universe.

Davey seemed ambivalent. It was somewhat annoying to realize that he might know far more than I was giving him credit for. Perhaps he knew the solutions already to some of the tough questions the cosmologists were tackling. I planned to talk a bit about these questions in a future dialogue, and I wondered if I could trick him into showing his hand at that time. In the meantime, I returned to the subject at hand.

– As you can see, the numbers associated with the study of astronomy are, well, astronomical! It takes our sun 250 million years to complete an orbit about the centre of our galaxy (by comparison, dinosaurs are thought to have appeared on Earth only about 230 million years ago). As previously mentioned, the average galaxy is estimated to contain one hundred thousand million (~10^{11}, or 1 followed by 11 zeros) stars, and astronomers believe that there are well over a billion galaxies in the universe! Over the past decade, it has been possible to observe the effects of large planets orbiting nearby stars. While not all stars have planets, clearly some of those that do, like our sun, have several planets.

At this time I know of no reliable estimate of the likely number of planets in the universe. It may be of the same order as the estimated number of stars (i.e. ~10^{20}), but even a number one thousand times smaller is still unimaginably large. Similarly, it is hard to find an estimate of how many, out of the total number of planets, could support life, but it would be exceedingly rash to imagine that our Earth is the only one amongst this huge number of candidates. A similar line of argument makes it seem extremely likely that more than one of these remote planets is home to life forms that are at least as intelligent as we are.

Most of us humans are still inclined to consider ourselves the wisest and most intelligent species in the universe. If there is a superior species one might reasonably expect that it would have been in touch with us by now, and left us with firm evidence of its (superior) existence. But is this really true?

Davey, you could really help me here. You say you don't even come from our universe, but rather from another universe altogether. Assuming that this is true, and that you really have wandered about our universe in ways we are not yet capable of, perhaps you could let me know what you have found in the way of other life forms in our universe and perhaps also, by the way, what you know about life forms in your own universe.

– Sorry, Peter. I thought I had already explained to you that I cannot possibly answer such questions without first understanding whether what I say could unduly harm life in your universe. I understand your impatience, but you will just have to be patient for the time being.

– It's all very well for you to ask for patience. You need to understand that I have to work hard to pull together the information you have asked for. Thus far I have received precious little from you for all my effort.

– I think you know that you have my sincere thanks as well as my appreciation. I was given to understand that most humans appreciate such gifts.

I rolled my eyes. As I did so I realized that he could not see the gesture. This realization led me to wonder how often he was misled by what he heard, as it would have been much more difficult for him to detect sarcasm than for the rest of us in the circumstances.

– Do please continue, Peter. What you are describing is a rather intricate method for determining the extent and nature of your universe. I admit that it seems rather awkward and unnecessarily involved to me, but given that you lack the tools that I have to understand my universe, I have to take my hat off to your scientists for what they have accomplished – and I am very interested to see how your story will end!

– All right. I suppose I should continue, at least for the time being.

In an attempt to introduce some order into what can at best be described as a highly speculative subject, Frank Drake, a physicist at the University of California in Santa Cruz, introduced in 1961 what has become known as the Drake Equation. (Carl Sagan, a well-known physicist and science writer of my time, gave the Drake Equation considerable prominence, to such an

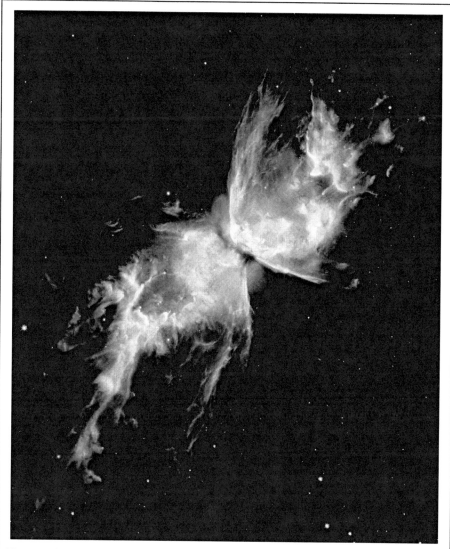

Figure 4.3 – *The Butterfly Nebula. This spectacular photo (NGC 6302) was taken in 2009 using the upgraded Hubble Space Telescope. It is representative of the great beauty frequently encountered by astronomers as they observe our universe.*

extent that it is also known as the Sagan Equation.)

The formula itself is simply the product of a set of fractions and average numbers multiplied by the rate of star formation in our galaxy; the equation addresses the likelihood of communication with intelligent life elsewhere in our galaxy only. If we wanted a likelihood for the whole universe, we might choose to multiply our result by the number of galaxies in the universe reduced by a factor that reflects the considerable difficulty of communication over distances greater than the diameter of our galaxy. If we sent a message to an intelligent being one million light years away, it would be at least two million years before we could receive a reply. And if we wanted to compute the likelihood of communication with intelligent life outside our universe, well, we probably would have to ask you, Davey, for an estimate!

Davey responded immediately.

– I've already given you an answer! Not today!

– That's disappointing. Perhaps you have a headache?

I paused to emphasize my displeasure, then I continued.

– Drake's formula looks like this:

N (the number of civilizations in our galaxy with which we might be able to communicate at any given time) equals:

R (the rate of star formation in our galaxy (about 6 per year), times

f_p (the fraction of those stars that have planets [current estimates are 20–50 per cent]), times

n_e (the average number of planets per star with planets that can potentially support life – a figure between 1 and 5 has been suggested), times

f_l (the fraction of planets in n_e where life actually evolves – fractions ranging from 0 all the way to 1 have been suggested), times

f_i (the fraction of f_l where intelligent life evolves – some believe that the fraction should be close to 1, because the survival advantage of intelligence is very large, hence intelligence is almost certain to develop; others see the development of brains in life forms as being highly improbable, perhaps as small as 1 chance in 10 million), times

f_c (the fraction of intelligent life (f_i), which develops both the means and the desire to communicate externally – most guesses are in the 10–20 per cent range), times

L (the average expected duration of the intelligent and communicating life forms calculated above – now there is a challenging number to estimate! We have only our own experience to go on. Some pessimists have estimated an average duration as low as 68 years. Drake himself suggested 10,000 years. Perhaps he was an optimist, but humankind in our present form appears to have been around for about 160,000 years, of which we have been attempting to communicate with extraterrestrial beings for only 50 or 60 years. The pessimistic estimate has us almost overdue for disappearance! Even the optimistic suggestion of 10,000 years is not very long except when viewed against the average human lifetime.)

At this point, to my delight, Davey intervened with a question.

– Why would you necessarily assume that if one intelligent life form ceased trying to communicate, another on the same planet would not immediately supplant it?

– I imagine it is because we on Earth find it difficult to believe that another life form, the chimpanzees for example, even assuming they survived whatever holocaust resulted in the extinction of *Homo sapiens*, would gain intelligence rapidly enough to engage in extraterrestrial communication in a reasonable span of time.

– I see. But by now you must have realized that in my universe this assumption is not valid. I am the living proof that one intelligent life form can be superseded by another. In due course I will explain to you how that came about.

After a pause to encourage further elaboration, I again continued.

– As you are likely already suspecting, estimates of the value of N vary very widely. If you are a relative optimist, and expect that the average communicating intelligent civilizations will last about ten thousand years, and that about 20 per cent of planets that evolve life will also evolve intelligent life, then you would expect that there are about a thousand civilizations in our galaxy trying to communicate with us. On the other hand, it seems that you don't have to be much of a pessimist to conclude that the likelihood of hearing from other civilizations in our galaxy is very small indeed, perhaps one chance in a million. Of course the likelihood of a two-way conversation is smaller still, since any communication we receive may easily have been transmitted halfway across the diameter of the galaxy, or from about 150,000 light years away. At that distance, any reply would be received at least three hundred thousand years after the original message was sent!

Davey, here is a place where you can be a big help to us! Let me ask you one more time. Have you run into intelligent life elsewhere in our universe? Does that life resemble us in any way? How many instances of intelligent life have you found? Are you even able to visit more than one planet in the universe at one time? And what about . . .

– Stop all those questions, please. Before I can say anything, I have to be aware of your understanding of your situation. Otherwise, what I know could be profoundly destructive. Think about it for a minute, and you will understand what I am saying.

What could he have meant? I wondered. Does he have bad news for us, such that life would lose all its meaning? Are we destined to be destroyed by extra-terrestrial life forms? Perhaps there is a big chunk of rock headed our way that will wipe us out? Or perhaps we are destined to be the instrument of our own destruction? I at once yearned for and dreaded the answer that my continued efforts might prompt him to provide.

After a long pause it was Davey who broke into my thoughts.

– So, what you are trying to tell me is that once your forebears understood that the sun and the moon were not living beings, but rather balls of matter, a ball of gas in the case of the sun and a ball of rock in the case of the moon, they decided not to worship the sun and the moon any more, and opened a search for gods elsewhere?

– Well, not quite. Religion in the developed world had already changed a lot before the truth about the sun and moon were known. The major factor influencing religious changes seems to have been the changes in organization of human societies, which went from being many small bands of hunter-gatherers operating independently to much larger societies with one person acting as the supreme leader. I suppose that in the circumstances it seemed natural that the heavens might be organized in the same way, so that the belief in a single omnipotent God became increasingly widespread. The Jewish faith is generally reckoned to have been the first successful monotheistic religion. It got its start in about 1000 BC, well ahead of the discoveries I have just reported.

– So, it looks as though science really does not have much influence on the direction your religions take after all. I thought you were trying to tell me that science is important!

– It is important. But the influence of science did not really become key until the discipline of science had itself evolved to the point where its influence and authority became widespread. This did not really happen until some societies became wealthy enough that some of its members could afford to devote themselves to scientific enquiry. But this did not happen until long after the founding of the major monotheistic religions.

– I see. But you must admit that this whole religious business is very strange. As far as I know, my friends in my universe were never religious.

– On the contrary. It is not at all strange. For us religion was and still is a logical activity closely connected with our desire to survive. Perhaps your people had little desire to survive. I would not be too surprised as you tell me they have died out already! But I shall be very interested to learn how they could have come to exist if they had no instincts for survival.

– You must be sure not to jump to conclusions too quickly. In any case, I assume you want to stop here and meet again in about a week's time. I wonder what you plan to talk about then. Perhaps you will tell me how humans finally came to understand their place in your universe?

Davey was getting under my skin! I responded in kind.

– You must be sure not to jump to conclusions too quickly.

I smiled grimly. How likely was it, I wondered, that Davey's people were really as intelligent and all-knowing as Davey seemed to think them to have been?

DIALOGUE 5

It's About Time

Rough winds do shake the darling buds of May,
And summer's lease hath all too short a date.
WILLIAM SHAKESPEARE, "Sonnet 18"

It was a sunny Thursday morning. My wife, Margaret, was out playing badminton as she did most Tuesday and Thursday mornings in the winter months. So far, I had successfully arranged my meetings with Davey to coincide with her absence. I was not keen to have her questioning my sanity as a result of a sometimes loud conversation in my workroom with an invisible being.

– You seem a little nervous, this morning, teacher.

I had not said a word, but was sitting patiently at my computer, a cup of coffee in hand, reviewing the briefing notes I had prepared. If Davey could not see, but could only hear, how could he know I was feeling a little nervous? Furthermore it was hard to tell from his tone of voice if his addressing me as "teacher" was meant to be humorous and to set my mind at ease, or, to the contrary, as a sarcastic reference to my feeble efforts to explain something he already understood much better than I. Later on I would learn that he rarely if ever tried to embarrass me. It was almost certainly a failed attempt to relax me and get me laughing. Not knowing this, yet, I bit my tongue and started in without further comment.

– I vividly recall being told in school that a butterfly only lived for about twenty-four hours. In reality the lifetime of a butterfly varies substantially by species but is more likely to be about ten days.[12] Such a short time span for a life, even if preceded by a few weeks spent as a caterpillar preparing for the metamorphosis, seemed incomprehensible to me at the time. I could

52

not begin to imagine how to pack a lifetime into one, or even ten, days.

If I were to live for a hundred years, that would be about 3600 times longer than the ten-day life of an average butterfly. The human sense of time must clearly be very different from that of a butterfly, assuming it has any sense of time at all. Take another leap by a factor of 3600 beyond a hundred-year lifespan, and you get 360,000 years. Now that is quite a long time. If we look backward in time, it takes us well beyond the appearance of people like us (*Homo sapiens*), generally reckoned to have occurred about 160,000 years ago.[13] These time frames must be totally beyond the ken of a butterfly and put a severe stress on our own imagination. I look on my great-grandparents, only three generations older than I, as having lived quite a long way back in time. 160,000 years takes us back more than six thousand generations!

The big dinosaurs such as *Apatosaurus* and *Tyrannosaurus rex* became extinct about 70 million years ago, or almost 440 times longer ago than the first appearance of *Homo sapiens*. As you have likely learned already, the first dinosaurs appeared quite late in the history of Earth, only about 230 million years ago. The first microfossil records of life on Earth date to about 3.45 billion years ago and are found in Apex chert (a variety of quartz) at Marble Bar in Western Australia. The fossil colonies are cyanobacteria (sometimes called blue-green algae), which built reefs. There are several other locations on Earth showing microfossils of a similar age.

Taking yet another step back, the ages of Earth and the solar system of which it is a part, are thought to be just over 4.5 billion years, in a universe thought to have started with the Big Bang about 13.7 billion years ago. Now there is some speculation about "other universes". Although the term "universe" was originally meant to include absolutely everything, it seems quite reasonable to speculate that other universes may also have started with a Big Bang at other times, and may exist as entities completely inaccessible to us and probably unobservable by us.

– Of course there are other universes! Do you still think I don't exist?

– Can you imagine me standing up at a conference of cosmologists and saying that I know there are other universes because I have been talking to a disembodied voice that hails from one?

– As a matter of fact, I can imagine it. Indeed I have been conversing with someone who did just that!

– And . . . ?

– You are right. He was not believed to be a credible witness.

– I assume I need say no more.

I said this rather too tartly, perhaps because I was guiltily aware that I had not answered his question. I had not yet fully settled in my mind whether or not Davey really existed, so I hastily picked up the thread of what I wanted to say.

– The philosophies and belief systems of my forebears are strongly biased by their understanding of time as well as by their understanding (or lack of understanding) of the world around them. For example, it is much easier to believe that humankind is extra special, and generally above the rules that apply to other mammals and to other life forms if you believe that the universe was created only about six thousand years ago (i.e. long after the evolution of *Homo sapiens*), than if you believe that our universe is about 14 billion years old. In the absence of any strong evidence for one position or the other, one must expect that humans would be naturally biased to believe that they were special.

I have thus found it helpful to learn about the gradual development of our understanding of time, not just because the history is interesting but also because it helps us to understand the origins of some of our belief systems.

For by far the greatest part of the history of *Homo sapiens*, the boundaries of time measurement have been set by oral or written records, passed down to succeeding generations. Many, if not most, of our great philosophers and religious leaders lived and died long before we had the benefit of understanding the history of other species as recorded in the fossil beds (an understanding that roughly dates from the remarkable work of William Smith in the early 1800s in England), and without the detailed knowledge of the ages of rocks and some cherished human artifacts. It is only during my lifetime, with the discovery of radioactive dating techniques, and even more recently the finding that some changes in genes over time can be used to measure the passage of time, that humankind has been able to learn and to wonder at the vastness of time from a human perspective. Of course the findings of Charles Darwin and William Smith in the nineteenth century certainly provided strong indications that Earth might be a lot more than six thousand years old, but precise measurement of the age of most artifacts was simply not possible. Had the prophets of yore had that knowledge, and had they shared the insights available to people of my generation on the workings of the human brain, and of the power and complexities of evolutionary theory, their writings would surely have been very different. To begin with, they would have been much less likely to place man at the centre of the universe with a prestigious role as the most-beloved species watched over by an omniscient and omnipotent god.

I have fallen heir to one of several family bibles passed down from clergyman ancestors. Like many old bibles, on the first page of Genesis, there is the annotation "Before Christ 4004" beside the opening text, which reads, "In the beginning God created the heaven and the earth". Behind this annotation lies a fascinating historical footnote.

Figure 5.1 – *Archbishop James Ussher, after a portrait by Sir Peter Lely, circa 1654.*

Archbishop James Ussher (or Usher, as his name is sometimes spelled) was born in Ireland in 1581, and died in England in 1656. He had been a professor and subsequently vice-chancellor of the then newly established Trinity College in Dublin before becoming primate (or head) of the (Anglican) Church of Ireland in 1625. These were adventurous (that is to say dangerous) times for a cleric in England and Ireland as there were ugly conflicts between supporters of the Church of England, Calvinists and Roman Catholics. Ussher, who leaned toward Calvinism, was heavily embroiled in the conflicts and published several tracts in defence of the Irish wing of the Church (which was Calvinist). As a result, the King viewed him with disaffection. Today, his best-known work, published late in life, is *Annales Veteris et Novi Testamenti* (Annals of the Old and New Testament). In this work, he sets out a time scale for biblical events, starting with the creation of Earth in 4004 BC (October 23 at high noon to be precise). Stephen Jay Gould has written a rather nice article on Ussher that appears in his book, *Eight Little Piggies: Reflections in Natural History*.[14] In it he scolds a number of writers of geology textbooks who pillory Ussher for being so narrow-minded as to use the bible (though it turns out he also relied on other historical sources) to come up with such a ridiculously low number for the age of the universe.

Gould points out that Ussher was actually a rather broad-minded man who used any sources available to him to help in his task. Gould claims that Ussher's estimate was simply considered to be the best of many competing estimates of the age of the universe in his day (and long after), primarily because he was prepared to examine all available sources. To be sure he made use of the copious data in the Old Testament about who begat whom and how old they were when the birth occurred, but this data still left some significant gaps. To close in on some of these gaps, Ussher referred to the works of secular historians, took account of the historical possibility that Christ was actually born in 4 BC (that is, before the death of Herod), and accounted for changes resulting from the adoption of the Gregorian calendar. Gould has written that Ussher's figure for the age of Earth was actually a good deal higher than almost all other contemporary estimates, but this contention is open to serious doubt.

In *As You Like It*, Shakespeare has Rosalind observe to Orlando that "[t]he poor world is almost 6000 years old". It is thought that the play was first performed in about 1600. Shakespeare died in 1616. Ussher was only nineteen in 1600 and didn't die until 1656. His estimate of the age of Earth was completed in 1654. It is just possible that Shakespeare learned of Ussher's estimate for the age of Earth some time after his play's first performance, and added the line in for a contemporary audience for whom Ussher's estimate was new, though as yet unpublished, news, but it seems a little unlikely. The facts as told in Wikipedia under the heading "Young Earth Creationism" strongly suggest that there were many estimates for a young Earth made between 1000 and 1700 AD. About fifty different estimates are listed in the article. These, in the main, were in the range 6000–3000 BC. Isaac Newton and Martin Luther, for example, weighed in with estimates of 4000 BC and 3961 BC respectively. Recall that Ussher was a senior prelate in the Church at the time that that wonderful literary masterpiece, the King James translation of the Bible, was being produced. It is hardly surprising that Ussher's estimate was the one used by the translators. Nor is it surprising that he should, as a result, be the target of opprobrium from a variety of scientists now in possession of much more accurate information about the age of the Earth. Shakespeare's reference to the planet's age doubtless came from one or more of the many similar estimates current at the time.

It is important to remember that in Ussher's lifetime, only about five hundred years ago, many tools now available for dating artifacts and events were unknown. In my generation, many people still either ignore or outright disbelieve the scientific evidence relating to the age of the universe and of Earth. The dissonance between the book of Genesis and what scientists have learned about our genesis is explained by most clerics by asserting that Genesis was always intended as a poem or fable of creation and was never meant to be taken literally. This assertion is a harmless example of historic revisionism, but it is certainly true that these days, most religious people have found a way to believe both the scientific discoveries and the religious texts, even when the two are incompatible. Those who have refused to do so have found themselves faced with some serious contradictions to resolve.

One of the falsehoods I was taught at school was that the oceans were getting gradually more and more saline, as Earth's rivers continuously carried small concentrations of salt to the sea. We were told that the salt gradually increased in concentration in the ocean, while the water that carried the salt from the land was recycled by a process of evaporation, formation into clouds and then precipitation back onto land as rainfall. It was not until quite late in life that I learned that this interpretation of events is wrong. Our oceans are not getting saltier. As it happens, again according to a perceptive and interesting chapter in *Eight Little Piggies*, this particular misunderstanding has been around for quite a few centuries. A distinguished British scientist, Edmund Halley (he after whom Halley's comet is named), tried to use this assumption as a way of estimating the age of Earth. Halley was born in the same year that Ussher died (1656) and is remembered principally for his contributions to astronomy and for his work on magnetic variation. He was Savilian Professor of Geometry at Oxford University, Astronomer Royal, and a distinguished member of the Royal Society.

Halley lamented the fact that measurements of the saltiness of the sea had not been made by the ancient Greeks and Romans so that he could compare the salinity he measured with that of ancient times, and thus derive a number for the rate at which oceanic salinity was increasing. Then, assuming that the rate of salinity increase was constant, and that the ocean started out as fresh water somewhere near the birth of Earth (both assumptions open to dispute), he reasoned that it should be possible to calculate the age of the oceans, and by extension, the approximate age of Earth. Lacking the all-important early measurements, he never made that estimate himself. However, in the late nineteenth century, Irish geologist John Joly did make an estimate using Halley's method and came up with 100 million years as the age of the oceans, and thus, he assumed, of Earth – still a long way from current estimates of about 4.5 billion years, but also over ten thousand times closer than James Ussher's attempt.

For reasons that are only fairly well understood, despite significant variations of salinity in the world's oceans, overall, the salinity of the oceans appears to have reached a stable state. While salt is continually carried to the oceans by rivers, it is also constantly being used up by various life forms and, it is now believed, at least partially stripped from the ocean water inside hot vents. These vents are located on the ocean floor where tectonic plates are drifting slowly apart, permitting ocean water to penetrate deep into the hot ocean floor before being expelled – usually sulphur rich but relatively salt-free – back into the ocean. It has been estimated that all the water in our oceans is flushed through these hot vents on average once every 10 million years.[15]

Gould draws the following insightful lesson from this classic case of drawing incorrect conclusions as a result of faulty assumptions:

> *The best signs of history are objects so complex and so bound in webs of*
> *unpredictable contingency that no state, once lost, can ever arise again in*
> *precisely the same way. Life, through evolution, possesses this*
> *unrepeatable complexity more decisively than any other phenomenon on*
> *our planet. Scientists did not develop a geological time scale – the*
> *measuring rod of history – until they realized that fossils provided such a*
> *sequence of uniquely non-repeating events.*[16]

While there are aspects of Gould's statement that I don't quite agree with (which I will discuss later), I was struck by the beauty of the insight. Most of the physical processes we observe have long ago reached some kind of natural equilibrium and so cannot be useful to demark time. The Halley ocean salinity test provides a nice example of this. However, life forms keep changing in unpredictable ways, and the odds of them reverting to a previous form are negligible. As a result, we can confidently associate an era with the existence of particular life forms, preserved as fossils. Perhaps there will be a time when human fossils will be useful to date soil strata from our era, in which case we had better not all opt to be cremated!

Gould's observation about the use of fossils to date the age of rocks refers to research pioneered by British canal builder William Smith in the early 1800s. Smith discovered that many of the layers of rock that he had to dig through to build canals contained fossils and that the fossils tended to be different in each layer. As he toured about England and Wales, digging more and more canals, he found that he could correlate the layers he encountered at different sites by the similarity of their fossil content. In short, he could determine a relative time at which a rock stratum was laid down by the type of fossilized animals it contained, and could thus correlate the relative ages of the various strata he encountered throughout England.

Smith's personal story is an interesting one and is ably told in a book by Simon Winchester, *The Map that Changed the World*. Over a period of twenty years, this observant and self-taught son of an English blacksmith compiled a large map of England and Wales depicting the relative age of the strata to be found near the surface. This map was published in 1815 and caused a small sensation. The credit he received for all his work was initially disappointing to say the least. Smith paid a huge penalty for not being a recognized member of the scientific establishment. His work was plagiarized, and he spent time in debtor's prison before finally, in his later years, enjoying the recognition he deserved.

Smith's discoveries did much to promote the scientific field of palaeontology – the study of life in the prehistoric past. There is now a large body of knowledge relating to long-extinct life forms. These studies demonstrate conclusively that humankind is very much a newcomer to the parade of life forms over the past few billion years. In common with most other life forms, many of which have disappeared, its future is by no means assured. This was a message that the Western world of Smith's time was ill-prepared to receive, since it was still fixated on the idea of spontaneous creation as outlined in the Old Testament, and buttressed by Archbishop Ussher's precise calculation of the time when that creation occurred.

Palaeontology has allowed scientists to determine the relative age of various rock strata, but it was a long time before non-specialists accepted its basic message about the existence of life long before 4004 BC. In part this was because palaeontology did little to help us determine the actual age of a given rock or relic. Our ability to make such identifications came much later (for the most part after the Second World War) with the discovery that radioactive isotopes of some commonly found elements could be used to determine the age of a variety of substances.

– Well, those were some quaint stories you told me. You humans seem to have had a lot of difficulty sorting out the age of things. It is not a problem I had thought about much before listening in on you and others. However, I now suspect it had much greater importance to my vanished friends than I had realized.

But I missed something in what you said. Can you tell me what radioactive isotopes are?

– I'm surprised you don't know! Some elements have variants (called isotopes) that are not stable and that decay into other elements.[*]

– How strange. You mean that the ninety-two natural elements, which I thought were the basic building blocks of your planet, are not always stable?

– That's right. Furthermore, the rates of decay vary widely from one unstable variant of an element to another. This turns out to be very useful for calculating time lapses. Don't you have radioactivity in your universe?

– Not that I am aware of. This is quite a surprise for me. I need to learn more.

– And so you shall.

I tried to mask my astonishment that Davey's universe had no

----

[*] For a longer discussion of isotopes, see Appendix 1.

radioactive elements. In fact, while I was prepared to talk about the application of radioactivity to the dating of old artifacts, I had not guessed that radioactivity itself would be a mystery to him. So, I began at the beginning with a quick summary of the discovery of radioactivity in 1896.

– Henri Becquerel was awarded the first-ever Nobel Prize in Physics in 1901 for his discovery of radioactivity. Marie and Pierre Curie and Ernest Rutherford were subsequently awarded Nobel prizes, too, for their follow-on research that yielded a basic understanding of the phenomenon. The world of the early 1900s was just as surprised and mystified as you apparently are today to encounter the phenomenon.

By their very nature, radioactive elements are unstable. They give up energy in the form of radiation and *decay* into the isotope of another element. That new isotope may also be unstable, in which case it, too, will decay. Ultimately, the last radioactive isotope in the chain will decay into a stable (i.e. non-radioactive) isotope.

The first watch I ever owned had the numbers on the watch face as well as the hour and minute hands painted with a special luminescent paint containing traces of radium. The radioactivity in the radium excited luminescence in the zinc sulphide with which it was mixed. As a young boy, I loved looking at the watch face in the dark, and wondered at this miraculous appearance of light without an obvious power source.

For health reasons, the use of radium-powered luminescent paints was banned in most countries in the two decades starting in 1960. The deleterious health effects were first noted in the women who painted the watch face dials, many of whom licked their paintbrushes to give them a sharp point and subsequently developed cancerous growths on their tongues. Today's luminescent paints are either powered by tritium, an isotope of hydrogen, which decays much faster and more safely than radium, or they are pumped by external radiation sources, such as ultra-violet light. The luminescent face on my watch today has only a very feeble glow as compared to that of my first watch!

Although clocks and watches with radium paint dials have now virtually disappeared, there is another kind of radioactive clock that has revolutionized our measurement of historical time.

As already mentioned, fossils provide a marvellous record of the comparative ages of different sedimentary rock strata. Using the simple assumption that a layer of sedimentary rock that is on top of another layer is therefore more recent in origin, it is possible to establish the relative ages of different fossils by correlating layers of similar fossils that appear around the world. As Gould so nicely pointed out, the very complexity of the evolutionary process means that the same animal has, for practical purposes, no chance of evolving twice in the same form. This makes comparison of fossils to other fossils found elsewhere in the world an excellent way of determining the comparative age of a rock or a fossil. The process is aided by the finding that in the early days of Earth's history there was only one, large, contiguous landmass, making it much easier for one animal to appear in many divergent places.

However Gould went too far when he wrote: "The best signs of history are objects so complex and so bound in webs of unpredictable contingency that no state, once lost, can ever arise again in precisely the same way."[17] For while the fossil record – the pre-eminent exemplar of this phenomenon – is very good at giving us relative time, it is of very little use in telling us exactly how old a given fossil might be. For that, we need a clock with a regular beat that leaves a record of how long it has been ticking away since the clock was "started".

Radioactive dating (or radiometry) provides just such a facility. Used in combination with the fossil records, it has enabled us to gain a spectacular insight into events that took place long before the arrival of humankind on Earth, providing some humbling news about our place in the world's history.

There were two main streams of radiometry development. Geologists were most interested in dating the age of rocks (whose ages are typically millions [and occasionally billions] of years old) and started work soon after the discovery of radioactivity in 1895. They tended to work with isotopes with long periods of radioactive decay, measured in millions or billions of years. Their estimates were rather inaccurate until after the Second World War, when much better instruments for measuring radioactivity were developed.

The second main stream of research focussed on isotopes whose radioactivity decayed much faster. These substances decay too quickly to be useful for measuring time scales as long as billions of years but very useful, for measuring much shorter time frames. Radioactive carbon-14 is the best known of the elements used for shorter-term dating, but it is not the only useful isotope applied to this task.

The undoubted pioneer in developing radioactive carbon–dating techniques was Willard Frank Libby (1908–80), an American chemist. According to Libby, carbon-14 and tritium dating had their origins in studies of the effects of cosmic rays on Earth's atmosphere.[18] In 1939, just before the Second World War, Libby developed an instrument that permitted him to detect very low concentrations of carbon-14 and tritium – respectively radioactive isotopes of carbon and hydrogen. This instrument allowed him to discover that interactions of cosmic rays with the upper atmosphere led to the generation of about two neutrons every second over each square centimetre of Earth's atmosphere. The neutrons thus created are trapped by nitrogen atoms, which then decay, for the most part into radioactive carbon-14 plus stable hydrogen. One per cent follow an alternative decay route that leads to the production of the most common (and stable) isotope of carbon (carbon-12) and an unstable isotope of hydrogen (i.e. tritium).

The atmospherically generated carbon 14 becomes mixed with stable carbon-12 and oxygen, and is incorporated into carbon dioxide in the air, where it is taken up by plants and, later in the food chain, by animals. The net result is that in living plants and animals, there are about 14 counts (or disintegrations of carbon-14) per minute per gram of carbon present.

But radioactive elements are elements in transition to becoming other elements, sometimes stable forms of other elements, sometimes yet more unstable isotopes. In the case of carbon-14, the radioactive element is on its way to becoming nitrogen once again, but it takes its time to decay. In fact, given a container full of carbon-14, it will take 5730 years (±40 years) for half of it to decay. Another half of the remaining carbon-14 will decay in the next 5730 years, and so on. Hence, 5730 years is defined as the *half-life* of carbon-14. (Incidentally, the term "half-life" was first coined by Sir Ernest Rutherford while a professor at McGill University.)

Stable carbon in rocks, which has generally been in place for time periods of the order of at least millions of years, will not show any detectable radioactivity since any carbon-14 present when the rock was formed will long since have decayed to undetectable amounts. However, remnants of organic material that date from times up to about fifty thousand years ago (i.e. about nine half-lives of carbon-14 decay) will still have detectable radioactivity. Life forms only take in carbon for as long as they are alive. Once the life forms die, carbon dioxide uptake from the atmosphere ceases, and the existing radioactive carbon decays at a fixed rate, hence providing a signature identifying its age.

One of the best-known applications of carbon dating that I am aware of was finding the age of the Shroud of Turin, which purportedly was used to cover the body of Jesus when he was taken down from the cross. Carbon-14 dating carried out in 1988 established that the shroud came from cloth woven between 1250 and 1400 AD, long after the death of Jesus. It needs to be said that this estimate has not been accepted by all experts on the grounds that the small samples of the shroud used to determine its age may not be representative of the age of the whole shroud. The carbon-14 dating technique is now in wide use by archaeologists and anthropologists. It will likely continue to be a workhorse for artifact dating for many years to come.

While carbon-14 is only useful for dating material up to about fifty thousand years old, experts are now using many different radioactive elements to measure time frames ranging from a few years up to the age of the universe, i.e. about 13.7 billion years.

I have already mentioned that the age of our solar system is about 4.6 billion years, without explaining how this number was derived. In practice, the age of the solar system is derived from radiometric dating applied to many meteorites, as well as to rock samples from the moon. Because the Earth is in a state of continuous change, it has to this date not been possible to find many rock samples older than about 3.5 billion years. However in 2001 a small fragment of the mineral zircon found in Australia was age dated at 4.4 billion years. The oldest large samples of bedrock have been found on the eastern shores of Hudson Bay in northern Canada. Some samples of that rock are as much as 4.28 billion years old.

Amongst other key dates established through radiometry are the disappearance of dinosaurs at the end of the Jurassic age 70 million years ago, and the first appearance of life on Earth about 3.45 billion years ago.

So, you can see that radioactive dating, when used in combination with the fossil records, has enabled us to gain a spectacular insight into events that took place long before the arrival of mankind on Earth, providing some startling news about our place in the world's history. Some humans find such a diminution of human centrality in the history of the Earth hard to reconcile with their religious beliefs. As a result these scientific findings quite frequently come under attack.

Davey broke in almost immediately. He was agitated and incredulous.

– But surely that is extraordinary! Why wouldn't everyone want to know as much as possible about their own origins?

– Actually, it is exactly what you should expect. Remember that we are the only animals in our planetary system with any curiosity at all about our origins. Our curiosity is almost certainly primarily motivated by an innate drive to seek knowledge that will enhance our pleasure and our odds of survival.

But what if that knowledge diminishes our pleasure, upsets our most hallowed beliefs, or threatens our economic or physical security? In such circumstances, our innate curiosity can evidently be easily vanquished by a strong urge to deny "inconvenient truth". Human history is full of examples of such denial.

– In that case, it is a wonder that humanity has survived so long. Surely it must be doomed if it doesn't change! It is one thing for a few astronomers to be imprisoned or killed because the authorities cannot adapt to the idea that Earth orbits the sun and not vice versa, but quite another to deny, for example, the real history of your species.

– You may be right. I wonder if your own long-lost friends in your universe have any useful lessons for us in that regard.

– We'll come to that in due course. In the meantime I must tell you that your words have been of great interest to me. I had not imagined a world where it is possible to date the origins of objects going back billions of years.

– And I cannot imagine a world where one has no idea how far back one's past extends. Did your superhumans really have no idea how they evolved? Did they not study history?

– Yes and no. I mean, I think so.

– What is that supposed to mean exactly?

Now I was really paying attention. I sensed the possibility of a chance to get a glimpse of life in Davey's universe that heretofore he had deliberately kept hidden. Davey spoke.

– Well, as I told you, we went through a horrendous period. First there was a huge amount of information lost beyond the possibility of recovery – a situation we had thought to be impossible, since we went to great lengths to preserve important information in many places. This disaster was very sad and very costly. But the truly horrendous result was that it led to the death of all superhumans. Not a single one survived. It was an unimaginable catastrophe!

I was stunned. The more so as I was just learning to detect Davey's emotions as he spoke, and I could feel how very deeply saddened he was. Then suspicious thoughts seeped into my mind.

– I don't suppose you or your colleagues were involved in any way in what you imply was an event of mass destruction?

Davey's response surprised me.

– How could you ask that? I told you I loved the superhumans more strongly than you on Earth can begin to imagine.

I immediately wondered how he could possibly know that his love for his friends was stronger than any love we experience, but I dared not ask. I remained silent, eyes downcast.

Davey was quiet for some time, as well. Eventually he appeared to rouse himself.

– Let's meet again at the same time next week.

I hesitated.

– Before you go, I wonder if a table of key dates would help you to get a feel for the critical time frames of the development of our universe, of life on our planet and of our civilization?

– Yes, but you will have to read out the columns and rows to me. You have no idea how frustrating it is to listen in on all those academic conferences where the participants apparently project something called a "PowerPoint" and then mumble out a few incoherent sentences to elaborate on what I cannot see in the first place.

And by the way, your references to useful books are very irritating to me. As you know, I cannot read them.

– But you can always talk to the authors if they are alive. Bill Bryson, for example. Why don't you go and talk to him? He is obviously intelligent, and he has a good sense of humour.

For the first time, I heard something resembling a hollowed out sigh fill the room.

– I suppose it is hard for you to understand, but I did start to listen in on Bryson shortly after he published his book about "nearly everything". I don't even want to think about how many interviews on radio and television he had. His voice was coming at me from all directions, but most of the time the questions he was answering were unhelpful and were posed by people who had not read his book. So, finally I dropped in on him much as I did with you. He was in bed at the time, and he was furious with me. He told me to get lost, go back to where I came from by way of a place called Hell, and never to haunt him again. He seemed not to have much of a sense of humour after all – though you will appreciate that it is difficult to appreciate a sense of humour when there are no shared instincts or culture involved. I believe that most humour arises from a conflict between instinctual behaviour and behaviour dictated by cultural imperatives.

– Sorry it didn't work out. I hope you weren't bored by my discussions of time and space. I am convinced that our evolving understanding of those two important concepts has been key to the development of religion and philosophy over the ages. An essentially static universe with Earth at the centre leads to very different conclusions about our place in nature from what we know today about our universe. Similarly, a world that is only six thousand years old necessarily lends a completely different hue to history than a universe which is over 13 billion years old, with intelligent humanity around to appreciate it only for the last 160,000 years or so.

– I appreciate your point, but what about mass, force and energy. Aren't they key concepts, too? Time and space have a certain aesthetic appeal, but without living mass, such as you, what meaning can they possibly have?

– Of course, you are right – and that is a really tough problem for we humans to get our collective minds around. In our next session I will give you a whirlwind tour of the subject, but don't expect that you will come away completely satisfied. Despite some early attempts by such brilliant scientists as Isaac Newton and Albert Einstein, both of whom provided theories that explained beautifully the world as it was known in their time, our expanded view of our universe, illuminated by a vast new store of astronomical data and buttressed by complex particle physics experiments, has shown us that their theories are inadequate to completely describe our universe as we now observe it. To make the data fit our theories, we have to "invent" (most scientists would say "postulate") great quantities of dark (i.e. unobservable) matter and energy, without any firm idea of their origin. Dark matter and dark energy stand as the two greatest beacons of human ignorance in our time.

– Perhaps, but just think how bored you would all be if you knew everything!

After I had carefully read out the contents of Tables 5.1 and 5.2 reproduced on the next two pages, we finished our dialogues on time and space, with the promise of a dialogue on matter and energy planned for the following week.

After listening to me read the tables, Davey's only comment was to say how fortunate it was that he did not visit our universe one or two hundred years ago, or he would have learned nothing of interest!

Table 5.1 - Some Perspectives on Time

Age of the Universe	~13.7 billion years	that is 13,700,000,000 years!
Age of the Solar System	~4.56 billion years	including the age of planet Earth
First appearance of life on Earth	~3.8 billion years ago	see Dialogue 15.
Period of rotation of the Sun in our Galaxy	250 million years	
First appearance of mammals on Earth	~125 million years ago	
Ages of bright stars in Orion's belt	6 -11 million years	
First appearance of man (*Homo erectus*)	~1.8 million years ago	cf. Dawkins, *The Ancestor's Tale*, p. 66
Earliest known figurines	800,000 - 220,000 years ago	Venus of Bereket Ram, Golan Heights, Israel and the Venus of Tan Tan, Morocco
First appearance of *Homo sapiens*	~160,000 years ago	cf. Dawkins, *The Ancestor's Tale*, pp. 62-65
First daughter of "Eve" is born	~143,000 years ago	see Appendix III - Everyone's Ancestors
Earliest known cave art	32,000 years ago	Chauvet, France
First migration of man across Behring Strait to North and South America	??	10,000 to 60,000 years ago
Earliest known pottery	13,000 years ago	Japan
"Dawn of Recorded History"	~7000 BC	Crete
"Dawn of Recorded History"	~3000 BC	Sumeria
"Dawn of Recorded History"	~3000 BC	China
Hinduism founded	~2500 BC	
Judaism founded	~1000 BC	
Buddhism and Taoism are founded	~500 BC	Buddha Siddhartha Gautama (563-483 BC) Taoism started about 500 BC, but was not formalized until about 500 years later.
Socrates dies in Athens	399 BC	
Aristotle's lifetime	384-322 BC	Tutor to Alexander the Great for 5 years, many of his writings were rediscovered in the late middle ages and have had a major influence on Roman Catholic Church doctrine.
Christ is born	0 AD	Possibly 4 BC (i.e. before King Herod died)
Mohammed is born	~570 AD	
Shakespeare's *Hamlet* first performed	1601-1602	
Bishop Ussher finishes estimate of the age of the Earth	1654	Ussher concluded that creation occurred at noon on October 23, 4004 BC (see Dialogue 3).
Isaac Newton completes formulation of Theory of Gravity	1686	Work started in 1666.
Darwin and Wallace publish Theory of Evolution by Natural Selection	1858	
Quantum Theory is born (Max Planck)	1900	Developed further over the ensuing 50 years with many different contributors.
Discovery of the Structure and Function of DNA	Feb. 28, 1953	see Dialogue 11
Preliminary Understanding of Long Term Memory function in the brain	2000-2005	see Dialogue 18

Table 5.2 - Some Perspectives on Distance Measurements		
Planck length	1.616199×10^{-35} m	The shortest measurable length. Approximate dimension of the vibrating "strings" in string theory. Not measurable in practice, the Planck length emerges from theory.
Diameter of the proton (hydrogen nucleus)	about 10^{-15} m	The diameter is measured by scattering electrons off the proton, and depends somewhat on the energy of the incoming scattering electrons.
Diameter of the hydrogen atom	50×10^{-12} m	i.e. 50 picometres
Diameter of a carbon nanotube	1×10^{-9} m	i.e. one nanometre
Diameter of DNA strand	$\sim 2 \times 10^{-9}$ m (2 nanometres)	Largest human chromosome (DNA strand) stretched out is 8.5 cm (8.5×10^{-2} m) long, or about 10,000,000 times its diameter.
Smallest transistor gate in commercial use (2012)	22×10^{-9} m (22 nanometres)	Estimate for the year 2020 is 5 nm (nanometres).
Typical virus	$\sim 10^{-7}$ m (100 nm)	Range from 20-450 nm (HIV virus 90 nm)
Wavelength of visible light	$(380 \text{ to } 740) \times 10^{-9}$ m	(380 nm-435 nm) violet, (625 nm-740 nm) red
Typical prokaryotic cell	$(1 \text{ to } 5) \times 10^{-6}$ m (i.e. 1 to 5 microns)	see Dialogue 13. A typical Eukaryotic cell (e.g. animal and plant cell) is 10 to 100 microns.
Average diameter of a human hair	100×10^{-6} m	i.e. 100 microns
Length of longest blue whale ever measured.	33 m	i.e. 108 feet long. Blue whales are the largest living animals.
Height of Mt. Everest	8,848 m	Deepest part of oceans (Mariana Trench) 10,911 m
Diameter of Earth	12,756 km (12.756×10^3 m)	Great Wall of China 6,400 km; Nile and Amazon Rivers about 6,600 km; planet Jupiter diameter 142,984 km.
Diameter of the Sun	1,390,000 km	Highest mileage recorded by a car - 4,200,000 km (a 1966 Volvo still going strong). That's three times the diameter of the sun!
Distance light travels in one second	299,792,458 m	i.e. approximately 300,000 km
Distance from the Sun to Earth	149,000,000 to 152,000,000 km	Light from the sun takes about 8 minutes to reach Earth.
Distance light travels in one year	9.46×10^{15} m (one light year)	Milky Way galactic disk is 100,000 light years across. Andromeda galaxy 2,500,000 light years distant.
Lower bound for the size of the universe.	2.4×10^{27} m	This is just a lower bound. It is likely that the universe is very much larger, possibly even infinite.

Sources: The two primary sources of information for this table are: (1) http://www.falstad.com/scale/, and (2)http://en.wikipedia.org/wiki/Orders_of_magnitude_(length), as accessed on 29/1/2013.

A Visit to the Boundaries of Human Understanding

Knowledge, a rude unprofitable mass,
The mere materials with which wisdom builds,
Till smoothed and squared and fitted to its place,
Does but encumber whom it seems to enrich.
Knowledge is proud that he has learned so much;
Wisdom is humble that he knows no more.
WILLIAM COWPER, *The Task*, Book VI, "Winter Walk at Noon".

It was time for my next meeting with Davey.

I had had a head-crunching week of preparation. Some of the models of our universe are too much of a challenge for most or perhaps even all of us humans to grasp. What, I wondered, would it be like for Davey? I was soon to find out.

He arrived, seemingly full of vigour.

– Well, well, well. Here I am back in the schoolroom and ready for my next lesson.

Who had he just been visiting? Someone in England I surmised, judging from the "well, well, well", which was a new turn of phrase for him.

– Shhhhhh! Not so loud, or my wife will come to find out who I am talking to. If she finds I am not talking to anyone, she may have me committed to an insane asylum!

– Ha, ha, ha! I wouldn't worry about that.

– All very well for you. You probably don't have a wife, or do you?

– Not really. At least not in your sense of the word. But don't worry. Margaret won't hear a word I say unless I open a talk channel with her.

– How on earth do you do that?

– Well, it is very easy if you know how. As I have already told you, I can locate any sounds in your universe with great precision. I can also direct my voice into your universe with similar precision, although there does tend to be a residual garble that can sometimes be heard by others.

– That's just great. So, Margaret will hear me talking out loud intermixed with a garbled mumble from you. She will be absolutely certain I have gone mad!

– So, all you have to do is to speak in a whisper so she doesn't hear you. I will still hear you quite distinctly. Or, if you like, you can go for a walk or a drive, and talk to me as we go along.

– That won't work. I have to refer to my notes quite a bit today. It is a tough subject for a mere mortal like me to teach. Let's just try the whispering routine. If it doesn't work, we will meet another day.

– Fine. Let's get started.

And start I did – in a whisper.

– There is an old story about a wise man from an Aboriginal tribe who was addressing a group of anthropologists. Like you when we first met, they were all anxious to know more about the "creation myth" of his tribe, so the man described in some detail how Earth as we know it is held up on the back of a turtle. "And what", asked one of the anthropologists, "holds up the turtle?"

"Well", the man replied, "the turtle is standing on the back of another turtle."

"Hmmm", said the anthropologist, who now had a bit of a glint in his eye, "and what do you suppose is holding up the second turtle?"

The man was quick to pick up the gist of the line of questioning.

"Oh", he said. "It's solid turtles all the way down!"

So, what does a modern "creation myth", based on extensive research by a lot of scientists, really look like?

We believe that our Earth is a satellite of our sun, and that our sun is a part of the Milky Way galaxy, one of several billion galaxies in the universe. That's a good start and is something we can see with our own eyes. But how was our universe created? Was it, as is currently believed, in one Big cosmic Bang? If so, might there be other universes? Are these other universes likely to obey the same physical laws as our own universe, or might they, for example, have a different value for the charge and mass of the electron, or of some other *fundamental constants* that have been the ultimate arbiters of the shape and structure of our universe? Some theoretical physicists believe that some fundamental constants were not so constant after all, having changed their values slightly over the life of our universe.

– Well, I told you last time we met that we have no unstable elements in our universe. From that you should have been able to conclude that some of our physical constants must be different, or we would have the same elements and the same chemistry that you have.

– I'm not yet totally certain that you exist. So, I can scarcely use your words as evidence!

You can tell I was feeling a little nettled! Davey noticed and immediately worked to calm me down.

– Quite right. I keep forgetting that the brains of you humans are subject to unbidden thoughts and visions, so you have to be very careful not to believe what you see and hear, unless it can be independently confirmed. Let's proceed.

– Very soon now I shall address Charles Darwin's monumental contribution to our quest to understand ourselves. As you almost certainly know already, Darwin believed that we have all evolved from the most primitive life forms through a process of natural selection, in which physical or mental capabilities that have species survival value are likely to be preserved, while changes that reduce the chances of survival have a tendency to die out.

It is difficult to see how understanding the origins of our universe will enhance our ability to survive. So, how could evolution through natural selection bestow on us, for example, the capacity to understand whether or not there are other parallel universes?

One answer is that it is well known that evolution frequently bestows properties that are peripheral to survival. Some genetic changes have no survival benefits, but neither do they materially reduce the odds of survival. Such changes simply alter our genetic makeup. It is to be expected that some such changes will subsequently prove to be advantageous. For example, it seems likely that our ability to write resulted from enhanced mental capacity more generally, combined with our manual dexterity, already a well-established inherited capacity. The upshot of our ability to record our thoughts and our progress has been nothing short of spectacular and has led many of us to view *Homo sapiens* as being in a class by itself relative to other animals. But we should not let this enormous success blind us to our probable limitations.

Let's think small for a few minutes.

Consider the ant. Edward O. Wilson, the famous Harvard entomologist, did for many decades and at some length.[19]

It has been estimated that there are at least a million billion (10^{15}) ants alive on Earth at any given point of time, and that there are thousands of different ant species, only some of which attend human picnics. Wilson describes ants as being "... in every sense of the word the dominant social insects". Almost all, if not all, ant species have several different castes in their societies, including a queen, some males, soldiers (which can be much larger than their sisters and brothers), and smaller and medium-sized workers. Wilson describes the morning hunting activities of one particular species of ant, *Eciton burchelli*, in the following way:

> When the light level around the ants exceeds 0.5 foot candle, the bivouac begins to dissolve. The chains and clusters break up and tumble down into a churning mass on the ground. . . . Then a raiding column emerges along the path of least resistance and grows away from the bivouac at a rate of up to 20 metres an hour. No leaders take command of the raiding column. Instead, workers finding themselves in the van press forward for a few centimetres and then wheel back into the throng behind them, to be supplanted immediately by others who extend the march a little farther. As the workers run on to new ground, they lay down small quantities of a chemical trail substance from the tips of their abdomens, guiding the others forward. A loose organization emerges in the columns, based on behavioural differences among the castes. The smaller and medium-sized workers race along the chemical trails and extend them at the points, while the larger, clumsier soldiers, unable to keep a secure footing among their nest-mates, travel, for the most part, on either side. . . . The smaller workers, bearing shorter, clamp-shaped mandibles, are the generalists. They capture and transport the prey, choose the bivouac sites, and care for the brood queen.[20]

Ants are not alone among insects in the highly developed innate nature of their social behaviour. Some species of termites build nests with special tunnels to provide air conditioning and heating as appropriate to keep termite larvae within a prescribed temperature range, and the dance of the honey bees to communicate the direction and distance to promising sources of nectar is now well known.

Change a few words here and there, and the quotation from Wilson could be a description of a raiding party by some remote human society, and yet experiments have conclusively established that almost all ant behaviour is inherited and not learned. It is enough to make us wonder just how many human activities are motivated more by innate behaviour than by our much-vaunted free will.

Ant societies have some definite boundaries to their understanding of the world about them. Exterior (and almost certainly, unimagined) forces can quite suddenly shatter the ant's world – a bulldozer blade or an elephant's foot on the nest, a sudden flood or a landslide.

So it is with humans and human societies, though our perception of just where those boundaries of understanding lie has undergone some very significant changes in recent years. We now think we understand how life evolved. We also understand the sources of earthquakes and violent storms. We know what fuels the sun and roughly how long the sun will continue as a benevolent source of energy. But there is still a lot that we don't know.

For me, as for many others, some of the most perplexing and complex boundaries of our understanding lie in the physics of the universe.

I am almost certain that if my science teacher had not taught me about Newton's law of gravity, it would never have occurred to me to wonder why it was I walked around on Earth and did not float up into the sky. Just like the vast majority of our ancestors, and, one supposes, all the other mammals and life forms, I would have lived my life without ever imagining such concepts as force and gravity. Furthermore, when I learned about Newton's equations, they came to me as just another item on a long list of things to learn about, and, if need be, parrot back during an exam.

Now I look at what Isaac Newton discovered in wonder. How is it that he looked about him – perhaps triggered by the famous apple he is reputed to have seen falling to the ground – and asked the simple question I would never have thought to ask: "Why did the apple fall down?" Of course, Newton went even further. He integrated the force of gravity into his laws of motion, identifying the gravitational force as identical in principal to the force felt in one's back arising from the acceleration of a plane down a runway, or, in his case, more likely a trotter moving a carriage swiftly away from a curb. What a stroke of genius!

Einstein, too, addressed questions it would not have occurred to most human beings to ask. Some famous experiments by distinguished scientists were showing that the speed of light is constant no matter what the relative motion is between the observer and the observed object. This is counterintuitive. If I throw a ball out of a speeding car in the direction of the car's motion, I expect that the initial velocity of the ball relative to the ground will be the sum of the velocity I give the ball by throwing it, and the velocity of the car. This is true for the ball I throw (though air friction rapidly slows it down), but, as you may know, not for the light in a flashlight beam I point in the direction of travel. In that instance, the speed of the light in the beam as measured by a stationary observer is the same as the speed measured by someone on a passing car going in the opposite direction, or indeed, as measured by me while speeding along in the car. In exploring the consequences of this very strange fact, Einstein evolved the Special Theory of Relativity in 1905.[*]

Figure 6.1 - *Two Geniuses, Two Theories of Gravity*
Sir Isaac Newton in middle age on the left; Albert Einstein aged sixty-eight at Princeton in 1947 on the right.

Eleven years later, he pushed much further into the realm of questions I would not have thought to ask, in the process adding new meaning and accuracy to Newton's gravitational equation. Gravity, he concluded, can be thought of as the curvature in space resulting from the presence of mass.

I beg your pardon? Space is curved? What kind of a concept is that? To

---

[*] Einstein's theory actually rested on two assumptions, the first being that the speed of light is constant. The second is that the laws of physics apply equally to any observer whether moving relative to other objects or not. It is conceivable that there may be instances when this last assumption is not true, though, as far as I know, no such instances are known in our universe.

most of us, space is space, stretching out in three dimensions. Curvature of space, unless in only one or two dimensions – such as the surface of a sphere, or the curvature of a line – has no obvious meaning.[21] Unnoticed by me, as well as by most others, was the unanswered question of what is really meant by the "force" of gravity. What is it that causes the force to be observable? We know it is not a piece of elastic tying two bodies together, so what can it be? And while we are on the subject, what causes oppositely charged bodies to attract each other, and similarly charged bodies to show mutual repulsion?[22] And what about the strange forces that hold the nucleus together? As you are quite likely aware, Einstein was only one of many who spent their lives endeavouring unsuccessfully to devise a coherent theory that would explain all these forces in what came to be called a "Theory of Everything".[23]

Now, in my later years, we are told that Einstein's four dimensions (three spatial dimensions plus time) are really only a convenient approximation of reality, although useful for almost all practical purposes. Some theoretical physicists now tell us that we live in a world of eleven dimensions, one of which is time, with the rest being spatial. Only three of the spatial dimensions are large, and thus visible to us. The other seven are very, very small *strings* (whose dimensions are of the order of the [tiny] *Planck length* of 10^{-33} cm) and are tightly curled up on themselves. In a wonderful book for the layman, *The Elegant Universe*, Brian Greene explains why he and his colleagues believe that ten spatial dimensions is the correct number, how the space can be described mathematically (Calabi-Yau space) and envisioned. He also explains how he expects pursuit of the very complex mathematics that are required to provide numerical rigour to the theory may lead us to be able to describe what the gravitational, electro-weak and strong nuclear forces actually are, thus at last realizing the dreams of Einstein, Newton and many others of explaining all forces using a single theory.[24]

Even a dim understanding of ten spatial dimensions is hard for most of us to imagine. I found one of Greene's many analogies especially helpful in this regard.

Imagine a hose stretched across a canyon that you can see off in the distance. The hose will appear to you as a line that is slightly curved.[25] This is a one-dimensional space. The position of an ant walking along the hose can be described by a single number, such as its distance from the left-hand end of the hose. Now imagine that you raise your powerful binoculars to peer at the hose, and you see that the ant can also walk around the diameter of the hose, although this second dimension is small and curled up, and not at all like the dimension along the length of the hose. You might even look again, and realize that ants are emerging from inside the hose. You might then conclude that there must be at least one more dimension curled up inside. Now, says Greene, consider the fundamental nuclear particles and imagine that in addition to the three spatial dimensions in which they observably exist, there are other tightly curled up and very small dimensions associated with these particles.

It still sounds pretty wild, doesn't it? But then these dimensions are so very small that none of our senses could possibly detect them, nor, if you believe that our senses are the product of an evolutionary development where only those faculties that had survival value predominated, is it to be expected that survival would depend on anything remotely close to an understanding of string theory and ten spatial dimensions. Just like the ants, for whom, from our perspective, there is so much of the world that is beyond their ken, is it not possible that there are some spatial dimensions that are likely irrelevant to our survival, and for which we have therefore not evolved any capacity to observe? Although these dimensions may be inaccessible to us by direct observation, is it possible that we could come across a few hints and clues occasionally providing scraps of evidence of dimensions we cannot properly ken?

This last proposition seems reasonable to me, based partly on some of the key non-intuitive discoveries of scientists, such as Einstein, Newton and Planck, but also based on what sensors one might expect humans and other animals to have evolved in their fight for survival over evolutionary time periods. Hence my admiration for those amongst us who have supplied verifiable answers to scientific questions I would never have thought to ask!

During my lifetime I have seen the theory of the universe evolve from an essentially static model in which stars were born, "lived" and died, to a model that posited our universe was spontaneously created about 13.7 billion years ago, has been expanding ever since and may or may not, at some time in the future, reach a point of maximum expansion, followed by a compression stage leading to annihilation – or, on the other hand (as is currently believed), may go on expanding forever. Spontaneous creation from a point source has always seemed to me to be a bit of a stretch, as it has to Arthur Eddington (who was quoted on this subject during our fourth dialogue) and to Einstein. String theory relieves some of the mental stretch implied by the Big Bang, by immersing our particular Big Bang in a "turbulent cosmic ocean called the multiverse," so that, in Eddington's (previously quoted) words, "the implied discontinuity of the divine nature" derived from the Big Bang can at least potentially be satisfied at a higher level as one of many Big Bangs in a "cosmic ocean".

String theory is by no means the only theory in contention to explain the workings of the universe at a fundamental level. Thus far the theory has been jigged so that it can explain most observed phenomena. However especially after Einstein's experience inserting a cosmological constant to explain what turned out to be an incorrect assumption that the universe is stable,* physicists are slow to accept a theory unless it can also make some predictions that can be tested. String theorists have so far been unable to meet this challenge in a manner that is widely accepted.

Some other leading theories do not require us to envision extra spatial dimensions that are inaccessible to us, but no theory has yet provided a widely accepted explanation of the observed properties of the universe. As our ability to explore the outer reaches of our universe has increased with the availability of scientific satellites, some challenging new phenomena demanding explanation have come to light.

Intense efforts to understand how galaxies formed have led to widespread acceptance of the existence of *dark matter*, matter which is undetectable by our available instruments, but which is needed to provide enough gravitational force to account for the formation of galaxies. Dark matter is believed to comprise 22 per cent of the total mass and energy in the universe. Recall that energy [e] can be equated to an equivalent mass [m] using Einstein's famous equation, $e=mc^2$, where "c" is the speed of light.

Even harder to comprehend is the likely existence of *dark energy*. Without it, the existing theoretical framework cannot otherwise explain the recently determined fact that the expansion of our universe is accelerating, when our best theories predict that gravitational forces should be slowing the expansion. Nor is dark energy a minor factor, as current estimates are that it comprises about 73 per cent of the total mass and energy in the universe. Between them, dark matter and dark energy thus comprise 95% of all matter and energy, leaving only 5% for the matter and energy that we know about and can explain!

One of the great attractions of pursuing research on dark matter and dark energy is that if we do succeed in understanding them, they lend further credibility to our current best theories about the known forces. These theories have gained a lot of credibility due to their ability to predict phenomena such as hitherto undiscovered particles and the mechanisms fueling the movement of stars, planets and galaxies in the universe. One of the very basic conundrums this research faces is that the Standard Model predicts the existence of a particle called a Higgs boson after a British physicist named Peter Higgs who was prominent amongst those who proposed in 1964 that such a particle was required if observed phenomena

---

* See Dialogue 4 for a brief reference to this occurrence.

are to be consistent with the Standard Model.

At the time, a major problem with this proposal was that the likelihood of ever seeing and identifying such a particle seemed very remote because of the huge energy output required of any particle accelerator that could make a Higgs boson identifiable.

However, in 2008 a new ultra powerful particle accelerator called the Large Hadron Collider (LHC) was commissioned at the European Centre for Nuclear Research (CERN) in Geneva, after more than ten years of construction. The single most important reason for the LHC was to search for the Higgs boson.

On July 4, 2012, two competing research groups, both based at CERN, announced that each had discovered a boson "consistent with" a Higgs boson. The wording was appropriately cautious. The Higgs boson is estimated to have a lifetime of about one zeptosecond (10^{-21} seconds) and would be accompanied by vast showers of unrelated particle interactions which obscure the event of interest. The discovery of the Higgs boson opens up new realms of exciting research. Some theories imply the existence of a whole family of Higgs-like bosons, some or all of which may interact with dark matter. Other theories implicate Higgs particles in a posited rapid expansion of the universe immediately following the Big Bang.[26]

An intriguing alternative explanation for the unexplained expansion of our universe is put forward by string theorists. They think that gravity may not be confined to the three spatial dimensions we are familiar with, and may be "leaking" into some of the other dimensions postulated by string theory, thus having a reduced effect in the dimensions we can observe. In this scenario, it is the reduced impact of gravity, rather than the countervailing force of the dark energy, that is driving the accelerated expansion of the universe. Yet others believe that our galaxy may inhabit a part of the universe that is much less dense than average, so that the velocity of expansion of the universe will vary from place to place, instead of being uniform throughout the universe, as most current theories assume. Clearly there is much yet to understand about our universe!

The Big Bang itself continues to be an intriguing enigma. While theories which explain the behaviour of the universe after the Big Bang are relatively well developed, the cause of the Big Bang itself is still unexplained.

One promising approach described by Neil Turok[27], the Director of the Perimeter Institute for Theoretical Physics in Waterloo, Canada, envisages the universe as a cyclic process which, after its expansion phase then contracts down to a very small (but not infinitely small) size, followed by another expansion phase. Whether or not there is more than one universe, the idea that our universe might be oscillating between two extremes has some intuitive appeal, though the complex mathematics involved in all these cosmological theories render further enqiry forbidding for all but the highly trained expert.

Another enduring challenge to our understanding lies in the field of quantum physics. Quantum theory seems like an essential component to our understanding of our world, since it is an integral part of the Standard Model, which provides an excellent explanation, accurate to many decimal places, of observed phenomena. In particular, it is brilliant in explaining the strong and weak nuclear forces, as well as the electromagnetic forces, but, as already mentioned, it has yet to be reconciled with what we know about gravitational force. In this latter domain, it is still Einstein's theories of Special and General Relativity that govern. Attempts to reconcile the two approaches into a combined "Theory of Everything" have failed miserably thus far.

For most of us, the most confusing aspect of quantum theory is the concept of *non-locality*, that is, the prediction, confirmed by experiment, that two widely separated particles can be "entangled" in such a way that changes in the state of one particle immediately affect the other, without there being any observable force to effect such a change. Furthermore, the changed states seem to take place simultaneously, which is to say that the communication between the two seems to be faster than the speed of light, which, according to relativity theory, is the fastest speed at which such communication can occur.[28]

This property of quantum entanglement is currently being vigorously investigated by researchers intent on developing, for example, ultra-high-speed quantum computers, but even the most brilliant minds (notably including Einstein and Niels Bohr) seem to have had the same difficulty the rest of us more intuitively experience in reconciling the observed fact that particle entanglements over very large distances really occur, perhaps even extending right across the universe, with the lack of any observable force to give effect to the entanglement.

In the circumstances, the idea that the force or forces from which the entanglement results might operate in one or more dimensions inaccessible to us has some attraction.

So, how about you, Davey? Do you have entanglement in your universe? And if so, do you understand it? Is it possible that entanglement can happen between particles in different universes, and if so, is that perhaps how you are able to communicate with me?

– That's a lot of questions to ask all at once!
    First of all, yes, I am familiar with the idea of entanglement.

Indeed I observe it all the time. It is an important phenomenon in my universe. I may lack your ability to review past history, since we lack radioactive isotopes, but we do have useful features in our universe that you seem to lack.

For example, you could not have known that in my universe we live in six spatial dimensions, and entanglement works in only three of them. It is very hard for me to imagine anyone living in only three spatial dimensions, and of course, since I can hear but not see you, my imagination gets no help from just listening to you talk. As far as I know, entanglement of particles between universes is possible between some, but not all universes. Indeed, I believe it is quite rare. However, it is possible between your universe and mine, a situation that has made communication with you humans possible for me.

Our universe is undoubtedly much more stable than yours, possibly as a result of having those three extra dimensions. I believe our past was very like our present.

– You come with very exciting news today! What is it like in your universe?

– What do you mean? It all seems pretty normal to me.

I was dumbfounded by his answer! Six dimensions! Everyday particle entanglement similar to what we call quantum entanglement! This hardly seemed normal to me. It took me a few seconds to see my question from his perspective, and even then, I was a little surprised that he had failed to see it from mine.

– Sorry, but you have to understand that what is normal for you is not at all normal for me.

– If only I could see you, it would probably be obvious, but sadly I cannot.

– Okay. Try this! Do you have weather? Trees? Sunlight? Nighttime? Animals?

– I think we have some, or possibly all, of these things. But it is sometimes hard for me to figure out what is our equivalent of those things you call weather, trees and sunlight, for example. It is harder still to understand how they might resemble what may be our equivalents. With no electromagnetic radiation and six spatial dimensions, when you have only three, resemblance is a hard word to define.

But I might be able to help you understand what you call non-locality. Let's go back to your example of the hose suspended in the distance on which, with binoculars, you are able to see ants crawling. Instead, let's assume that you see two trains approaching each other from opposite ends of the suspended hose, and apparently running on tracks on the hose. A head on collision seems certain to be about to occur.

But what you are looking at (essentially) is a one-dimensional hose. However, unseen by you are two pairs of tracks on a flat ribbon, rather than what looks to you like a (one-dimensional) hose. To your amazement, the trains roll right past each other. Ah, you may say to yourself, this is amazing. The trains just passed right through each other without any impact! The second spatial dimension is unobservable by you, though it can be inferred as one explanation for the fact that the trains passed each other without colliding. I can tell you that the quantum non-locality observation is the result of you not being able to envisage other dimensions through which a local force can operate.

– An interesting idea. I wonder if what you say is correct. Maybe our two universes have no relevance at all to each other.

– I totally disagree. Several things you have said already have helped me to understand the plight of my superhumans, as I have already told you. That is what makes what you are telling me so very important!

I can tell that this subject was a difficult one for you to try to explain to me. From listening in on conversations on Earth I have learned that many humans do yearn to know more about their origins.

And now you even have me wondering how I came about, and why our universe is the way it is! Alas we don't seem to have any easy way of finding out about our origins.

By this time I was tired. All I could think of to say was, "How very odd!"

But then I realized that I needed to summarize for Davey what conclusions I thought he should draw, even though they may not be the conclusions he would wish to draw.

– Well, Davey, this pretty well concludes what I wanted to tell you about our knowledge of the physical world we live in. We have two underlying theories that explain the actions of the physical forces we are aware of. The Standard Model very accurately describes the actions of the strong and weak nuclear forces and the electro-magnetic force, Einstein's theories of Special and General Relativity describe the way that the gravitational force behaves. The Standard Model, for the most part, does an excellent job of describing nature in the small, while Einstein's theories, again for the most part, do an excellent job of describing nature where the force of gravity predominates, i.e. over large distances. Neither approach, however, explains everything even within its own domain. Perhaps most notably, in the domain of the small, we have yet to comprehend and explain the phenomenon of non-locality. In the domain of the large, perhaps the greatest remaining mysteries are the apparent existence of dark matter and dark energy. Despite extensive efforts by some of our brightest minds labouring over many decades, we have yet to arrive at a satisfactory "Theory of Everything" that might unify these two theoretical approaches and clean up the mysteries at their fringes.

In his summary at the end of *The Elegant Universe*, Brian Greene wrote:

> *Already . . . we have seen glimpses of a strange new domain of the universe lurking beneath the Planck length, possibly one in which there is no notion of time or space. At the opposite extreme, we have also seen that our universe may merely be one of the innumerable frothing bubbles on the surface of a vast and turbulent cosmic ocean called the multiverse. These ideas . . . may presage the next leap in our understanding of the universe.*[29]

Eight years after Greene wrote those words, the beginnings of what may turn out to be the next leap were posited by a physicist named Lisi who used some advanced geometry to construct a plausible family of new particles and forces that could explain dark matter and dark energy and encompass both the Standard Model and Einstein's General Relativity theory. He calls his theory E8 Theory, where E8 refers to an *E8 Lie Group*. It is particularly of interest that the theory predicts that the newly constructed European particle accelerator, the LHC, will be able to generate some of the new particles that the theory predicts.[30]

Davey, I cannot tell you what an E8 Lie Group is, but I can tell you that it is vital to know that there appears to be a way to test whether or not it has promise. You might be surprised to learn how many theories in this domain are essentially untestable.

Perhaps one of the most wonderful phenomena of our history as a species is that we humans have been able to adapt the brains that evolved to enhance the survival chances of our forebears when they were hunter-gatherers into engines that have not only allowed many of us to experience a much better, longer and pleasanter life, but also has allowed us to contemplate the laws that govern our universe.

It is a source of wonder that scientists have been able to formulate testable theories about the origins of the universe and thus create a credible knowledge base about worlds undreamt of only a hundred years ago.

However, an important part of self-knowledge lies in an understanding that humans, like ants, may never be able to see and understand all of the real world.

There is a useful corollary to this statement. The history of science clearly demonstrates that it is neither useful nor helpful to accept without question unverifiable explanations for those things we do not understand. If a satisfactory science-based explanation is not available, wait a while. There may be one in the making!

– Bravo, Peter! I know it was a lot of work for you to prepare for this, but that wasn't so bad was it? I do hope though that your brilliant researchers will continue to explore the use of additional spatial dimensions to help explain what they do not now understand. I can assure you from personal experience that it is a promising avenue to pursue.

Let's meet again at the same time next week.

– Well thank you Davey. I was afraid that you might find my discussion either much too simple or totally incomprehensible! It is very helpful to get your reactions to what I say, especially as I know you have a habit of talking to others who may know a lot more than I do about any given subject.

You should find our next discussion much easier to understand, though many humans get a headache as soon as the word "statistics" is mentioned.

In much of human experience we encounter potential events that either cannot happen (i.e. have zero probability) or that are certain to occur. Of course we often have to deal with situations that might or might not occur, though we seldom feel impelled to understand in numerical terms just how likely it is that a particular event will occur. Our distant forebears didn't have the tools available to make such calculations, and we clearly have little innate drive to calculate probabilities. In many fields of science it is important to place a value on the likelihood or probability that a particular event will take place.

So, next week I will try to give you a feel for the human fascination with lotteries, and other statistical marvels, and show you how knowledge of statistics can help us to better understand our origins.

Before I could say another word I sensed that he had withdrawn to another universe.

DIALOGUE 7

The Universal Lottery

Our wisdom and deliberation for the most part follow the lead of chance.
MICHEL DE MONTAIGNE, *Essays*, Book 3, 1595

– Well, Davey, how much do you know about lotteries and statistics?

– Interesting that you should ask! How did you know that I was finding the idea of lotteries a little mystifying? You humans seem to like them for some reason. You keep buying tickets so that you can make someone else rich. Why don't you just give the ticket money to a friend?

– But you don't understand! By buying a lottery ticket I buy a chance to become rich!

– Hah! Have you ever even met a lottery winner?

– I did once know someone who won.

– Lucky you!

Where, I wondered, had he learned to be so sarcastic?

– Has it ever occurred to you that if you buy enough tickets you are bound to win? If that were not true, we humans would not be here today.

– Tell me about it.

– Okay, I will.

About four or five times a year I buy a lottery ticket. Almost always the ticket I buy is for what is today called Lotto 649. Basically the buyer selects (using any selection "system" he or she likes), six different numbers between one and forty-nine. No duplication of numbers is allowed. The lottery draw is held twice a week, when the six winning numbers are randomly selected. Some weeks no one wins and the pot simply gets bigger for the next draw. Other weeks there is more than one winner, in which case the pot is shared by the winners. Since my odds of winning, small as they are, are not affected by the size of the pot, nor by the number of people buying tickets, I, like many others, prefer to participate only when the pot becomes large.

What do you suppose my chances of winning are?

The answer is actually quite easy to calculate. When the first number is drawn, there are six chances in forty-nine that one of my six numbers will be a winner. In the next round, there are five chances in forty-eight that I will have another winner. In the third round, my chances are four in forty-seven, and so on, until the sixth and final draw, when my chances will be one in forty-four. Thus the chances of being lucky six times in a row, the kind of luck I need for a big win, are:

$$(6/49) \times (5/48) \times (4/47) \times (3/46) \times (2/45) \times (1/44) = 720/10{,}068{,}347{,}520$$

This is equivalent to about one chance in 14 million of winning (more precisely, it is one chance in 13,983,816). With this sort of odds, my expectation of winning is not very high! Perhaps because I know this, I sometimes forget to check whether or not I have the winning number for several weeks.

If I were a bit more of a gambler than I am, I might buy one ticket for every draw, i.e. twice a week, or 104 times a year. My chances of winning would then be 104 times improved to roughly one chance in 134,500, and if I kept up this habit for fifty years, my lifetime chances of winning would improve to one chance in about 2,700, still a bit of a long shot! But if I should, by some off chance, discover the elixir of youth, and live for 134,000 years, faithfully buying a lottery ticket twice a week the whole time, I would very likely have won the lottery at least once, and quite possibly more than once, by the end of my life.

The lesson here is quite clear. Even what look like wildly improbable events can become near certainties if the "experiment" is repeated often enough.

Right now we don't know what the odds are that in the chemical and thermal conditions found on Earth half a billion years or so after our planet came into being, a primal strand of ribonucleic acid would form and commence the process of reproduction essential for the evolution of life as we know it We do know though that even if the odds were millions of times worse than the odds of winning a lottery, life would still almost certainly have evolved.

In my years as a graduate student at Oxford, the department was led by a man named Denys Wilkinson, later Sir Denys Wilkinson. He was rumoured to have been one of the youngest persons ever inducted into the Royal Society. In his leisure time, he studied and published papers on the means used by migratory birds to determine direction so seemingly unerringly.

Wilkinson had a knack for making difficult subjects sound easy – the true mark of someone who understands his subject in depth.

He has accomplished just such a feat in a little book published in 1991 entitled *Our Universes*. In it he explores our limits to understanding the universe and what we know about its origins with the deft touch of a master, as well as with a fine sense of humour. His writing spans philosophy and metaphysics as well as physics. It will be a good read for many years to come.

As you are probably aware, many critics of evolution believe that life is simply too complicated to have evolved without the intervention of an *intelligent designer*, a being who is usually designated as God. Although the intelligent designer hypothesis would be difficult to disprove, it is at least as difficult to prove, and, moreover, the hypothesis does not appear to be needed. Given enough time, even extremely remote possibilities become virtual certainties. If there is a finite probability that life started spontaneously, then it is a certainty that given enough time, it will start. Of course, it is important to note that at this stage of our understanding we don't know what the probability actually is that life started spontaneously. Perhaps we will know one day. For now, it is the most probable explanation for the existence of life.[*]

In his book, Wilkinson discusses a number of other unlikely events that had to occur before life on Earth, or anywhere else in the universe, could arise. Part of his introduction to this subject reads as follows:

> There is an astounding goodness of fit between us and our own Universe. If, as I shall relate, the Universe had been only slightly different in any of many ways to do with the laws and constants of Nature and to do with the properties of the substances to which those laws give rise, we would not be here to wonder at it; our existence seems to be due to the delicate interplay of a large number of individually incredible accidents. This bundle of considerations goes under the general heading of the anthropic principle, which has been developed in a multiplicity of forms by Brandon Carter,

---

[*] Davey and I discussed this topic in greater detail in Dialogue 15.

Robert Dicke, and many others. I will not attempt to survey this ramification of forms but just state the principle in the crudest possible way: our Universe has to be as it is because if it were not, then we would not be here.[31]

Wilkinson goes on to suggest that it would be very arrogant of us to suggest, in such circumstances, that there has been only one Big Bang – perhaps designed by someone to make it possible for humans to develop. Rather, he favours the idea that the *"concatenation of unbelievable coincidences on which we . . . depend in a unique Universe might . . . lead one to suppose that we are indeed an accident and were not in mind when the Universe was constructed . . ."*

There are over twenty *"unbelievable coincidences"* listed in his book. Included on the list are items such as: the fine structure constant; the mass of the electron; the lifetime of stars; the fact that our universe, at least for everyday purposes, has three spatial dimensions and one time dimension; the density of the universe; the production and retention of carbon; the nature of the electrical force; and the fact that the electrical charge carried by the electron and the proton is identical to within one part in 10^{21}. It may be that the various physical constants that help describe the laws that these phenomena obey are all mutually related by some as yet undiscovered theory, but this seems to be unlikely. If indeed God did intervene to create us and our universe, the current state of scientific knowledge suggests that he would not have been needed any later than 10^{-10} seconds after the Big Bang, when "electroweak symmetry broke".[32]

All this leads some of us to conclude that there have likely been (and may still be) many universes, though it is hard to imagine how we may find out. [I heard Davey clear his throat rather ostentatiously at this point.] As we will discuss later, our senses as well as our capability to reason were almost certainly developed and honed as a part of our struggle for survival on Earth. It must come as no surprise that there are limits to our ability to understand our own universe, let alone any others that may exist.

If there is only one universe, then the odds that it would have just the properties needed for life to begin and humankind to evolve are very much smaller than the odds of winning a 649 lottery. But if there are billions of billions of billions of universes then the chances of one coming into existence that suits the evolution of creatures like us may be very high indeed.

We may never know for sure, but by calculating the statistical probabilities of possible outcomes wherever such calculations are possible, we can gain a much better feel even for very challenging topics, such as whether or not the universe we inhabit is one of many or is the only universe that exists!

When we come to examine the very challenging mystery of the origins of life, we shall find that we are again encountering a need to understand statistical probabilities, such as whether and in what circumstances certain critical chemical reactions could have occurred.

– Thanks for this Peter. I'm glad you kept the discussion short, since I neglected to tell you last time that I actually have a deep understanding of statistics. I routinely carried out statistical analyses to help my Envergurian friends, who were not as good at statistics as they should have been.

PART 3

INHERITANCE

My next encounter with Davey came about two weeks later. I was feeling a little miffed that he seemed unwilling to provide any help in unravelling the mystery of the creation of our universe or even of his own. Perhaps he was as ignorant on this topic as we are, but if so, he might have said so. Instead, he implied that he knew but could not tell us just yet – the sort of excuse one gives to children dying to know where the cookie jar is hidden.

Because I was feeling put out, I started our conversation with something of an ultimatum.

– Perhaps we should stop this exercise. It seems to be all one way, more a monologue than a dialogue. You could be a real help to us, yet you hold back and simply absorb the information I give you. What will you use it for anyway? What possible use can it be to you other than to pass your time of day, unless of course you have some nefarious scheme in mind . . .

Even this provocation got no immediate answer. I was beginning to wonder whether Davey really was one of those classic invaders from outer space so popular in TV entertainment. Could it be that "beneath his suave and friendly exterior beat an evil, scheming heart!"? How could I possibly know?

At last, he responded.

– Right! I think I understand your problem!. I wonder if I can help you to understand mine. Even allowing for the fact that time has a rather different meaning in our universe than in yours, our society is hugely older and thus, not surprisingly, more advanced than yours. In the process, we have lost track of our origins, and no amount of research seems able to recover that history satisfactorily. Our best opportunity of learning about our own origins seems now to rest in understanding the development of intelligent beings elsewhere. We have scoured our own universe – a task much easier for us than it would be for you – and found the exercise largely unrewarding. Almost by chance we found that we could gain a partial presence in another universe, namely yours. From our observations, the planet Earth seemed to be the best (but not the only) place to research the history of development in your universe.

My task is to understand all I can about your origins so that I can better understand my own. All reasonably intelligent life forms are overlain with a culture by means of which behavioural norms and ideas are transmitted. This is why the history of knowledge development is so important, for it is the hidden hand of cultural development that determines so many peculiar features of any civilization. We, in our universe, are ignorant of some critical early phases of our cultural development. You are providing us with some helpful insights – as are others with whom I am having dialogues. The way in which you humans gradually obtained an understanding of the size of your universe and of the time scales involved is fascinating, especially when viewed as the backdrop against which changes in your widely accepted belief systems occurred.

I understand that you also want feedback from me. You want to know how our life forms evolved, and whether that evolution has any relevance to you. I will tell you right now that I believe it does have relevance to you, but I am convinced that what I have to say will be more useful to you when I have finished hearing you out. For as important as the underlying physics of your universe is, it seems to be your chemistry that is the most remarkable feature of your universe. Before telling you much about us, I need to understand much better than I now do what impact that knowledge could have on you and your fellow human beings.

I therefore beg you to continue with our dialogue. Let me assure you that even if we had evil intent toward your universe, there is nothing we could do to harm you, with the sole exception that by revealing our own history of development we could influence yours, perhaps in ways that ultimately are not beneficial to you. Thus, just as your space explorers have to be very careful not to contaminate the new planets they wish to study with detritus from your planet, so we have to be very careful that we do not unduly influence life on Earth.

So, I operate under strict rules about what I can and cannot tell you. I suspect that these rules are far stricter than they need be. Even those few humans who actually believe that I come from another universe are very sceptical of everything I say, and most of you seem to be convinced that I am some kind of hallucination or fraud.

It was now my turn to let silence reign while I thought over the implications of what Davey had said. You will appreciate that my task was made immeasurably more difficult by the fact that I had only his voice as a measure of his reliability and truthfulness. I could not see a shifty glance, a smile or a smirk, a wrinkled brow or the wink of an eye. It was a bit like downhill skiing in thick fog. I felt more or less balanced and somewhat in control, but I had no idea where or when the next bump or dip was coming. In the end, I decided to continue on.

– All right for now, but I may change my mind later on.

I take your point about the importance of chemistry to understanding human development and behaviour, but I have decided to postpone that part of our dialogue until I have given you an overview of earlier scientific efforts to understand the development of life on our planet, and to comprehend the drivers of human behaviour. You will understand, I think, that these attempts took place in the absence of any understanding of the chemical mechanisms in place to implement inheritance of both form and behaviour from one generation of life to the next.

Charles Darwin is of course pre-eminent in this part of the story, but following on from him were some crucial observations that resolved a vigorous debate about the degree to which human behaviour is preprogrammed (i.e. inherited), as opposed to arising from a blank tablet that gets filled in by life's experiences. This *nature vs. nurture* debate, which was still raging when I was a young man, is clearly of the greatest relevance to obtaining an understanding of oneself. We can usually change behaviour we have learned; we don't (yet) know how to change behaviour we have inherited other than through sheer willpower. In my day, it is common to hear people say, "Clearly our behaviour is influenced by both nature and nurture". While this is true, it is not very helpful in determining how much freedom of action we really have, and how we should endeavour to nurture our behaviour to have satisfying lives.

We can do better than that, as I hope to convince you in ensuing dialogues. But for now let's discuss that eminent and eminently likable man, Charles Darwin.

– I hope you don't plan to talk about Darwin for too long. After all, he died well over a hundred years ago, and you keep telling me that the discoveries that have taken place in your lifetime are key to understanding the place of mankind in nature.

– A good point. Here's my answer.

First of all, you probably realize already that there have been some extraordinarily wise people in all periods of recorded history, and almost certainly going back long before recorded history. These people had a deep understanding of human nature. Some of them became religious leaders (such as Jesus Christ and Mohammed), some became writers (such as Shakespeare and Molière) and some became scientists. These people still have something of relevance to say to all of us, especially since we don't think that human nature has changed materially for at least tens of thousands of years. They are still worth listening to, but we would be very foolish if we did not filter their thoughts through the screen of new knowledge, just as they would have done had they had the benefit of knowing what we know today.

But of greater importance in Darwin's case is the value of understanding how knowledge developed, what caused the major steps in human understanding to come about and how these new ideas were received and adopted, for these facts all tell us something about human nature. Darwin's case is particularly interesting since the discovery of evolution by natural selection challenged cherished religious dogma head on, much as Galileo's confirmation that Earth orbited the sun and not vice versa had done two hundred years previously.

You haven't yet told me much about the extinct species of "superhumans" of which you were so fond, but it is very likely that they encountered similar struggles in the course of their development, as perhaps did your ancestors.

I paused for a brief interval, in case Davey wanted to say something, but he made not a sound. Perhaps he was asleep or absent doing loop-the-loops in his or our universe! I plunged ahead anyway. This was a discourse I had been looking forward to.

Charles Darwin and Evolution

As many more individuals of each species are born than can possibly survive, and as consequently, there is a frequently recurring struggle for existence, it follows that any being, if it vary however slightly in any manner profitable to itself, under the complex and sometimes varying conditions of life, will have a better chance of surviving, and thus be naturally selected. . .
CHARLES DARWIN, *On the Origin of Species*, 1993 (First edition 1859)

– To put it mildly, he was not a good student. The son and grandson of prominent doctors, Charles Darwin abandoned medical studies at Edinburgh University after two years. Like me, he felt faint at the sight of blood! He then moved on to Cambridge where he took much pleasure in attending shooting parties, and in parties more generally. Given his academic record, he would almost certainly not have been admitted to Cambridge these days. He found course work very boring, but he only needed a passing grade in order to become a clergyman and find a quiet position as a vicar somewhere in the English Midlands, close to his parents' home.

However, he was passionate about his hobby – natural history. As Alan Moorehead wrote in *Darwin and the Beagle*: "Everything in the fields delighted him. Flowers, rocks, butterflies, birds and spiders – from boyhood he had collected them all with the sort of absorption that belongs only to the besotted amateur or the true professional."[33]

While at Cambridge, Darwin befriended the Rev. John Stevens Henslow, Professor of Botany, who included him in his inner circle of botany enthusiasts. Botany was, after all, a perfect hobby for a country parson, and Henslow liked the young man. As things turned out, he provided the key impetus to Darwin's future.

At Henslow's suggestion, the twenty-two-year-old Darwin, recent graduate of Cambridge with a (bare) pass degree, received an invitation from Captain Robert FitzRoy, the twenty-six-year-old captain of HMS *Beagle*, to meet him in London to discuss his candidacy for the position of ship's botanist on its forthcoming expedition. Different as they were from each other, there was an almost instant mutual liking, and it was not long before arrangements were made for Darwin to join the expedition.

The principal tasks of the *Beagle* were to continue charting the South American coast and to endeavour to improve the measurements of longitude (a perennially difficult task until the invention of highly accurate chronometers) both on the South American coast and around the world.[34] The voyage was planned to take at least two years (starting in 1831), and possibly as much as four years, and was scheduled to leave

Figure 8.1 – HMS Beagle *in Sydney Harbour (detail).*
From a watercolour by Owen Stanley, circa 1835.

within weeks of Darwin's meeting with FitzRoy. The *Beagle* was a very small ship for the task by today's standards, only ninety feet in length, yet she carried a ship's complement of seventy-four people!

Darwin and FitzRoy shared a strong belief in God and in the biblical account in Genesis of the creation of Earth and of man. Both looked on the expedition as a valuable opportunity to find evidence of the first appearance of life on Earth and of the changes brought about by the flooding of planet in Noah's time. This common objective of finding evidence to confirm the religious teachings of the day – teachings widely believed in the Christian world at the time – adds some poignancy in hindsight.

After numerous delays, the HMS *Beagle* finally sailed out of Devonport on the southwest coast of England on December 27, 1831. She was not to return until October 2, 1836, almost five years later, after having lingered on the east and west coasts of South America for about three and a half years, and then returned to England via Tahiti, New Zealand, Australia and the Cape of Good Hope (with a final brief return visit to the northeast coast of South America on the way home). Though grievously afflicted with sea sickness that never let up if any kind of a sea was running, Darwin proved to be an indefatigable and energetic explorer on land. He frequently mounted major treks inland or along the coast, to be picked up several weeks later by the *Beagle*, which, in the meantime, was carrying out its primary mission of hydrographic surveys. Darwin regularly shipped cases of specimens and notes back to England, and he eagerly received from England the latest news from John Henslow.

While Darwin was away, the last two volumes of Sir Charles Lyell's groundbreaking three-volume treatise, *Principles of Geology*, were being published. Lyell was a Scottish geologist (and lawyer) whose books led to widespread acceptance of the view that the observable surface of Earth is the result of a slowly acting series of chemical, physical and biological processes. He concluded that these forces for change must have been acting over a much longer time period than the six-thousand-year horizon so carefully calculated by Archbishop Ussher almost two hundred years before.

Darwin had received the first of Lyell's volumes hot off the press as a parting gift from Henslow in 1831. He received the second in Montevideo about a year later, while the third volume caught up with him in Valparaiso in July 1834. Darwin would later say, "The greatest merit of the *Principles* was that it altered the whole tone of one's mind, and therefore that, when seeing a thing never seen by Lyell, one yet saw it through his eyes."[35] Lyell and Darwin were to become great friends after the *Beagle* returned to England, and Darwin was to have at least as profound an influence on Lyell, whose firmly held Christian belief in spontaneous creation was eventually overcome, but only after a long struggle, by the evidence presented by Darwin.

While a student at Cambridge Darwin had little time for geology, but aboard the *Beagle* he devoured Lyell's volumes, and they influenced his thinking greatly, perhaps starting with his discovery of some fascinating and very ancient bones at the foot of a cliff in Bahia Blanca on the southern shores of Argentina in September 1832.

He was struck by the similarity of these bones to those of living animals, except that the ancient animals were much larger. For instance, there were parts of a giant sloth, much larger than the extant very small version, a giant armadillo and a llama as big as a camel. As Darwin wrote at the time, "This wonderful relationship in the same continent between the dead and the living will, I do not doubt, hereafter throw more light on the appearance of organic beings on earth and their disappearance from it."[36] The issue for Darwin and Captain FitzRoy, to whom Darwin gave a full report, was to explain how this finding squared with their belief that God created the world as we know it in six days, that man was created in God's image and that all extant beasts were survivors of the flood, their ancestors having accompanied Noah on the Ark as described in Genesis. Already Darwin was beginning to agree with Lyell that the animals had evolved over a very long period of time, much longer than Archbishop Ussher's six-thousand-year time horizon. FitzRoy was unconvinced. Not all animals made it onto the Ark, he suggested; some got left behind and drowned.

It was in the Galapagos Islands, visited by the *Beagle* as it finally left South America in September 1835 and headed toward Tahiti on the way home, that Darwin collected the last pieces of evidence needed to convince him that the biblical account of creation could not be literally correct. Here,

once again, Darwin observed unique species very similar in many ways to those he had seen in South America, and yet strikingly different in others. Most surprisingly, he even found significant differences in the same animal living on different islands in the Galapagos group, yet many of these islands were separated by only eighty to 120 kilometres of water. The Galapagos are one thousand kilometres west of their nearest landfall in South America. As we now know, they are volcanic in origin (like the Hawaiian islands) and rose out of the South Pacific Ocean perhaps as long as 80 million years ago, so all original life forms either drifted, swam, flew or blew onto the islands at some time since the Galapagos emerged from the

Figure 8.2 – Charles Darwin in the late 1830s. Reproduction of a watercolour by George Richmond.

sea. As Darwin later wrote, "It was most striking to be surrounded by new birds, new reptiles, new shells, new insects, new plants, and yet, by innumerable trifling details of structure, and even by the tones of voice and plumage of the birds, to have the temperate plains of . . . Patagonia, or the hot dry deserts of northern Chile, vividly brought before my eyes."[37] The familiar birds of the South American mainland had reappeared as similar, but different animals one thousand kilometres offshore!

It was Darwin's discoveries about the finches of the Galapagos that later attracted the most public attention. Finches on different islands of the Galapagos were different. As Darwin's biographer, Alan Moorehead wrote: "[For example, on] one island they had strong thick beaks for cracking nuts and seeds, on another the beak was smaller to enable the bird to catch insects, on another again the beak was adjusted to feeding on fruits and flowers. . . . Clearly the birds had found different foods available on different islands, and through successive generations had changed accordingly."[38]

Before going to Cambridge, as Darwin later wrote, he did "not in the least doubt the strict and literal truth of the Bible". In the years following his return from South America, Darwin's thinking underwent a gradual metamorphosis, or, as Darwin himself later wrote: "Disbelief crept over me at a very slow rate . . ."[39] Within a year after his return, he started a series of notebooks on the mutation of species. A few years later, he outlined his theory of evolution, and left it with his wife to publish in the event he died before he could publish his ideas himself.

Darwin later described his heretical theory as follows:

> As many more individuals of each species are born than can possibly survive, and as consequently, there is a frequently recurring struggle for existence, it follows that any being, if it vary however slightly in any manner profitable to itself, under the complex and sometimes varying conditions of life, will have a better chance of surviving, and thus be naturally selected. . . .
>
> Although much remains obscure, and will long remain obscure, I can entertain no doubt, after the most deliberate study and dispassionate judgement of which I am capable, that the views which most naturalists until recently entertained, and which I formerly entertained – namely, that each species has been independently created – is erroneous. I am fully convinced that species are not immutable; but that those belonging to what are called the same genera are lineal descendents of some other and generally extinct species, in the same manner as the acknowledged varieties of any one species are the descendants of that species. Furthermore, I am convinced that Natural Selection has been the most important, but not the exclusive, means of modification.[40]

One of the more difficult tasks for historians is to recreate for the contemporary reader the intellectual and cultural environment of a previous era. Charles Darwin died on April 19, 1882. Even today there are a few vocal minorities who believe that the world was "spontaneously" created by God in six days in 4004 BC, in the manner described in Genesis.

However most Christians today view the book of Genesis as a kind of poetically licensed description of reality, not to be taken literally. The accumulated body of scientific evidence for the evolution of life on Earth over several billion years is too convincing for most people to think otherwise.[41]

This was clearly not the case in Darwin's Christian world. The vast majority of the population took the description of the creation of the Earth as described in Genesis as the literal truth. Heretical beliefs were not well tolerated and had many negative social consequences. As we shall see, there certainly were those, even within the Darwin family, who doubted that Genesis was literally true, but such doubts were usually aired amongst a select few chosen friends. The situation was made doubly difficult for Darwin by the fact that his wife was a devout Christian and a believer in spontaneous creation.

One of the strongest strands of evidence for the fierceness of the belief in spontaneous creation in Darwin's day is the fact that Darwin, though clearly immensely excited by his discovery of evolution through natural selection, and though he shared his belief and his evidence with a few selected scientific friends (most notably Sir Charles Lyell and botanist Joseph Hooker), he did not make his theory and its accompanying evidence public until his hand was forced by Alfred Russel Wallace, another (younger) British naturalist. In 1858, twenty-two years after the return of HMS *Beagle*, Wallace sent Darwin a copy of his essay entitled *On the Tendency of Varieties to Depart Indefinitely From the Original Type*. It was nothing less than a treatise on evolution through natural selection. Wallace asked Darwin to show the essay to Lyell, whose *Principles of Geology* had influenced both men. Darwin duly forwarded the essay with a warm recommendation, adding parenthetically, "So all my originality will be smashed."

In one of the nicer chapters of honourable scientific endeavour, Hooker and Lyell convinced Darwin that a simultaneous publication of the theory by both Wallace and Darwin would be an appropriate resolution of the matter of scientific precedence. The joint paper was presented in July 1858, and the next year Darwin published his book *On the Origin of Species by Means of Natural Selection, or the Preservation of Favored Races in the Struggle for Life*. By 1860 evolution and natural selection had become a national, even an international issue. This was the year of the famous debate held in the library of what is now the Museum of Natural History at Oxford between the Bishop of Oxford, Samuel Wilberforce and T.H. Huxley, a distinguished Oxford naturalist. From that day forward, the creationists, though often loudly heard, have been on the defensive.

In my day, Darwin is generally considered to be the father of the theory of evolution. Yet this is not strictly true. He is certainly the father of the theory of evolution through natural selection (though not, strictly speaking, its first proponent), but the idea of evolution itself has a longer history. In 1815 William Smith published a remarkable geological map of England that clearly demonstrated the layering of rock strata from different time periods and showed that life on Earth, as preserved in fossil form, must have existed long before the presumed creation of Earth and the heavens in 4004 BC. Even before 1815, Jean-Baptiste Lamarck of France had concluded that "... by the action of the laws of organization ... nature has in favourable times, places, and climates multiplied her first germs of animality ... and increased and diversified their organs."[42] As well, Darwin's grandfather, Erasmus Darwin, a respected physician and a keen naturalist, published one of the first formal theories of evolution in 1794–96. As already mentioned, Sir Charles Lyell, the renowned geologist, had also come to believe that current life forms had evolved from more primitive forms, though as already recounted, he clung to creationist beliefs for a long time.

If evolution is true, then what was the mechanism by which species evolved? In hindsight, knowing what we now know about DNA, the answer, at least in general terms, is easily supplied. However the early proponents of evolution came up with a variety of answers.

Jean-Baptiste Lamarck (1744–1829) is perhaps the best known evolutionist proposing an alternative to natural selection. As already stated, he believed that changes in the environment in which an animal lived led to changes in behaviour, and that greater or lesser use of an organ could lead to organ enlargement or atrophy respectively. According to Lamarck, these changes are inheritable. Charles Darwin's grandfather, Erasmus Darwin (1731–1802) held almost identical views, and most other proponents of evolution at the time believed in variants on this general theme. The key idea that *natural selection* was the arbiter of the success or failure of natural variations within species simply did not surface in scientific circles.

It is a telling mark of the quality and humanity of Charles Darwin that it is his own roster of evolutionists and their contributions that is the most frequent source of quotation for those who write on the history of evolutionary thought. Darwin is generous in his assessment of the contributions of others (even including Aristotle), and he clearly went to considerable lengths to find and assess their works. He even uncovered two prior proponents of the idea of natural selection, a Dr. W.C. Wells, who in 1813 and again in 1818 produced papers enunciating the principle of natural selection as it applied to mankind, particularly with respect to the superior adaptation of certain races to certain habitats. The second was a Mr. Patrick Mathew, who, in 1831, "published his work on 'Naval Timber and Arboriculture', in which he gives precisely the same view on the origin of species as that... propounded by Mr. Wallace and myself.... Unfortunately the view was given by Mr. Mathew very briefly in scattered passages in an Appendix to a work on a different subject, so that it remained unnoticed until Mr. Mathew himself drew attention to it in the 'Gardener's Chronicle', on April 7th, 1860."[43]

Wallace, too, is now known primarily for his role in inducing Darwin to go public with his ideas on natural selection. His brilliant work as a naturalist, especially his work in the Malay archipelago, has faded in public perception and understanding, while his later attempts to introduce spiritualism into evolutionary theory have diminished his stature in the halls of science.

You might ask, "How is it then, if others thought of evolution through natural selection ahead of Darwin, that he is acclaimed as the great father of the idea?"

In my view there are two good reasons for this.

The first (and lesser) reason is that Darwin was such a fine human being, and was so much liked by his fellow scientists, many of whom were influential. Stephen Jay Gould, the noted Harvard biologist writes of Darwin, "[H]e was a person whose basic

Figure 8.3 – *Darwin caricature from* Vanity Fair, *September 30, 1871.*

kindness and decency defy the numerous attempts of detractors to demean or defame him.... Darwin's humanity, with all its foibles, shines through in his life and writing."[44] Most humans prefer to support people they admire and respect, and in this respect scientists are no different from the rest of us.

Good scientists do however go to extreme lengths to be fair, and to try to discount their natural biases. The real reason for Darwin's pre-eminence is that Darwin's predecessors reached their conclusion without a lot of supporting evidence. Even Wallace, who shared the initial honours with him, first explained his case in the form of an essay. Darwin, it turns out, had already spent half a lifetime collecting supporting data. *On the Origin of Species* was published only eighteen months after the Darwin–Wallace paper was presented to the Linnaean Society. Near the beginning of the 649 pages in which he states the case for natural selection, he starts by calling his book an "Abstract" and apologizes: "I can here give only the general conclusions at which I have arrived, with a few facts in illustration . . ." This "Abstract" includes the chapters: Variation under Domestication; Variation under Nature; Struggle for Existence; Natural Selection, or the Survival of the Fittest; Laws of Variation; Instinct; Hybridism; On the Imperfection of the Geological Record; Geographical Distribution; Mutual Affinities of Organic Beings; and more. It was obvious that the man had done his homework – in spades. Thus it was his book that convinced many doubting fellow scientists and laymen that the doctrine of spontaneous creation was little more than a fallible theory invented by humans who had little or no access to the scientific facts needed to refute it. Hence too it was his book that drew the wrath of many clergymen and their adherents.

There can be no doubt that Charles Darwin really is the father of the theory of evolution through natural selection.

I cannot think of any discovery that has done more to knock man off his conviction of his own superiority[*] over all other life forms, and the related conviction that man has been especially chosen by God and created in his image. It clearly came as a shock to the authorities in Galileo's time to contemplate a universe that did not have Earth at its centre, and many years later the similarity of the genomes of mice and men was also a bit of a surprise. The discovery that Darwin had in store for us is likely the greatest of all time. It is not an exaggeration to state that Darwin provided the framework into which the work of thousands of later scientists – including animal behaviourists, biochemists, psychologists, and even economists – would fit so nicely.

Thus did my contribution to the dialogue come to a close. I was especially proud of my discourse on Darwin. It had required a lot of work on my part. Moreover, in the process, I had come to see him as both a scientific giant and a modest and thoroughly likable human being. What, I wondered, would Davey have to say about Darwin and about my

---

[*] Of course, humankind is superior to other life forms on a scale of intelligence but on little else, as Davey and I had already discussed. While belief in one or more gods has been characteristic of the majority of humankind since well before historical times, the existence of God has been questioned by sceptics at least since Greek antiquity.

presentation?

A preliminary reaction was not long in coming.

– I find what you said today particularly interesting, but I think you have a lot more to cover. It is one thing to conclude that all species alive today have evolved from predecessor species, but if you keep going back in time you have to come to . . . what? So far, I have heard a wide range of answers to this question. It is strange for me to see that you humans are still a long way away from a consensus. You clearly still have a lot of evolving left to do! But you can have no idea how useful it is for me to learn how you think you evolved. The kind of evolution you talk about is not applicable to me, though what is exciting to think about is that it may have been applicable to those great superhumans whose demise is still so devastating for me.

What Darwin uncovered is really quite exciting – but I don't understand at all why you think it important to mention that he was a nice man. He has been dead for over 125 years. Who cares any longer whether he was nice or nasty?

– I do, for one. Darwin is regarded by many as the greatest scientist of all time. Yet many people seem to think that "nice guys finish last". Perhaps you think so, too? The evidence I have seen suggests that, on the whole, nice guys have happier lives than the rest of us – so I am naturally hoping that my grandchildren will choose to be nice. It is important that they understand that being so does not disqualify them from making important contributions to human welfare.

– Perhaps you are right. But quite a few of your great men whom others have described seem to have been arrogant and uncaring of other people.

– You are right there, but in part that is because we are still busy trying to structure society so that it pays to be a nice guy. Too often there are still rewards for being nasty – but every year we get better at it.

By the way, you surprised me when you admitted that you have no real knowledge of how you evolved. What with fossils and DNA analysis, we seem to have a wealth of evidence about evolution here on Earth. How could your ancestors have destroyed all the evidence so completely?

– I have to go now. I'll be back next week at the usual time.

Just when our conversation started to get really interesting, Davey departed! Ask him a question he doesn't want to answer and he simply ignores it.

Just like a man!

DIALOGUE 9

Learning from Animals

Although we are still far from understanding the mechanisms which maintain life in its multitude of forms, it is obvious that a living organism is in an extremely unstable state. One of the main characteristics of the living organism is that it possesses an overwhelmingly complicated system of mechanisms that protect it against adverse influences of the environment and enable it to maintain itself as a living organism. It is these mechanisms that the biologist studies, whether he is an anatomist, a physiologist, or an ethologist. Causal study of these mechanisms is not complete unless their contribution to the primary activity of life is demonstrated.
NIKOLAAS TINBERGEN, *The Study of Instinct,* 1951

I had known from the start of my discussions with Davey that I was not his only source of information about ourselves and our universe, but since Davey supplied only the occasional hint about what he was learning from others, it was often difficult for me to decide what I should tell him and what would be wasted effort on my part.

My concern was particularly acute at this juncture, for it seemed to me that it would likely never have occurred to him that in the 1960s and 1970s, there was intense debate about the degree to which we humans are shaped by our culture as opposed to being shaped by behaviour that is inherited in much the same way that the contours of our nose and the colour of our skin are inherited.

Most of the heat has now gone out of the debate. It is generally acknowledged that we humans (and many other animals as well) all inherit some neuronal structures that generally drive our thoughts, desires and actions, but that these are subject to modification as life experience and circumstances dictate.

Did it matter that Davey might not know about this debate?

In the end I decided that the issue was too important to leave out. If Davey wanted to play Mr. Know-it-all in the face of what I had to tell him, so be it.

Nonetheless, I was a little nervous as I waited in my office on that chilly November morning. The insipid cloud-filtered sunshine was no match for the cold west wind gusting in from the North Pacific Ocean. I realized too late that I had no alternative material prepared should Davey reject out of hand what I wanted to tell him, but brazenly, and only half-convincingly, I told myself that it didn't matter. After all, it was me doing him a favour, not the reverse. If he didn't like what I had to say, he could just wander, fly or spirit (whichever was appropriate) back to his own damn universe!

By then it was ten in the morning, and right on cue he was suddenly present, though how I knew he had arrived before he actually spoke, I never did find out.

– How I wish I could see you and your surroundings! I feel it would help me a great deal to understand you and your species.

– Perhaps so, but I don't know how I can help you with that.

– Is there not a way that you can convert what you see into sound that I can interpret?

I immediately thought of ultrasound machines, which can show us pictures of the insides of our bodies.

– Have you tried hanging about by an ultrasound machine?

– Yes, of course! The problem is that ultrasound machines tend to broadcast all sound at the same frequency, and the changes in sound volume as the beam encounters different material don't really tell me very much.

– Then maybe your best bet is to talk to a good artist and get him or her to describe what they paint. I would guess that at least some of them are pretty good at it. But in the meantime, how about you giving me an idea of what you look like, and the superhumans you admired so much, and the general scenery where you live? Do you have plants and animals? Oceans and mountains? Ice and sunshine? And why can't you detect our light and our radio waves and our sunshine?

– Well . . . some of your questions are hard to answer. Basically, I cannot detect your electromagnetic radiations, such as sunshine and radio waves, because we don't have electromagnetic radiation in our universe. So, we have never developed eyes or other organs to detect what we didn't even know existed until I chanced on your universe. We do however have the equivalent of your acoustic waves, where our materials are alternately compressed and expanded – hence our ability to hear the noises you make. We do not, in fact, have days and nights in the way that you do. Our eyes can detect what is going on all the time, since our universe is bathed in our own unique radiation, the equivalent of your light, on a continuous basis.

I really cannot describe myself or any of the other many life forms in our universe. I don't know how to start. Perhaps when I have found an artist who can describe you and your species to me I shall be able to describe my universe to you – but so far I have to confess that I just don't know where to begin.

However, I would very much like to hear what you have planned to tell me this week. Is it asking too much to suggest that we get started?

– All right, but you may not like what I have to say. You may think it is painfully obvious, or you may have heard it from someone else before.

– Well, if it is painfully obvious, why do you feel driven to tell me?

– There are two reasons. First, while it may now be obvious to you and me, there are still a lot of people who don't think it obvious at all. Indeed, they are still busy denying the obvious. The second reason is because it is so important. Unless you accept the truth of what I am about to tell you, you will never fully understand human behaviour.

– This sounds serious! Perhaps you had better tell me what on earth you have in mind!

– Okay, I said, heartened by his evident interest. Here goes.

Before Charles Darwin provided a detailed description of the primary mechanism by which we came to look and behave as we do, there was no end of speculation about how we came about and why we behave the way we do. Most religions, and therefore most people, seem to have concluded that we look the way we do because we were made in the image of God or of gods. And then there were many suppositions about why we behave the way we do. Most of these explanations involved some kind of tussle between good and evil influences, with us hapless humans strenuously urged to listen only to the good influences. No further explanation of human behaviour seemed to be needed.

Darwin changed all that. The theory of evolution directed us to think in terms of humans and all animals inheriting behaviour patterns that were survival enhancing. But that was not the end of it. Many people argued that the survival strategy most likely to succeed for we humans would be to learn everything we needed to know about survival-enhancing behaviour from our parents and their friends, as well as more directly from experience. In other words, human behaviour was all learned, not inherited.

The best-known advocate of the view that human behaviour derives entirely from learned (or culturally provided or nurtured) behaviour was B.F. Skinner, a psychologist at Harvard. He was the originator of the study of *operant behaviour*, defined as behaviour that "is shaped and maintained by its consequences for the individual".

As one commentator wrote of Skinner:

> *Skinner expressed no interest in understanding the human psyche. He . . . sought only to determine how behavior is caused by external forces. He believed everything we do and are is shaped by our experience of punishment and reward. He believed that the "mind" (as opposed to the brain) and other such subjective phenomena were simply matters of language; they didn't really exist.*[45]

If the beliefs attributed to Skinner in the above quote are accurate, then Skinner was surely wrong.

When thinking about the inheritability of behaviour, the first question you might ask yourself is: if the major physical attributes of animals (including humans) are inherited (for this the evidence is incontrovertible), and if Darwin and his colleagues are correct that inherited traits are largely the result of a long-term selective process where the fittest survive (for this the case is very strong indeed), and if animal behaviour is a very strong factor in the ability of animals to survive (which is clearly the case), then why wouldn't behaviour be just as inheritable as physical traits for humans as well?

Skinner might answer: Because learning is the way that best guarantees the animal's survival. The animal learns how to survive by a combination of operant behaviour conditioning and being taught by its parents or their substitutes.

If he did answer that way, Skinner would be partially right, but only in a very restricted sense.

In 1973 two Austrians and a Dutchman shared the Nobel Prize in Physiology and Medicine. One (Karl von Frisch) was a zoologist, the other two, Nikko Tinbergen and Konrad Lorenz, were ethologists, or specialists in the study of animal behaviour. All had conducted a variety of fascinating experiments on insects, fish and mammals, and all had concluded, contrary to the widely held belief at the time, that animal behaviour was very substantially inherited or innate, though with many behavioural traits modifiable by learning. Their Nobel Prize was awarded primarily for developing a unified evolutionary theory of animal and human behaviour, a theory that essentially showed that Skinner and his associates were wrong. In the process, they clearly demonstrated that there is much we can learn about human behaviour by observing the behaviour of other animals.

Some of the experiments these men carried out amaze me both because of their ingenuity and because of the immense patience and love of research that they evidence. Ask yourself, for instance, how you might go about observing bees in the wild. How, for example, do you suppose they find good flowers for making honey? Do they communicate a good find to others? How do bees navigate? Are all bees in a colony alike? If not, how are the jobs in the hive shared? What causes bees to swarm? How do they survive the winter (or is it only the eggs that survive)?

Don't ask me! Ask Karl von Frisch. Dr. von Frisch (who died in 1982), listed the following as his most important publications.

> *(1914–15) (The bee's sense of colour and shape.)*
> *(1919) (The bee's sense of smell and its significance during blooming.)*
> *(1923) (Bee's 'language'- an examination of animal psychology.)*
> *(1932) with R. Stetter. (Examination into the position of the sense of hearing in small insects.)*
> *(1934) (The bee's sense of taste.)*
> *(1941) (On the repellent substance on fish skin and its biological significance.)*
> *(1946) (The bee's dances.)*
> *(1949) (The polarisation of skylight as a means of orientation during the bee's dances.)*
> *(1950) (The sun as compass in the life of bees.)*
> *(1965) (The Dance Language and Orientation of Bees.)*[46]

Imagine if you can the hours at a time over a fifty-year span that Dr. von Frisch spent in the field observing bees, and the ingenuity he brought to bear in discovering how the bee, which has discovered a good source of nectar, communicates that information to her coworkers through an elaborate dance.

Amongst other behaviours, the bee's complex "dance" by which she informs her colleagues of the distance, direction and quality of her nectar discovery is largely inherited, not learned. The behaviour pattern is passed to her through her genes.[47] While this may not strike you as in the least peculiar, it was, as already mentioned, totally contrary to some widely and strongly held beliefs about the plasticity of the behaviour of humankind (in particular) and animals (in general).

Of course, insects are very different from humans and have much smaller brains. The fact that bees inherit some very complicated behaviours is certainly cause for astonishment and wonder, but it hardly proves that humans and other higher animals also inherit behaviour.

Of the three Nobel laureates just mentioned, Lorenz was the best interpreter of ethology for the non-expert. His best known work is *King Solomon's Ring*, published in 1952. It is a delightful, humorous and informative work, in which he does not hesitate to draw parallels between animal and human behaviour where he thinks it warranted. I only wish, Davey, that you were able to read the book. I am sure it will still be both entertaining and informative in a hundred years! Have you tried using talking books?

– No, I haven't, though you are not the first to suggest it. If you can find a talking version of this book you like so much and can play it for me, I would be very glad to listen.

– I'll see what I can do.

I paused to pick up the thread of my discourse, and ploughed on.

– One great advantage those who study animal behaviour have over those who study human behaviour is that they are much freer to experiment. For example, it is possible to raise animals without any contact with other members of their species, and, by observing their behaviour, learn which behaviours are substantially modified by learning and which are essentially innate.

I particularly enjoyed reading about Lorenz's experience with the jackdaw,[48*] an abnormally intelligent bird with a lifespan comparable to ours, and which habitually mates for life. Through constant contact over many years, he learnt how they communicate (as with humans, much more than just their roster of different vocalizations), saw that betrothal took place well ahead of sexual maturity, observed what happened in the only case he saw of a married male being seduced by an unmarried female, realized that the bird calls with an important meaning were all purely instinctual, and learned what young jackdaws need to learn from older ones.

If I succeed in finding a talking version of *King Solomon's Ring*, you will laugh with Lorenz as he describes tourists peering over a fence to see him sliding along on his behind while quacking like a mallard as he leads some mallard chicks for a stroll. This action came about because mallard chicks, in common with the young of some other species, *imprint* on, or adopt as their mother, the first animal they see when they open their eyes after hatching as long as the animal in view has certain innately prescribed characteristics. In order for Lorenz to continue as their "mother", he had to keep a very low profile and quack almost incessantly. He thought the tourists might have been doubly mystified as the mallard chicks he was leading were in tall grass, and thus were likely invisible to them!

Lorenz spent a lot of time studying animal communication. You would enjoy his description of the behaviour of his dog, Tito, who had an uncanny way of understanding if his master had a guest who got on his nerves. In such circumstances, Tito would walk slowly over to the guest and nip him or her in whatever part of the anatomy was convenient. Try as he might, Lorenz was unable to fool the dog, so that even important (but boring) guests could not be saved from the Tito nip!

Through all of Lorenz's writing comes the inspired vision (though not, I think, the ultimate proof) that higher animals' actions are controlled at a fundamental level by inherited behaviour patterns. These basic behaviour patterns can be nurtured, varied, enhanced or even suppressed by external factors, which in effect superimpose *learned behaviour* on this underlying web of innate behaviour.

While Lorenz was the great communicator of the findings of ethology in its early days, Niko Tinbergen was the careful experimenter whose disciplined approach gave the subject its strongest scientific foundations.

Listen to what Tinbergen has to say near the conclusion of his classic text *The Study of Instinct*, first published in 1951:

> *Man is an animal. He is a remarkable animal and in many respects a unique species, but he is an animal nevertheless. In structure and functions, of the heart, blood, intestine, kidneys, and so on, man closely resembles other animals, especially other vertebrates. Palaeontology as well as comparative anatomy and embryology do not leave the least doubt that the present-day primates have only recently diverged from a common primate stock. This is why comparative anatomy and comparative physiology have yielded such important results for human biology. It is only natural, therefore, that the zoologist should be inclined to extend his ethological studies beyond the animals to man himself. However, the ethological study of man has not yet advanced very far. While animal neurophysiology and animal ethology are coming in touch with each other, there remains a wide gap between these two fields in the study of behaviour of man.*
>
> *One of the main reasons for this is the almost universal misconception that the causes of man's behaviour are qualitatively different from the causes of animal behaviour. Somehow it is assumed that only the lowest building-stones of behaviour, such as impulse flow in peripheral nerves, or simple reflexes, can be studied with neurophysiological or, in general, objective*

methods, while behaviour as an integrated expression of man as a whole is the subject-matter of psychology. Somehow it is assumed that, when, in investigating behaviour, one climbs higher and higher in the hierarchical structure, ascending from reflexes or automatism to locomotion, from there to the higher level of consummatory acts, and to still higher levels, one will meet a kind of barrier bearing the sign 'Not open to objective study; for psychologists only'. It is of fundamental importance to recognise the utter fallacy of such a conception.[49]

Over fifty years later I believe that most of us still grossly underestimate the degree to which we are beholden to our innate behaviour patterns, and grossly overestimate our "freedom" to take independent action. Darwin and his colleagues have demonstrated our close affinity with other animals. Lorentz, Tinbergen and their colleagues and successors have shown us how much of animal behaviour is innate. The available evidence

Figure 9.1 – *Nikolaas Tinbergen and Konrad Lorenz in 1978.*

indicates that we, too, are very much beholden to our inherited traits. The degree to which this is the case will likely always be a matter of debate. In later dialogues, we will discuss evidence from other disciplines that supports my contention, shared by many others, that our instincts are very strong forces, and that free will is essentially an overlay of freedom of action that extends and modifies defined inherited behaviour patterns. The degree to which humans are able to suppress and modify instincts is likely unsurpassed in the animal kingdom, but it is increasingly clear that prolonged suppression of instinctual behaviour, even for we humans, can be extremely stressful.

It is hard to overestimate the importance of this finding. For example, it is said that the great Chinese Communist dictator, Mao Tse-Dong believed that human nature was infinitely malleable. Accordingly, he believed that the Great Leap Forward, which he instigated in the late 1950s, would demolish the remnants of the old Chinese regime and permit him to fashion a new China, like painting pictures on a clean sheet of paper.[50] It is estimated that the famine resulting from the Great Leap Forward killed as many as 40 million people! If Mao had realized the extent to which human behaviour is governed by instinct, might he not have implemented a different and much more humane initiative to reform his country?

It seems at least possible, too, that had the pontiffs who decreed that priests in the Roman Catholic Church must be celibate failed to realize the extent to which the edict would put the hapless priests at war with their instincts. If they had known, perhaps they would have devised a safer and more effective way for their priests to demonstrate their devotion to God.

– You probably don't realize that what you have just said is extremely significant to me.

Davey's voice betrayed rare emotion. At last I was getting a reaction! When he continued, his speech was suddenly strangely precise and clipped.

–You see, for complicated reasons, we have lost the history of the early development of life in our universe. Yet what you seem to be saying is that to fully understand behaviour one needs to be able to parse that part of behaviour that is innate and that which has been learned.

I suppose it is only natural – Darwinian you might say – that each species would focus in on itself and regard the rest as supporting actors. But before today I had not understood the extent to which even moderately intelligent beings, such as humans, would allow themselves to be trapped into thinking of themselves as in some sense the "chosen" ones, different in kind from other life forms. What you have told me so far all adds up. Even you humans are near-slaves to your inherited behaviour. Of course, I now wonder if the dreadful, dreadful end of the superhumans in my universe may not have been due, at least in part, to similar underlying inherited behaviour that was invisible to me at the time.

Davey's voice changed again as he spoke the last few sentences. There was a muted echo to it, and the pitch was higher. Even though it was definitely not human, I could tell that his emotions were heavily engaged.

– I have to go now, as I am meeting with an Austrian psychologist, but I hope we will continue with this theme next week. And I would like to listen to *King Solomon's Ring* if you are able to find it in an spoken version.

– Don't go now! This conversation is just getting interesting! What makes you think that your superhuman friends might have been influenced by inherited behaviour? And how is it even possible that you lost so much important historical information about their early development?

But even as I spoke these words, I somehow knew that he had already left.

DIALOGUE 10

Learning from Our Human Forebears

> *. . . when men can neither read nor write, when they have nothing*
> *external to distract their minds, they can spend their lives in*
> *minute observation, and if they have thousands of years in which*
> *to accumulate folk wisdom, it can become in time wisdom of a very*
> *high order. Such people discover plants which supply subtle drugs,*
> *and ores which yield metals, and signs in the sky directing the*
> *planting of crops, and laws governing the tides.*
> James A. Michener, The Covenant, 1980

I had been convinced for some time that Darwin's contribution to the task of humankind developing a realistic view of itself was crucial. However the years between his discoveries and the decryption of the chemical code of life in 1953 saw important contributions from others. In my previous dialogue with Davey I had highlighted the important contribution of the animal behaviourists. There is a problem with this approach. A person convinced that humans are infinitely superior to animals will not be easily swayed by evidence collected about animal behaviour. But that same person should reasonably expect that evidence about human behaviour as collected by anthropologists and psychologists will enlighten his or her understanding of both self and fellow humans.

In this dialogue I intended to demonstrate to Davey how some key individuals in these fields had developed understandings of human nature that greatly enhanced our evolving understanding of ourselves.

I was chomping at the bit and ready for Davey when the time for our next meeting rolled around and was pleased to note that Davey seemed to be in a good mood when he arrived.

– Good morning, Peter.

– And good morning to you. I have some important questions for you before we get going.

I scarcely paused to take a breath, as I was afraid he would interrupt before I got my questions out.

– Last week you deliberately avoided my questions and bailed out on me! That does not augur well for a growing spirit of co-operation between us. What I wanted to find out before you disappeared is how it could possibly come to pass that you lost all your early historical records? Surely you stored spare copies in some remote places. I would have expected that without even trying to back up important information, it would be stored and used in so many different places that it would be impossible to experience such a wholesale disappearance of your historical records. It just doesn't add up!

There was a long silence – not even the sound of breathing – for almost thirty seconds! I waited impatiently but was careful to make no sound myself. Davey began in a monotone, then as he warmed to his subject his voice took on more expression.

– Yes. I probably do owe you a bit of an explanation, though if you thought about it a bit, you might have supplied part of the explanation yourself.

You see, lacking radioactive elements to use to help us assign ages to the artifacts that surround us, our early history is difficult to disentangle at the best of times. Our world does resemble yours in that it is undergoing constant change. If we know the rate of change we can sometimes work back to calculate the approximate age of something, but these calculations are complicated, in part because the rate of change can vary over time.

– That much I can understand. But surely a civilization as sophisticated as yours must have the equivalent of our written records, and likely the equivalent of our photographic and digital records, as well. Time may well, as you once said, have a very different meaning for you (whatever that may mean), but you clearly do think in terms of age to some extent.

– Yes, of course, we have a sense of time, but for me time is not as brief as it evidently is for you – and as it was for my superhumans. Oh, they resembled you humans in so many unfortunate ways. Like you, they had a lot of disagreements. Thus it was that at one particularly difficult time, some superhumans nearly accomplished the impossible, that is, they nearly caused all of the accumulated knowledge of our civilization to be wiped out. It was a very close call indeed, and did a massive amount of damage, damage that would be unimaginable to you. As a result we now have huge gaps in our knowledge of how the superhumans developed, and indeed even of how I myself developed. It is extremely frustrating.

Even though your civilization is not nearly as attractive as ours was, and though it seems to behave very stupidly at times, we think we might learn from you a little about how our more advanced world came about.

– In my universe, we much prefer to learn from geniuses than from the stupid.

I spoke a little tartly and not entirely correctly, but Davey's air of superiority annoyed me more than a little at the time.

– What are you anyway? You keep telling me you miss your superhumans, so I presume you are not one yourself. Did you evolve from them by being more fit to survive? Or did you just drop in from elsewhere? Or perhaps you suddenly happened out of thin air as it were?

– If I told you that indeed I did just appear out of thin air would you be any happier?

– Certainly not!

– Fine. Then let's continue with our dialogue. But first let me thank you for finding a talking version of *King Solomon's Ring*, the book you recommended so highly last week. I did find it entertaining. It reminded me of the sense of fun that my closest friends amongst the superhumans used to have.

– You're welcome, of course.

– Quite.

Davey was clearly at his enigmatic best, but I chose not to point it out this time. He quickly continued.

– Last week you told me that despite all the evidence to the contrary, many, perhaps even most, humans seemed to believe all their behaviour was learned – none was inherited. The human mind, these silly individuals seemed to think, was a blank slate just waiting to be written on by life's varied experiences and the accompanying lessons of what those experiences demanded of human behaviour in order for the mind's owner to survive.

– Who are you to call them silly? You seem to speak with what we often label as "the wisdom of hindsight". You may recall from our fifth meeting that many of my contemporaries have similarly labelled the calculation by Archbishop Ussher that the universe began in 4004 BC as silly. Yet Ussher was simply making the most of the information he had available, and his estimate was in fact at least as good as most of the other "expert" estimates made during his lifetime.

I shall show you that B.F. Skinner and his friends saw a lot of evidence to back up their beliefs. They were certainly wrong, but they were not silly.

– Oops! I am very sorry. I did not mean to offend you. I thought that you said last week that they had missed some obvious information, and should have known better. You were so convincing that I felt I had to agree with you.

– Well, try looking at it this way.

First of all, all the available evidence tells you that humankind is in a different league from all other life forms. For example, no other animal has a language that comes close in sophistication to human language, and none have a written language. Few use tools, and those that do, use very low-tech, unsophisticated tools.

Secondly, humans have achieved superiority over all the animals. Although one could say that the superiority simply made humankind the "king" of the animals, but not different in kind, the difference is so great that it is easy to believe that the difference is indeed a difference in kind.

Finally, most religions portray humans as being in the image of God or gods, and hence, *ipso facto*, superior to all the animals and imbued with at least some God-like characteristics, perhaps the most notable of which is a free will.

Until Darwin came along, there was not much obvious evidence to refute these contentions. For about a century after publication of *On the Origin of Species*, there was still lots of room for doubt. However, after the discovery of the genetic code in the 1950s, and the surprising similarities of the genetic codes of, for example, all mammals, including humans, it became really clear, at least for those whose minds were not closed by a blind adherence to current religious doctrine, that humankind is just a sophisticated animal. In practice, of course, it takes a good deal of time, as well as strong evidence, for such a radical idea to be widely accepted. Such conservatism is doubtless just another mammalian characteristic, time-tested as a useful survival strategy.

In practice, the debate was further beclouded by the fact that all behaviour is not inherited. Many species, especially the mammalian ones, have developed the capacity to adapt behaviour to circumstances, and we now know some of the mechanisms in the brain by which that occurs – so there was and is lots of evidence that behaviour can be adapted to enhance survival, and is thus nurtured (or learned).

But the picture gets even fuzzier and more complex, as you likely know.

Early humans seem to have found a survival advantage in forming into smallish groups or tribes. Although not always the case, it seems often to have been true that almost by definition, those in your tribe were "friends", while those outside it were "enemies", who tended to compete for essential resources, such as the territory necessary to provide satisfactory sustenance and shelter. Enemies, of course, rapidly became objects of scorn.

It was thus a small step for "civilized" humankind to view those inhabitants of less advanced societies as being lesser human beings, and almost certainly possessed of lesser intelligence. There were not many facts about to counter the prevailing view, especially since most considered the greater material prosperity of "civilized" societies as *prima facie* evidence to confirm their instinctual reaction to members of other tribes as being inferior.

However, those from the more advanced societies who studied these strange Native societies, especially the anthropologists, gradually discovered otherwise.

Perhaps the best-known anthropologist of his day was a German-born American named Franz Boas. He led the questioning of the "superior race" assumption and made the term *cultural relativity* familiar in households around the world.

Shortly after completing his PhD in Germany, Boas joined an expedition that left in 1883 to study the Inuit on Baffin Island. He returned convinced that all peoples were by and large equal in intelligence and that their varying social organizations were simply the result of differing historical and geographic conditions. (In more recent times, this argument has been further developed by Jared Diamond, who, in a book entitled *Guns, Germs and Steel*, convincingly demonstrated the important roles of climate and geography in the development of civilization, especially as they related to the availability of plants and animals suitable for domestication.) "If we were to select the most intelligent, imaginative, energetic, and emotionally stable third of mankind", Boas wrote, "all races would be present."[51]

This is hardly a surprise given what is now known, but it has taken a very long time for Boas's proposition to become generally accepted, and even more recently extended to an understanding that any racial differences in average intelligence are so small as to be insignificant for all practical purposes when compared to variations in intelligence within a particular group or tribe.*

---

* The concept of race is itself controversial. Luigi and Francesco Cavalli-Sforza treat *(footnote continued)*

A lot of our knowledge about how humans behaved in prehistoric times comes from studies of "primitive" tribes still extant in recent times. In the latter part of the nineteenth century and the early part of the twentieth, there was a great rush to study their social behaviour as much as possible before they, too, became "civilized" and started to live more like the rest of us. Indeed when I was a young man, there was just such a rush to study the Inuit in Canada's North. It was only half-jocularly declared by many familiar with this state of affairs that the average Inuit family unit of six included 1.5 grandparents, 1.8 parents, 1.7 children and one anthropologist!

It is widely believed that "pre-civilization" humankind lived in groups ranging in size from ten or so up to several hundred. In order to prevent excessive inbreeding, small groups evolved methods of mixing their gene pool with neighbouring groups. There is evidence that at least some early societies deliberately divided after reaching a size of several hundred individuals, with one group moving out to establish itself elsewhere. This is known to have happened, for example, amongst the Haida people in northern BC and southern Alaska. The fact that most of us can recall only several hundred names and faces is sometimes quoted as evidence that in prehistoric times we had evolved to societies of that order of size in at least some cases. Evidence from studies of primitive societies in the nineteenth and twentieth centuries generally supports this conclusion about the range of populations of prehistoric human societies.

Consider, for example, the San people in the south of Africa about thirteen thousand years ago. They were hunter-gatherers. A viable clan might consist of about twenty-five people. More than twenty-five was too difficult to feed, while fewer would make the clan too vulnerable to attack from wild animals.[52] We know that the San had developed some very sophisticated weapons with poison tips used to kill prey. We also know

the subject at some length in their book *The Great Human Diasporas: The History of Diversity and Evolution* (1995). They conclude that in practice, race is defined much more in cultural than in genetic terms, for which differences between groupings of humans are minimal. For example, skin colour differences are controlled by differences in only three or four genes, whereas humans generally differ in their genetic makeup by about 0.1 per cent or three million genes. As is the case with skin pigmentation, diseases prevalent in certain regions have led to the predominant appearance of genetic forms that enhance resistance to the diseases in that region. A systematic study of the genetic differences of races has lead to a dead end insofar as definition of race in genetic terms is concerned. This finding is consistent with the fact that the spread of *Homo sapiens* out of Africa occurred in recent times, leaving insufficient time for significant genetic differences between humans to have developed.

that some clan members had significant artistic talents, as evidenced by cave engravings of rhinoceroses and elands dated to that period. The survival value of their undoubted intelligence seems, in their case, to have lain at least partially in their ability to use it to learn how to kill prey much stronger and/or more agile than themselves. The complex actions needed to prepare the poison for their arrows, and the ingenious design of the arrows themselves provide good supporting evidence of their considerable intelligence.

James Michener writes cogently of the early uses of human intelligence as follows: "... when men can neither read nor write, when they have nothing external to distract their minds, they can spend their lives in minute observation, and if they have thousands of years in which to accumulate folk wisdom, it can become in time wisdom of a very high order. Such people discover plants which supply subtle drugs, and ores which yield metals, and signs in the sky directing the planting of crops, and laws governing the tides."[53]

Madhusree Mukerjee writes, "[Hunter-gatherers] simply have a different strategy for survival: rather than using technology to mould their surroundings, they use an extraordinarily subtle understanding of their environment to live in harmony with it."[54]

Clearly, we humans should never mistake a lack of formal learning for a lack of intelligence or wisdom. Rather, it is useful to speculate on all the learning and wisdom we have missed, which was common knowledge to our forebears of thousands of years ago before writing was invented – for it is certain that they knew much that would be of interest to us today.

The foregoing brief journey back in time has, I hope, helped illuminate the survival value of intelligence in those times, and perhaps gives pointers to an inherited sense of well-being that might derive from physical strength and agility, from being a good hunter, from constructing a satisfactory shelter and from begetting and raising children. However it does not really help us to delineate in specific terms the nature of our network of inherited drives, and the relative strength of the components.

For the most part, anthropologists have tended to be descriptive of life in the various hunter-gatherer societies they studied and have not drawn too many conclusions. In part, spurred on by Boas, this was a reaction to the widespread belief in the inherent superiority of one race over another. Theirs was a doctrine that each culture tended to have different objectives and to live in different circumstances from others. Different customs, therefore, did not result from differences in intelligence but rather from differences in culture and customs.

The zenith of this approach was achieved by one of Boas's star students, Ruth Benedict.

In 1934 Benedict published a bestseller entitled *Patterns of Culture*. In her book, she describes three very different, widely separated hunter-gatherer cultures. She concludes:

> *The three cultures of Zuñi, of Dobu, and of Kwakiutl are not merely heterogeneous assortments of acts and beliefs. They each have certain goals toward which their behaviour is directed and which their institutions further. They differ from one another not only because one trait is present here and another there, and because another trait is found in two regions in two different forms. They differ still more because they are oriented as wholes in different directions. They are travelling along different roads in pursuit of different ends, and these ends and these means in one society cannot be judged in terms of those of another society, because essentially they are incommensurable.* [55]

In other words, culture is relative. One culture cannot be judged as better than another, it is just different because it likely has different goals, and it certainly has a different history of development.

If true, this is an important finding. It would mean, for example, that actions enhancing the general welfare in one society could have negative consequences in another society. Indeed, we know this to be true in many circumstances. For example, child daycare facilities can be essential to the smooth-working of some societies, nearly useless (and hence a waste of valuable resources) in others. But does cultural relativity necessarily mean that one cannot be judgmental about the relative efficacy of two societies?

– I'm darned if I know the answer to that! Where I come from, we didn't have a lot of different "societies" of superhumans; we had only one large society. From what I have heard, this is one of the most important differences between you people and my superhumans. It makes life so complicated for you. Why can't you all get together and act as one coherent, sensible society?

– Maybe you're right, and maybe we will some day. But it may never happen, and that may even be for the best.

– Really, Peter! You must have slipped a cog! How could it possibly be better to have your situation of wars, poverty, human degradation, cruelty, hunger, rampant disease and a host of other problems that come with having disparate societies, than the well-functioning society created by our superhumans?

– I don't really know yet. I have to learn a lot more from you before I can tell you with confidence. Besides, until you learn more about how we humans are put together, I don't think you would understand. You may just have to wait!

I smiled inwardly at having given him back some of his own medicine. I had concluded that there had to, be a lot that he had not yet told me about his superhuman society. I was sure there must be a key somewhere to their disappearance, and I wanted to find it before our conversations came to an end.

Then, almost without thinking about it, I added:

– I think I can safely give you one clue, though: we are still alive and healthy. Your so-called superhumans have disappeared. Why do you suppose this is so?

I thought I heard a hollow sigh.

– All right. Let's get on with the discussion. I have other meetings to get to quite soon.

– A feeble excuse. You have already told me that you can listen in on many conversations at once. Why do you keep holding back important information from me? If you told me more, I could probably help you more.

– The fact is that although we did have one homogeneous society of superhumans, as I have told you already, they were not all in agreement all the time. Toward the end there was a large minority, led by what you would call intellectuals, who violently disagreed with the majority on how to deal with an important medical issue. The upshot was that all the superhumans perished. I still wonder what I could have done to prevent such a disastrous outcome.

Davey's voice was wretched and haunting. I imagined that his face must have been downcast and nearly tearful. But at the time, I was not feeling as sympathetic as I would be now were the scene to be repeated. This voice had been scornful of we humans, and had lauded his friends as having been superior in every way – but now, at last, it appeared that perhaps his superhumans had their flaws after all! I was determined to press for more information.

– Ah ha! So, these superhumans of yours were not so perfect after all! They had disagreements, even fatal disagreements. Did they have wars? Poverty? Perhaps even pestilence? What did they look like? How did they reproduce? How did they communicate with each other? And . . .

– Enough! Please stop asking questions. The time will come when I can tell you much more. In the meantime, I must ask you to trust me, and to continue to help me to understand you better.

His voice was uncannily piteous, as though he had learned the correct tone of voice by listening to many of my fellow humans being piteous, and somehow averaged and merged the result.

I succumbed at once. My sense of remorse was eerily all-encompassing. Through it all, I sensed that I was being manipulated, but I seemed to be powerless to resist. With wrenching effort, I barely managed to murmur "I'm sorry".

– Please don't worry. All will come clear in due course. In the meantime, I hope we can continue with our discussion. You were telling me that your anthropologists had concluded that all cultures were equally valid and laudable, but you hinted that that view was about to change. Is that what happened?

– Yes. And this was an unusual case. At least amongst us humans, a major change in accepted wisdom is usually initiated by a new player who challenges the accepted norm.

In this instance, it was Ruth Benedict herself, whose popular book laid out the case for cultural relativity so convincingly, who initiated the argument for its rejection.

For Benedict, the intellectual pillars of the theory of cultural relativity weakened as she observed the rise of Nazism in Germany prior to the Second World War. For it was clear to her that Nazi Germany was not a place she would want to live in, just as she felt she would much prefer life amongst the Zuñi, to that amongst either the Kwakiutl or the Dobu, as she described those societies in *Patterns of Culture*. Most, if not all of us, would agree with her after reading the book and learning how pleasant and friendly the Zuñi way of life was.

Benedict's change of mind was disclosed to only a few people in her lifetime. Her hesitance and her secrecy are easy to understand. Her mentor Franz Boas, she herself and many colleagues had spent decades fighting the facile sense of superiority, felt especially (but not exclusively) by the white man, with respect to other peoples, and particularly with respect to hunter-gatherer societies. The Rousseauian *noble savage* was understood by most "civilized" societies to be more ignorant than noble and certainly mentally inferior. The arguments to be found in *Patterns of Culture* were aimed at demolishing this false sense of superiority.

In Benedict's search to find a scientific rationale for her discomfort with the theory of cultural relativity, she gradually, very hesitatingly, developed the concept of *social synergy*. Her hesitation was in some ways similar to the hesitation Charles Darwin had in publishing his concept of evolution. The world, perhaps especially her world of anthropologists, was already sold on the idea of cultural relativity. She herself was one of its best-known advocates. Her discomfort did not seem adequately scientific to her, so, like Darwin, she shared her new ideas with a few close associates, most notably Abraham Maslow, but she did not publish them. They only came to light after her death in 1948.

As she described her concept in a paper published a full twenty-two years after she died: (Benedict, 1970),

> *I shall need a term for this gamut (of cultural configurations), a gamut that runs from one pole where any act or skill that advantages the individual at the same time advantages the group, to the other pole, where every act that advantages the individual is at the expense of others. I shall call this gamut synergy, the old term in medicine and theology to mean combined action . . . greater than the run of their separate actions. I shall speak of cultures with low synergy, where the social structure provides for acts that are mutually opposed and counteractive, and of cultures with high synergy, where it provides for acts that are mutually reinforcing.*[56]

Reduced to its simplest terms, a high-synergy society is one in which virtue pays; that is, a society whose rules of conduct are structured so that the interests of society as a whole are best served when individuals pursue their own self-interest. In a low-synergy society, pursuit of self-interest tends to disadvantage others. Maslow described societies with high synergy as those "in which the social institutions are set up to transcend the polarity between selfishness and unselfishness [and] between self-interest and altruism."[57] At last, Benedict could understand why Zuñi society appealed to her so much more than those of the Kwakiutl and Dobu. The Zuñi had structured their customs and laws so that synergy was higher than in either of the other two.

It is unlikely that there is any philosophical concept of merit that is entirely new. The concept of social synergy is no exception. A variety of philosophers have come close to enunciating the idea, but most if not all turn out to be ethical admonitions that, for a variety of reasons, individuals should act in the common interest. I have not been able to locate explicit sources that touch on the essentials of the Benedict thesis, namely the importance of shaping the customs and laws of a society so that, to the greatest extent possible, individual and common interests are synthetical rather than antithetical.

Adam Smith's *invisible hand* proposition suggests that by pursuing self-interest an individual is automatically benefiting his fellow man. This converse of the usual moral precept will only be true in a society where synergy is high. Smith, in effect, was maintaining that the free-market system is high in synergy.

Indeed, most modern societies are quite high in synergy. Productive enterprise by individuals and by companies is usually rewarded; governments help to redistribute wealth, so that the gain of some individuals is at least partially shared by all; those who are generous and serve society well receive recognition and sometimes rewards from society. On the other hand, it is clear that some, such as drug dealers and other criminals, receive huge monetary rewards for their anti-social activity, and it is not uncommon for some of the most socially useful professions to be underpaid relative to other, less directly beneficial work.

Franz Boas (1858–1942)   Ruth Benedict (1887–1948) Abraham Maslow (1908–1970)

Figure 10.1 – *Franz Boas, Ruth Benedict and Abraham Maslow*

Maslow's own research had led him to acknowledge the strong role of inherited behaviour patterns in dictating the way we humans behave. This conclusion was most graphically communicated in his well-known *hierarchy of needs*.

This was a crude but effective attempt to calibrate the relative strengths of human instinctual drives. Maslow's idea was that the lack of satisfaction of any need at a higher level on the scale trumped any otherwise-felt need at a lower level. Thus, for example, the lack of food to eat or air to breath, if not satisfied adequately, would leave a human indifferent to such lower-level needs as freedom from fear and higher social status.

According to Maslow, the hierarchy runs as follows:

1.   *Physiological Needs*: oxygen, liquids and the food required to keep the body stable and healthy.

2.   *Safety Needs*: "security; dependency; protection; freedom from fear, from anxiety and chaos; need for structure, order, law, limits; strength in the protector."[58]

3.   *Belongingness and Love Needs*: perhaps correspond to the animal behaviourists' reference to herd instincts, and the importance of being an "accepted" member of the herd.

4.   *Esteem Needs*: divided into a need for self-esteem and a need for esteem from others. The latter is closely tied to our tendency to form social hierarchies, of which esteem is an important measure of place in the hierarchy. Self-esteem is something rather different. It seems to be at the core of a subjective sense of well-being, although an excess of self-esteem is more closely allied with excessive pride, both of which most of us regard as unattractive attributes.

(Some of the difficulties of this scheme are becoming apparent!)

5. *Need for Self-Actualization*: "the full use and exploitation of talents, capacities, potentialities, etc."[59] Maslow was particularly interested in the concept of self-actualization as being key to having a strong sense of well-being. Intuitively, the idea is very attractive, and indeed it does appear that striving for the kind of excellence that self-actualization implies is very satisfying to human beings. But it is hard to know how much of the satisfaction derived from self-actualization comes from the status obtained from recognition of the specific actualized competence by friends and colleagues, and how much is independent of peer recognition and status-gaining behaviour.

In the years since Maslow first enunciated his hierarchy of needs we have gained a much more nuanced view of the complex structure of innate drives, but I find the Maslowian hierarchy is still useful as shorthand for understanding human actions. It helps us understand, for example, why such important concepts as democracy and equity are so quickly abandoned by otherwise civil societies when they are stressed by fear of survival.

But to my mind the most important conclusion to draw from this brief excursion into social anthropology is that by observing how a wide variety of societies have been structured, anthropologists have been able to shed light on innate behaviours we all share, as well as on the limits of our adaptability to differing circumstances. While anthropologists lack the ability to conduct experiments to test hypotheses in the manner of animal behaviourists, they have the advantage. The animal under observation is us, and not some other animal whose instincts may or may not resemble our own, and with whom it is difficult to communicate.

I wonder what relevance you see, Davey, in what I have told you over the past three weeks to your search for an understanding of your superhumans.

As I see it, the century or so between the publication of Darwin's *On the Origin of Species* in 1859 and the discovery of the mechanism that makes species evolution possible (i.e. DNA) was constructively used by humankind to digest the hitherto unpalatable thought that humans were not after all created in the image of God or gods, who graciously bestowed them with free will to do as they wished. Rather, humans were highly evolved mammals whose free will is circumscribed – some would say overseen – by innate behaviours that evolved over millions of years. Furthermore, these behaviours tend to evolve further at a very slow pace unless forcefully modified by selective breeding or a crisis (perhaps a fatal disease) that causes death to those not possessing a crucial inherited characteristic.

This was a very dramatic change in perspective for us. One that evidently requires more that a hundred years to permeate the world's population of humankind, since many if not most of us still cling to our ancient beliefs, including the rather bizarre belief (in my view) that humankind is created in the image of God or gods. But anthropologists and social psychologists like Boas, Benedict and Maslow, and animal behaviourists like Tinbergen, von Frisch and Lorenz, and their students and successors played crucial roles in transforming our view of ourselves from demi-god to mammal.

If further proof were needed of our mammalian status, it was provided by the unravelling of the structure of DNA and the subsequent growing understanding of the mechanisms for cell reproduction and development, which I will try to describe for you during our next four or five meetings. Before I have finished, I shall hope to convince you, Davey, that this understanding leads us logically to a much improved foundation for knowing not just how we came about but also what we need to do to lead happy lives and promote the survival of our species.

– Well, to be honest, I am not very interested in learning what makes you happy and how to ensure your survival. I cannot even see what you look like, and I don't believe I ever will be able to see you. Even if I could, I am not sure I would care about you humans very much. As I have told you, my goal is to be able to understand how my superhumans came into being and why they disappeared. You would really get me excited if you could teach me how to recreate them. You see, I miss them more than you can imagine!

This last comment of his really took my breath away. It had not occurred to me that he might think he could create a new, very sophisticated form of life. If that was what he hoped I could help him achieve, he had another think coming!

I was on the point of telling him so when I realized that he had gone.

PART 4

BIOSCIENCE'S BIG BANG

My last session with Davey had led to a real change in our relationship, at least from my perspective. To begin with, I was becoming a lot more comfortable with him. Familiarity had not bred contempt, though his reticence did annoy me from time to time. As Davey himself had remarked when talking about us humans, it is hard to warm to a voice alone, bereft of any visual cues, but it is possible nonetheless. Did he ever frown or smile? Or wrinkle his nose? Gesture in the heat of an argument? Pace about the room? I wondered. I imagined, as time went on, that he did all of those things, mainly because his voice was so naturally human most of the time and contained its appropriate quota of inflections to connote the range of human emotions. I also wondered how he dressed. Was he clad in rumpled trousers, sandals, a faded cotton shirt and sweater? Or was his dress more formal? Perhaps he wasn't dressed at all!

In short, just as most of the old religions had assumed that man and God(s) resembled each other, so I tended to assume that Davey, being a nice guy, most likely looked like me!

But the previous dialogue had brought in a new element. Davey had let slip, deliberately or not, that he was thinking of recreating his beloved superhumans.

What did he mean by that? Had he preserved some DNA (or his universe's equivalent) of his friends, so that he might be able to clone them from their own hereditary map, much as we on Earth had been speculating about the possibility of recreating extinct species, such as dinosaurs and woolly mammoths? Or did he think he might be able to re-invent the superhuman equivalent of the human chromosomes? Or perhaps fiddle with animal chromosomes so that the progeny would resemble his superhumans more closely, much as we could, in theory, but certainly not yet in practice, imagine monkeying with monkey chromosomes so that they became a human genome? If so, there was not much help I or anyone else on Earth could give him. He would have to hang about for at least a few more generations before we could begin to teach him how to do that!

The timing of his remark about resurrecting his superhumans was perfect from my point of view, as I was on the point of telling him about the discovery of DNA and the immense boost it gave to our understanding of life. But only Davey could determine what relevance our knowledge about life in our universe might have for him.

I was as convinced as ever that a sense of the history of humankind's search for a proper understanding of humanity and of our place in the universe is essential to understanding the strengths and limitations of our current state of knowledge about ourselves, as well as to predicting where this knowledge might lead us. But having treated most of what I wanted to say about the early development of our knowledge, I was now starting to broach very active fields of ongoing research, thus increasing the possibility that what I told Davey was out of date or even incorrect.

My situation reminded me of Sir William Osler, the famous Canadian medical educator of the late nineteenth and early twentieth centuries, who used to remind his students that there was still much to discover in the field of medicine. He would tell them that half of what he was teaching them was incorrect. The problem was to know which was the incorrect half!

I wondered how important it was for Davey to understand that the discovery of the structure of DNA was nothing less than a scientific sensation. Suddenly we could see not just how life forms reproduced but also, perhaps more important, how the many, very different forms of life were related to each other. Millennia of debates about the special status of humankind relative to other animals and many decades of intense discussion about the validity of evolutionary theory – about whether or not mankind really is related to monkeys, about whether behaviour could be inherited, and even about human morality – were hugely impacted by this discovery and the subsequent elaboration of the biochemistry of the reproductive process.

Darwin's theory emerged almost unscathed from the bright illumination cast by the discovery, though not without some tweaks and refinements. The "tree of life", purporting to show the evolutionary path of life forms, and encapsulating the painstaking work of a small army of palaeontologists and biologists, also prospered, though a number of corrections, some of them major, had to be made arising from a detailed analysis of differences in the DNA of different species. By examining similarities in ribosomal RNA* and the genes that code for it, the common ancestry of all known life forms on Earth was placed beyond any reasonable scientific doubt (see Appendix 2). Furthermore, because a lot has been learned about the natural rate at which genes mutate, scientists found that by comparing the differences in certain genes between different life forms it was possible to estimate the time that had elapsed since they had a common ancestor!

How could I best clarify and explain key developments in this important field? I already knew full well that many, perhaps even most people on Earth, find the idea of our common ancestry with all life forms both unpalatable and unbelievable. Might Davey, too, find it impossible to believe that he is related to other life forms in his universe? If so, how would he likely react to what I had to say?

And how much detail did I need to go into? I had hoped he would be content with an outline, but I already knew from his comments at our last encounter that Davey would want to delve into some details. But how far would he want to go? What about his other interlocutors? Perhaps he already knew from them all he wanted or needed to know.

It was in this somewhat querulous state of mind that I began the introductory dialogue on the biosciences.

By this time, I knew Davey would start listening to me at 10:00 a.m. sharp, exactly a week from the previous dialogue, so, for a change, I initiated our discussion.

– Good morning, Davey.

I suppose you remember our discussion of the many early theories about the origin of our universe, and the discovery that it almost certainly started with one big cosmic explosion (or bang). You may also recall how this

---

* Note that a definition of ribosomal RNA and other bioscientific terms can be found in the Glossary.

discovery was greeted at the time with surprise and a good deal of incredulity by scientists, clerics and laymen alike.

– Of course I do.

– Well, the discovery of the way in which the characteristics of living beings are preserved from one generation to the next was another transforming moment in human history. For the first time we could begin to understand the great mystery of life. What most had assumed must be divine (since it apparently surpassed human understanding) suddenly opened up to detailed examination. Much of life was (and still is) a mystery, but suddenly we had a key to gain admittance to that vault of many mysteries.

So, I am going to call this part of our dialogue, where we'll discuss some of the advances arising from this discovery, as "Bioscience's Big Bang." I hope you appreciate the analogy.

– Certainly. I have been looking forward all week to this discussion! I have just been talking to a very nice gentleman. He tells me that life arrived on Earth from space. I don't suppose you are going to tell me that one day a little man from Mars or perhaps a piece of genome from some distant galaxy landed on Earth with a bump and caused all the life forms visible on Earth today?

– In this case, you don't suppose entirely correctly. Available evidence makes such scenarios very improbable, as I hope to show you.

I was surprising myself by the relaxed way I was able to parry Davey's challenges. I must have been getting used to his sudden off-the-wall remarks. But I also did not want to disappoint him, a situation that could quickly lead to an irritating lecture on how much more intelligent and capable his superhumans were than are we humans.

– Before I go on, however, you should be aware that we have no easy formula for re-establishing vanished species. Frankly, we just don't know how to do that yet, and even if we could, it would be unlikely to be useful to you, since presumably your life forms are entirely different.

And by the way, I have been meaning for some time to suggest that you stop referring to your erstwhile good friends as "superhumans". Clearly, as inhabitants of a different universe, they must be totally unrelated to humans, and, since you tell me they no longer exist, I have to wonder how truly "super" they were, even at the best of times!

– Well, I am sorry if you feel jealous of them. To me, they seem superior to you humans in so many ways, and yet also like you in quite a few ways, too. I thought by referring to them as superhumans you would more quickly come to understand this. However, it is not hard for me to come up with another name for them. How about Envergurians?

– What a strange name! Where did you ever get that from? It sounds a bit like a French word.

– I know that.

– I suppose it's interesting that you like French, but that does not explain why you might choose to name your superhumans Envergurians.

– As you may know, the French use the word *envergure* to describe a woman or man who has a great breadth and depth of personality and of interests, dimensions that make him or her admired by most others. The primary meaning of *envergure* relates to the depth and breadth of the sails on a sailboat. Later, the word was used to describe the wing span of a bird or a plane. From there, the French started to use the word as a descriptor of people as I have just mentioned. I am told that there is something wonderfully poetic and descriptive in the image of a ship underway, its sails spread and full under the pressure of a strong breeze, and the use of this image to describe a person possessed of a broad understanding and a profound depth of knowledge, coupled with the ability to use these attributes well. Don't you agree?

In any case, it seems to me that such a word would perfectly describe my superhumans.

Of course, this explanation almost knocked me out! Where had Davey been? How did he or could he acquire a sense of what is beautiful to humanity? Was he, perhaps, only aping what he knew was what we humans think of as beautiful, without himself having any feeling for beauty as we know it? How would or could I ever find out? I found that I had no answer for Davey at this juncture, no reply to his astute coupling of the French term with his superhumans, whom he obviously loved and admired. I could only force myself to say:

– Okay. I suppose we call them Envergurians from here on in.

– Peter, I'm disappointed! I thought you might congratulate me on finding such a descriptive name. Were you not just a little surprised?

– Very surprised. When will you cease to amaze me?

– Never, if I have my way!

Davey sounded a little on the smug side to me, but I refrained from saying so.

– Well, I suppose we should get started. Perhaps you could begin by giving me an outline of what you intend to tell me.

– All right.

There was a note of strain in my voice, for I had found it difficult to decide on the best manner of relating the fascinating tale of this exciting phase in the development of human self-knowledge. I wished I could look him in the eye and judge his reaction to what I was saying. After a pause to collect myself and refocus my thoughts on what I had previously prepared, I started off.

– There can be no useful debate about the importance of genetic research to gaining an understanding of ourselves and our origins. The challenge for me is to present the key findings in a comprehensible and interesting manner, while keeping my comments brief!

Davey, of all the subjects we have discussed, the science underlying our understanding of genetics, accrued since the discovery of the structure of DNA in 1953, is probably the subject of the greatest interest to you. It is also undergoing rapid development with important new findings published frequently.

It has been no easy task for me to determine the order and content of our dialogue on the discovery of the shape and form of the human genome, and the many fascinating developments that flowed from that understanding. In part this is because comprehension of one part of the field is greatly enhanced by prior knowledge and understanding of others. Yet all aspects cannot possibly be assimilated simultaneously – or at least we humans cannot do so.

I propose the following order for our dialogue:

1. Historical Background: An account of the development and early understanding of heredity at a molecular level, including a selective discussion of some early experiments and conclusions. This is the story of how we at last came to understand the coding system nature uses to preserve and pass on from one generation to the next the information needed to create another very similar life form, be it a germ, a puffball, an elephant, a tree or a human.

2. The Life of a Cell: An elaboration of some of the remaining mysteries concerning the manner in which DNA and RNA shape the form and functions of life's ever-changing stream of species.

3. Small Animals and Monkeys: A focus on unicellular i.e. single-celled life forms and their relationship with us multicellular plants and animals.

4. Adaptation by Gene Selection: A brief look at how the environment we experience, particularly in early life, can influence what genes are actually used (i.e. expressed) and which ones are shut down. Studies in this area are hugely important in making choices about early-childhood education.

5. Emergence: A short summary of what we think we know about how the first-ever life forms on this planet came into existence. To my mind, this is where the biggest unresolved "riddles of life" reside. Once unicellular life existed, it is now, at last, relatively straightforward to see one's way clear to the evolution of humankind. However, the manner of the jump from a "primordial soup" to cellular life is still very uncertain.

I have also prepared a Glossary of Bioscientific Terms especially for you. For the non-expert human being, microbiology is a difficult field to address, not least because there is a whole new set of terminology that experts writing on the subject tend to forget may present difficulties to the uninitiated. I suggest that I read the glossary to you. It is true, isn't it, that once you have heard what I will read, you will not forget it?

– True enough. You might as well read your glossary to me right now. Then we can proceed with your background.

At this point I picked up my notes and read out the glossary of terms, which is reproduced in this book immediately following the Appendices. I had the distinct impression that every word was assimilated by him, and was readily accessible should he ever need the information. How I wished I had a memory like that! Davey thanked me at the end of the reading, but there was no reaction, not even a grunt, during my recitation. At the same time, he never did subsequently ask me for a definition of terms.

The stage having been set, I wondered if he might have had enough already for one sitting, but Davey seemed keen to carry right on, so carry right on I did.

DIALOGUE 11

Crick, Watson, Franklin and Wilkins

It has not escaped our notice that the specific pairing we have postulated immediately suggests a possible copying mechanism for the genetic material.
J.D. WATSON AND F.H.C. CRICK, from the original paper announcing the discovery of the structure of DNA, "A Structure for Deoxyribose Nucleic Acid", 1953

– As you know, Davey, Darwin's work was accomplished without having any clear notion of how life originated and even of how, once originated, the replication process for the creation of new life copied so faithfully the pattern set by its predecessors (even though accompanied by minor but significant variations). Just think how excited he would have been to learn about the discovery of the code of inheritance! Perhaps it is better for us that he developed his Theory of Evolution in ignorance of the structure of deoxyribonucleic acid (DNA), ribonucleic acid (RNA) and perhaps as many as five million proteins that these molecules generate. Absent this knowledge, Darwin, in *On the Origin of Species* treats us to a wonderful spectrum of arguments based on a near-encyclopaedic knowledge of both wild and domestic animals and plants, and of their many variations, evidence that nicely complements that provided by our relatively new-found understanding of genomes. Imagine the glow of pleasure on his face had he learnt that humans share 80 per cent of their DNA in common with the lowly mouse, and a full 99 per cent in common with the chimpanzee. "Yes, of course!" he would have said. "All those similarities of structure which got used by different species in so many different ways, such as 'forearms,' which may see service as flippers, fins, wings and legs in different species, are all crafted from DNA, which is similar in very many respects. How I wasted my years in demonstrating what has become so

135

obvious now we can compare DNA records!"

It was while I was a high school-student in Grade 10 that two scientists at Cambridge University at last discovered the structure of the DNA molecule, and hence the manner in which (in principle) innate characteristics of plants and animals are passed from one generation to the next. It may well be hard for you to imagine a time when we had no idea how inheritance worked much beyond what the European monk Gregor Mendel[60] had published in 1865 as a result of his study of pea plants. Now, less than sixty years after the discovery of the structure of the DNA molecule in 1953, knowledge of DNA has revolutionized the production of effective drugs and medicines, and has been routinely applied to criminal investigations. But for our purposes, the primary impact was that it set us on the course to solving one of the greatest mysteries facing humankind, namely, how we and other life forms, including apes, trees, fungi and fish, pass on our attributes to the next generation.

As you may well imagine, the manner in which inherited traits were passed from parents to progeny was the subject of a good deal of speculation over the years. In *On the Origin of Species*, Darwin wrote: "When a character which has been lost in a breed, reappears after a great number of generations, the most probable hypothesis is, not that one individual suddenly takes after an ancestor removed by some hundred generations, but that in each successive generation the character in question has been lying latent, and at last, under unknown favourable conditions, is developed."[61]

James Watson, who with Francis Crick and Maurice Wilkins won the Nobel Prize in Physiology or Medicine in 1962 for the discovery of the structure and code of DNA, credits Erwin Schrödinger, the Nobel prize-winning physicist and father of the Schrödinger equation (which describes the wavelike properties of matter) with inspiring a renewed search for the secret of inheritance through the 1944 publication of a book entitled *What is Life?* Watson wrote: "Schrödinger argued that life could be thought of in terms of storing and passing on biological information. Chromosomes were thus simply information bearers. Because so much information had to be packed into every cell, it must be compressed into what Schrödinger called a 'hereditary code-script' embedded in the molecular fabric of chromosomes. To understand life, then, we would have to identify these molecules, and crack their code."[62] With the wisdom of hindsight, we can appreciate that Schrödinger's conjecture was spot on, but to many people at the time it must have seemed like wild speculation.

New knowledge frequently comes to us in the wake of the development of new scientific instruments. The invention of the telescope, when put in the hands of Kepler and Galileo, led to revolutionary changes in man's understanding of the universe. The microscope, which came into use at about the same time, led to the discovery of a world of very small animals, hitherto unsuspected. Louis Pasteur used this tool to better understand the sources of some dangerous human diseases. Darwin and many others used microscopes to better understand the variety of Earth's life forms.

But in Darwin's time, the best available microscopes had a magnification of only about one thousand times and were capable of resolving lines separated by about a micron (or one millionth of a metre). A key factor in determining the limit of resolution is the wavelength of the light that illuminates the specimen. In the nineteenth century, the wavelength in question was that of the light we see by.

The critical tool needed to unravel the mystery of DNA turned out to be X-ray diffraction. If X-rays, which have a much shorter wavelength than visible light, are shone through crystals whose atoms have a regular spacing comparable to the wavelength of X-rays, then those waves will be diffracted to project a regular pattern on a film set up to detect the emerging rays.

In 1913 William Henry Bragg and his son Lawrence first developed the technique of X-ray diffraction. The X-rays the Braggs used had a wavelength about a thousand times shorter than visible violet or blue light, the shortest wavelengths visible to human eyes. Using X-ray diffraction, Bragg and others were able to see two-dimensional diffracted images of the atoms arrayed in crystals of materials such as common salt or diamonds. By rotating the sample in an X-ray beam, a three-dimensional image could be reconstructed. It turns out that both salt and diamonds have quite simple crystalline structures, so the multiple two-dimensional images viewed were relatively easy to decode in order to derive the three-dimensional crystal structure. Organic molecules, however, often consist of very long chains of (simpler) molecules assembled in regular patterns, but also often replete with bends, twists and turns. As you can well imagine, the two-dimensional interference pattern of the tangle of atoms making up such a molecule leads to a very confusing picture, rather like trying to unravel a messy ball of twine by looking at its shadow projected on a screen, as the ball of twine is rotated.

Despite this complexity, by the 1950s, scientists were busy refining Bragg's X-ray diffraction tool so as to be able to better understand complex organic molecules. The interpretation of the X-ray images was critically supplemented by the building of models out of cardboard, tin, wood and plastic at a large scale so that scientists could start to envisage what the substance under study might look like at sub-microscopic scales. Computers (at the time very expensive, new and bulky machines) rapidly became indispensable tools to help sort through and make sense of the diffraction data.

Maurice Wilkins and Rosalind Franklin were the first to apply X-ray diffraction techniques to DNA successfully. They were also the first, with Alec Stokes, a colleague of Franklin and Wilkins, to surmise that DNA might exist in the form of a helix. Wilkins was awarded the Nobel Prize with Watson and Crick in 1962 for his contributions. Sadly, Rosalind Franklin, whose artful X-ray diffraction images of DNA provided the ultimate clues to the decipherment of the DNA molecule, died of ovarian cancer in 1958, four years too early to share in the prize.

Figure 11.1 – *Francis Crick with a DNA model.*

The final piece of insight to crack the DNA structure came from Francis Crick and James Watson on Saturday, February 28, 1953 at the Cavendish

Laboratory of Cambridge University while analyzing Rosalind Franklin's latest X-ray diffraction image. They concluded that DNA comprised a double helix (and not a triple one as postulated by Linus Pauling, and as previously seriously considered by Crick, Wilkins and others). They also concluded that the basic elements of the code (analogous to "dit" and "dah" of the Morse code used by wireless operators of their day) are four chemical compounds that they called *nucleotides*.*

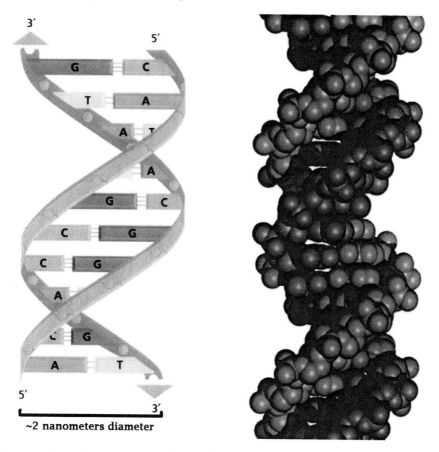

~2 nanometers diameter

Figure 11.2 – *Two representations of the DNA molecule. The DNA double helix is about 1/50,000 the diameter of a human hair, but, if unwound, would be about two metres in length! Within the cell it is wound up on molecules called histones. The outside support structure of DNA consists of sugar and phosphate molecules, while the four coding nucleotides (adenine, guanine, thymine and cytosine) are inside the structure.*

---

* See Table 11.1 on the next two pages to pursue the Morse Code analogy.

So you want to write a book in Morse Code!

Table 11.1 - The Genetic Code and Alphabet Compared to English Language Coding

Human Language coding	Genetic coding
Morse Coding for Alphabet (26 letters)	**Base Coding for Genetic alphabet (20 amino acids)**
Morse Code has two elements, i.e. dot and dash. These elements are used in various combinations of from one to five elements to provide a code for each of the 26 letters of the alphabet. For example, the letter A is [dot dash], E is [dot], S is [dot dot dot] and Z is [dash dash dot dot]. There is also coding for punctuation and numbers (for example the number 2 is [dot dot dash dash dash]).	There are four different elements (called bases) in the DNA code, i.e. G, C, A, and T (U replaces T in certain circumstances). Specification of a given amino acid requires a string of three bases. This string is called a codon. For example, either UGC or UGU specify the amino acid cysteine (most amino acids have one or more alternate codon sequences). Together the 64 possible codons specify the 20 amino acids, which are the genetic equivalent of the letters of the alphabet. There is also a punctuation code, in this case a "stop" code which indicates where the code for a particular protein ends. (See Table 11.2 for more detail.)
Use of Alphabet to spell words	**Use of Amino Acids to "spell out" protein structure.**
Letters of the alphabet are combined in various ways to form words. Estimates of the number of words in the English language range from a low (by the Oxford English Dictionary [OED]) of about 250,000 distinct words "excluding inflections, and words from technical and regional vocabulary not covered by the OED", to a high by Google and others of just over one million words.	The 20 amino acids are the "alphabet" used to spell out genes, and the genes in turn specify the structure of proteins in a two step process by which the RNA coding for the protein is assembled beside the gene, and the RNA is later used as a template to produce the protein. All proteins consist of interconnected chains of amino acids. Some of these chains are very long. Proteins are the basic components of all body structures in all life forms. The "dictionary" of known proteins used in life forms is in excess of two million, but a much smaller number are to be found in any given cell. Humankind is thought to use at least 90,000 different proteins, some researchers believe we use as many as five million! Most proteins have a large number of variants, making a proper estimation of the number of proteins at least as difficult as estimating the number of words in the English language. Part of the difficulty is that some proteins are exceedingly large molecules. In theory the number of possible proteins goes up exponentially as the number of amino acid components increases. One of the largest known proteins in the human body, titin, contains about 27,000 amino acids. Titin is found in muscle tissue.

Available words stored in Dictionary	Available genes stored in the genome
A dictionary is a compilation of words and definitions bound into a book with a spine and covers. Some dictionaries are very comprehensive, while others specialize, for example in technical terms or geographical areas, such as American English or Medical terms.	The genetic equivalent of the dictionary is the genome, which is the entirety of genetic information for the organism in question. For multicelled life forms, the same genome is used to specify many different cell types, for example neurons, liver cells and blood cells. Any given cell type will use a subset of all the genes in the genome, just as a book will use a subset of all the words in a dictionary. The chimpanzee genome has about 99% of its contents identical to that of mankind. The mouse genome shares only about 80% of its contents with mankind and chimpanzees. But the really striking fact is that all life forms use the same language. It is not the case that mice use a French dictionary and cedar trees a Greek dictionary. We all make use of the same 20 amino acids (i.e. the same alphabet), as well as a lot of the same proteins (i.e. words).
Author produces output	**Organism both regenerates and maintains itself**
A selection of the words in a dictionary are ordered by a writer to produce plays, poems, technical papers, essays, etc.	All organisms speak the same language. But as with human language, the order and timing of use of the words/genes, has a huge impact on the result. We know that there are a host of special RNA sequences and special function proteins which control the timing of cell development and the selection of proteins used. Together they dictate the form and function of each cell and ultimately of each organism, much as word order and punctuation often differentiate particular literary forms. Also, events outside the cell can and often do influence cell development. (See, for example, Dialogue 14)

Each nucleotide is made up of a sugar-phosphate backbone combined with one of four simple organic chemicals, namely adenine, thymine, guanine and cytosine. These chemicals, called bases, are internal to the backbone structure and not external to it as initially thought. The chemical nature of these four bases is such that the adenine from one strand naturally bonds with the thymine of the other; the other two nucleotides, guanine and cytosine, similarly form interlocking pairs. The beauty of this pairing system is that it enables two essential methods of replication. Firstly, for purposes of cell reproduction, the entire DNA sequence (i.e. the chromosome) that comprises many individual genes, can be copied by the division of the molecule into two mirror-image halves, each of which attracts its pairing nucleotides complete with base and attached phosphate group, so that, bit by bit, the two half-DNA segments are restored to yield a pair of DNA molecules.

But each cell also relies on its DNA to manufacture the proteins it needs to sustain itself. Thus there is a requirement for a second replication system that causes a particular gene in the total DNA sequence of thousands of genes to be "activated" on command to produce an RNA (ribonucleic acid) sequence, which provides the template for protein manufacture. We'll take a look at this process later.[*]

The Watson-Crick epiphany must surely rank as one of the great discoveries of all time. It has unleashed armies of researchers who, to this day, are making life-saving and life-enhancing discoveries, with much more yet to be learned.

However, it is one thing to know that we inherit our looks and much of our behaviour through genetic codes in sets of long molecules containing four nucleotides. It is quite another to understand how the coding works. For example:

- How can you be sure that the DNA does split neatly into two pieces as described above, as the first essential step in the replication of DNA?
- The DNA molecule is a very long string. How do you suppose that the "genes" show up on the DNA string? And how do we know when one gene begins and another ends on the long

---

[*] There are some nice animated sequences showing the DNA and RNA replication and transcription processes. If the reader is not already familiar with the way this works, he or she might find it interesting to do a web search on "DNA replication" and/or "RNA transcription" to get a better idea of how these beautiful processes work.

strand of a DNA molecule?

- The genes mostly code for proteins, which are made up of many different combinations of only twenty amino acids. How do four different nucleotides code for twenty different amino acids?
- Once a gene has been identified, how do you know what it does, and what happens if you make minor modifications to it?
- What are the roles of RNA, and how are they carried out?
- How do genes "express themselves" to form the flesh and blood we are made of?

Over the sixty odd years that have elapsed since the structure of DNA was discovered, much has been learned toward answering these and other related questions, but there is still a huge amount to learn. Watson's book *DNA: The Secret of Life*, published in 2003, provides some interesting insights into the ingenuity of some of the experiments devised to answer such questions. Let me tickle your imagination by discussing a few of these experiments, but first a few reminders to provide background.

A human has 24 chromosomes, each consisting of one long (double-helix) DNA strand and some proteins that provide the backbone to support the base pairs. The chromosomes are numbered from 1 (the largest) to 22 (the smallest), plus the X and Y chromosomes. Y chromosomes are uniquely inherited from one's mother, X chromosomes from one's father. The rest come from both parents. Chromosomes vary quite a bit in size. Chromosomes 1 and 2, for example, have about 255 million base pairs, while chromosome 22 has only about 33.4 million. (These estimates are still subject to change, as more is learned about the makeup of our genes.) Within each DNA strand there are a number of genes, where a gene is defined as a functional and physical unit of heredity passed from parent to offspring. Genes are thus pieces of DNA. Confusingly, for many multicellular life forms, including humans, these pieces are usually not even contiguous pieces of DNA, but have seemingly random strings of base pairs (*introns*) interspersed in the gene sequence. Most genes encode the information needed for making a specific protein. Proteins are molecular chains built up from 20 different known building blocks, the amino acids. The number of permutations in the ordering of these building blocks along the protein molecule is very large, thus allowing for the encoding of millions of different proteins. For many years, humankind was thought to make use of about ninety thousand different proteins. By 2012, some researchers believed that this number was more like five million different proteins. (I shall return to this subject later.)

The matters of the roles of RNA and the manner in which genes express themselves will be addressed at our next meeting.

The first bulleted question above – How can you be sure that the DNA does split neatly into two pieces that in turn cause replication of more DNA? – was a topic of vigorous debate in the 1950s. It was a nice idea that the two intertwining strands would split into two identical pieces, which would then assemble the appropriate nucleotide constituents to make themselves into two new DNA molecules. But how could one prove that this was actually happening? The proof came in 1958 after Matthew Meselson and Franklin Stahl raised bacteria in a medium containing a heavy isotope of nitrogen (^{15}N). Thus ^{15}N was incorporated in both strands of the DNA. They then transferred some of this DNA to a new culture, which contained a lighter isotope of nitrogen, so that the next round of replication took place using ^{14}N. If the DNA really did replicate as predicted by Watson and Crick, i.e. by "unzipping" the two helical strands, then the next generation of replicated DNA should contain one strand with heavy nitrogen and one with light nitrogen. Subsequent generations would contain some DNA with one strand each using light (^{14}N) and heavy (^{15}N) nitrogen, and some that used 100 per cent light nitrogen. When they put the new mixture of DNA into a centrifuge, a device that separates out molecules of different weight, they found that the DNA did indeed cluster into three bands corresponding to both strands heavy nitrogen; one strand of each, and both strands light nitrogen — a neat and ingenious demonstration.

Since this initial discovery, a lot more has been learned about the process of DNA replication. Special families of workhorse proteins called enzymes facilitate and guide the replication processes. For example, a series of enzymes (called topoisomerases) initiate the gene replication process by causing the DNA strand to unwind, while still other enzymes recruit new nucleotides to bond to the unwound and unbound strands. RNA has an extra hydroxyl group in its nucleotides, which inhibits it from forming a stable double helix, which in turn frees it to bend into a variety of different, stable, three-dimensional forms. For example, the fact that in RNA uracil (U) substitutes for the thymine (T) found in DNA permits the formation of "hairpin loops", which are short sections where segments of the RNA fold over and form a series of pair bonds with the preceding segment.

By means of another ingenious experiment published in 1958, Crick and Leslie Barnett determined that the twenty different amino acids that make up proteins are coded using triplets of nucleotides. Each such triplet, which specifies a unique amino acid, is called a codon. Further work by Marshall Nirenberg established what the codes were for one amino acid (phenylalanine).[63] This knowledge unleashed several years of feverish code-breaking activity to determine the codes for the remaining nineteen amino acids. Nirenberg won a Nobel Prize for his leadership in uncovering the coding system. By 1966 the world knew what the triplet codes (i.e. codons) were for each of the twenty amino acids [listed in Table 11.2, which I had to read to Davey].

The Genetic Code

The Basic Four-letter Alphabet for DNA and RNA

Base

guanine	G (pairs with cytosine)
cytosine	C (pairs with guanine)
adenine	A (pairs with thymine [DNA] or uracil [RNA])
thymine	T (in DNA, pairs with adenine)
uracil	U (in RNA, pairs with adenine)

The four letters of the DNA (G, C, A, T) or RNA (G, C, A, U) alphabet code for the 20 amino acids used to make up millions of different proteins in plants, animals, and other life forms. The codes for each of the amino acids are shown on the right. Each code comprises three bases. There are thus 64 different codes (i.e. 4 x 4 x 4) available to specify only 20 amino acids. As can be seen, some amino acids are uniquely specified by as many as six different three-base combinations.

The Twenty Amino Acids Used to Manufacture All the Many Thousands of Proteins

Amino Acid	RNA Codon
alanine	GCA GCC GCG GCU
arginine	AGA AGG CGA CGC CGG CGU
asparagine	AAC AAU
aspartic acid	GAC GAU
cysteine	UGC UGU
glutamic acid	GAA GAG
glutamine	CAA CAG
glycine	GGA GGC GGG GGU
histidine	CAC CAU
isoleucine	AUA AUC AUU
leucine	UUA UUG CUA CUC CUG CUU
lysine	AAA AAG
methionine	AUG
phenylalanine	UUC UUU
proline	CCA CCC CCG CCU
serine	AGC AGU UCA UCC UCG UCU
threonine	ACA ACC ACG ACU
tryptophan	UGG
tyrosine	UAC UAU
valine	GUA GUC GUG GUU
STOP CODONS	UAA UAG UGA

Stop codons mark the end of the coding segment of the gene.

Table 11.2 – *The Genetic Code Used in DNA and RNA*

These twenty amino acids are the building blocks for the likely millions of very complex proteins that give us and all plants and animals our unique properties. The proteins often contain thousands of amino acids and tend to bend and fold in characteristic ways. A single missing amino acid can have very serious consequences, at least in part because the shape of the resulting protein will change, so that it will no longer fit correctly with other proteins with which it has to interact, or it may interact with chemicals it should not interact with. One of the largest proteins discovered so far amongst those coded in the human genome comprises about 27,000 amino acids. It is a substance called *titin* (also known as *connectin*) and is found in muscle tissue, where it functions as a spring.

A good example of the possible catastrophic effects of a missing amino acid is sickle cell anaemia, a disease caused by the displacement of a single nucleotide in the DNA sequence. As a result of this displacement the manufactured haemoglobin molecule is altered, with the unfortunate consequence that it is unable to carry oxygen properly. This genetic alteration is carried by about a third of indigenous sub-Saharan Africans. The defective gene is much less common elsewhere. Sickle cells provide the owner with partial protection from malaria, which explains its prevalence in Africa.

Starting in the late 1960s, researchers learned how to cut DNA strands in well-defined locations, for example, immediately after the nucleotide sequence GAATTC (or: guanine, adenine, adenine, thymine, thymine, cytosine), using naturally occurring restriction enzymes, each of which cuts the DNA when it encounters a specific sequence of nucleotide bases. Equally important, and at about the same time, scientists learned how to make many copies of these cuttings so that they could produce them in useful quantities. Replication was enabled through a process of "gluing" the strip of DNA to be studied onto a plasmid, a small loop of DNA that lives within bacteria, and, in normal circumstances, is replicated and passed on during bacterium cell division. This process of cutting, splicing and replicating rapidly became a standard laboratory procedure for genetic research.

Determination of the actual sequence of bases (e.g. GAATTC) was quite another problem. Think of a string of 250 million combinations of the four bases (A, T, G and C), and your task is to identify the order in which each of the bases appear on the string – without looking at them, because the bases are too small and too entangled to be seen with enough precision to identify the order or sequence of the individual bases. It was a daunting challenge, to put it mildly.

In 1980 Wally Gilbert of Harvard and Fred Sanger of Cambridge University independently arrived at two different methods for base sequencing. As a result, both won a Nobel Prize for Chemistry – Fred Sanger, for the second time! Ultimately, it was Sanger's base sequencing method that proved to be the most practical.

In order to save time, Davey, I shall not provide a full description of the procedure now. If you want to learn more about these methods, I will procure a talking book version of Watson's 2003 book for you to listen to.

– Thank you, Peter. What you are describing is absolutely

fascinating, but already a little difficult to follow even for me. I think I will take a pass on listening to Dr. Watson's book for the time being.

– Well, I'm glad to know you are still listening in. You have been quiet for a long time. It is unlike you, Davey.

– I wouldn't want you to get bored by my company. It seemed to me that a little variety in my behaviour might help – besides, I find I have to concentrate on what you say or I lose track.

– Just hang in a little longer if you can. I have almost finished what I wanted to tell you today.

– Okay. Let's keep going. But I can tell you that I am very glad that the key to my existence is not contained in long strings of DNA – but don't even think to ask me how the design of me is stored; that will come later.

I was too tired by now to remonstrate. I just wanted to finish what I had to say and go for a long walk, so I plunged ahead.

– As you have probably realized already, just understanding the sequence of base pairs is not enough to be very useful. We also need to be able to determine where on the long DNA chain the code for construction of one protein ends, and where the next begins. By 1990 a massive ($3-billion, 15-year) international Human Genome Project (HGP) was launched with the primary objectives of:

mapping the entire human genome of some 3.3 billion base pairs,

locating the 20,000 to 30,000 genes thought to be contained therein, and

determining the function of each of those genes.

This project succeeded magnificently, as did that of its private-sector competitor, the US-based Celera Corporation, which was founded by Craig Ventner, a prominent American biologist and entrepreneur. Celera launched its venture capital–funded $300 million project in 1998, and benefitted from access to earlier work of the HGP. On the other hand, the HGP was moved to adopt the much more efficient method of gene sequencing used by Celera in order to stay competitive.

In announcements roughly spanning the time period 2000–2006, both groups announced various stages of completion of the gene mapping, accompanied by steadily improving estimates of accuracy.

Theirs was a monumental accomplishment, and one that generated much more than the usual drama surrounding massive science projects, for it was during this time period that the issue as to whether specific genes could be patented was decided after vigorous debate. As might be expected, Celera was intent on patenting genes it discovered, which had significant commercial potential, whereas the Human Genome Project, a tax payer–funded initiative, was required to make its findings public on an immediate basis, with no plans for patenting. This situation generated intense competition between the two groups and contributed substantially to earlier than planned project completion.

In March 2000, US President Clinton announced that gene sequences would not be eligible for US patents. Celera's shares nose-dived along with those of many other biotech stocks, which combined lost about $50 billion in market capitalization in a two-day period.[64]

Quite early on in the gene-sequencing endeavour, it was discovered that only 1.5–2 per cent of the base pairs on a human DNA strand are components of genes! In the DNA of most plants and animals (but not bacteria) there are long sequences that, until fairly recently, appeared to have no vital use. As I have mentioned already today, the base pairs in these regions are called introns. Until quite recently, the 98 per cent of our DNA that comprise the introns were often referred to as "junk DNA", and presumed to be largely the left-over rubbish of previous genetic alterations.

Unlike unicellular creatures, for which each gene has a specific role, multicellular creatures need to replenish and replicate a variety of different cells, all using the same set of genes. Yet a nerve cell in the brain has very different properties from, say, a cell in the kidneys or the blood stream. Multi-celled creatures have evolved methods of reusing the same gene to produce different proteins in different cells. It soon became evident that some intron sequences help orchestrate this process.

Furthermore, researchers discovered that not all genetic information comes through the DNA⇒RNA⇒protein classical sequence. Several other replication processes seemed to be possible, including one in which RNA acts directly to inhibit or enable reactions, without the formation of a protein. It is thought that such processes could be particularly important to brain development. If so, this could go a long way to help explain why humans have far more introns than other species but have about the same number of genes as most other mammals, including mice, rats, cows, dogs and horses.[65] I have prepared some tables that provide comparative numbers relating to the genes of different multi-celled species. I can read these to you in a minute or two, Davey. I will be saying more about genes over our next two meetings.

– Peter, I think it is safe to say that you should forget all about the phrase "junk DNA". If I may say so, it is a piece of junk phraseology left over from the end of your twentieth century. While it is true that introns may contain some garbage sequences, it is now very clear that many intron sequences are vital to the reproduction and health of multicellular life. Introns are currently the subjects of very active research. For example, it now seems likely that intron modification will prove to be an important method for detecting and curing health problems, though probably not in the immediate future.[66]

– Well! Isn't that interesting? It sounds to me as though you have been talking to some experts again, and kept those contacts secret from me. I don't want to be rude, but you might save me a lot of time and effort by letting me know in advance if you have become an expert in the subject we are about to discuss.

– Frankly, I think not. I don't want to be rude either, but I find it very helpful to hear several speakers on very important subjects. Luckily everyone has a somewhat different approach to their favourite subjects, so in the end I come away with a much better understanding if I listen to several experts discuss the same subject. DNA is to me a very important subject. Listening to experts on the subject is helpful, but it can often be confusing. You are able help me sort through their more detailed explanations and to view them in a broader context.

– I suppose I should take that as a compliment. I'll have to think about it. For the time being I shall just carry on and talk a bit about the next large international scientific project related to the human genome, which is called ENCODE (short for ENCyclopedia Of DNA Elements). In essence, ENCODE addresses the likely functions of the 98 per cent of the DNA strands that are not genes.

On September 6, 2012, in the co-ordinated release of thirty papers, six of which are published in a special issue of *Nature* magazine of that date, the research group of over four hundred scientists demolished the already fragile myth that the introns were essentially junk DNA. ENCODE project members have identified functions for 80 per cent of the human genome as part of a process that has added enormously to our understanding of gene expression mechanisms

Project researchers used comparative data on gene transcription in 147 different human cell types, as well as cells drawn from some other life forms, for example mice, flies and worms, to identify groups of introns associated with important control functions, such as enhancers (which regulate the rate of gene replication), promoters (i.e. the sites at which DNA replication is initiated) and a variety of other RNA transcripts whose detailed functions have not yet been identified.

Instead of envisaging a long ribbon of DNA with genes and parts of genes located at rare intervals down the length of the DNA strand, we are instead urged to envisage a long ribbon partially wound around a number of biochemical reels called histones. Gene transcription is heavily dependent not just on the location of the gene on the histone reel (which can make access to the chemicals needed for transcription easy or difficult) but also by what (RNA) enhancers and promoters are nearby on the reel, even though they may be relatively far away on the DNA ribbon. Other RNA sequences can and do affect the configuration of the histone reel itself, which introduces yet another control variable. As the ENCODE summary paper put it: "Beyond the linear organization of genes and transcripts on chromosomes lies a more complex (and still poorly understood) network of chromosome loops and twists through which promoters and more [distant] elements, such as enhancers, can communicate their regulatory information to each other."[67]

Journalist Stephen Hall summarized the new state of our understanding of the DNA molecules rather nicely. He wrote, "the ENCODE group has produced a stunning inventory of previously hidden switches, signals and sign-posts embedded like runes throughout the entire length of human DNA."[68] By the way, the word "rune" may not be familiar to you. In the sense used here by Hall it refers to marks of mysterious or magical significance included in an early Germanic alphabet used in Scandinavia.

The research has opened up new realms of the unknown for further investigation. For example, one group of researchers observed: "genes are highly interlaced overlapping transcripts that are synthesized from both DNA strands. These findings force a rethink of the definition of a gene and of the minimum unit of heredity."[69]

In one broad stroke, our understanding of the complexities of the human genome has massively improved, and our ability to tackle cures for genetic diseases has been infused with a new life and new challenges.

It is helpful to understand the key role of computers in many aspects of DNA research. Recall that in examining the human genome we are talking about twenty-four human chromosomes each with from 33 million to 255 million base pairs. Humans are not at all good at dealing with such huge quantities of data. Fortunately, computers are very good at such tasks. Without them to help with data analysis, our ignorance about the structure and function of DNA would be immeasurably worse.

For example, computers have proven to be very useful in helping to isolate defective genes. One can compare the DNA sequences of groups of people with and without a particular disease and look for differences from the healthy group that are also common to the group with the disease.

Given the huge number of total sequences of bases, it helps a lot if the researcher has a clue as to which chromosome likely houses the gene being searched for, and roughly where on the DNA sequence it might lie. The search is often complicated by the fact that many diseases result from a combination of genes being defective, so finding one gene often does not provide a solution. Despite these difficulties, several defective genes associated with disease have already been found, and more are being uncovered as we speak. However researchers found out early that about 90 per cent of the genome anomalies they identified in this manner were not in the gene itself but elsewhere in the maze of introns. The work of the ENCODE group in finally making sense of the role played by most of the intron sequences lends new life and meaning to this research.

So Davey, in a nutshell, I am glad to say that I completely agree with your suggestion that the term "junk DNA" should be junked.

– Holy Maloney!

This interjection came out of the blue!

Where, I wondered, had Davey ever learned to say "Holy Maloney"? Was he consorting with an Irish biologist? A teen-aged genius in a rural high school? A Hollywood comedian? Furthermore, I thought I detected a scent of delighted triumph in his voice, as though he knew before he said it that I would be surprised by his comment. Of course the scent was every bit as elusive as the "hints of blackberry and orange juice" that professional wine tasters claim to find in some quite ordinary wines, but it was there nonetheless. Furthermore, I thought I could almost hear him grinning from ear to ear. I took the bait.

– Hold on a minute, just where did you get "Holy Maloney" from? Have you taken up with an Irish girlfriend?

– I thought you'd appreciate my continuing improvement in the use of colloquial English, and I felt I had to interrupt. You are coming very close to losing me here – something I never thought would happen! You seem to be implying that your shape and behaviour are defined by the ordering of only four particular chemicals dangling on a strand, and that the ordering of these four chemicals in turn codes for only twenty more complex chemicals, which in turn are the components of up to five million different proteins. What you haven't said anything about is why these particular four and then twenty chemicals out of all the millions that could have been involved, are directly responsible for all life in your universe . . . er . . I mean, on your planet.

– Which do you really mean? What do you know about life elsewhere in our universe?

– Oh, not much really.

Davey's response came quickly. It was clearly not a subject he wished to pursue.

– But I would like to know how you have evolved such complex life forms from such a simple basic structure. And by the way, if you can, please leave out some of the complicated details of the protein chemistry. As I have already told you, our chemistry is very different from yours, which is perhaps why I find it a little difficult to follow.

Our discussion had already lasted longer than I had intended, but I told him not to despair, but rather to keep working at it. Ensuing discussions would address most, if not all, of his concerns. In the meantime, I told Davey I would like to close our session off with some numbers that provide a bit of a perspective on how the human genome compares with the genomes of some other plants, animals, yeasts and viruses. I read out two tables, taken in the main from Watson's 2003 book *DNA: The Secret of Life*.[70]

Species Name	Number of Genes
Human	**Estimated 35,000 in 2003** **Estimated 20,244 in 2012**
Mustard plant	27,000
Nematode worm	20,000
Fruit fly	14,000
Baker's yeast	6000
Gut bacterium (*Escherichia coli*)	4000

Table 11.3 – *Sample Gene Counts*[71]

It is indicative of the rapidity of developments in this field that estimates of the size of the human genome have continued to decline from Watson's estimate of 35,000 genes, to Gil Ast's estimate of "less than 25,000 genes" in 2005,[72] and on to an estimate of about 23,000 genes in 2009, and by early 2012 to an estimate of 20,244 genes. You can bet that this is not the final number either, given that the ENCODE group have indicated (as previously quoted) that new evidence about the operation of DNA forces "a rethink of the definition of a gene and of the minimum unit of heredity." The first table I read to Davey (Table 11.3) shows the estimated number of genes for a limited selection of species, while the second (Table 11.4) lists the approximate size of the genome in millions of base pairs for another short list of species. For most of us, these numbers are rather bewildering. Does it really only take 20,244 genes to define human beings, including behaviour, bone structure and shape of nose? Watson reports that the relatively small total number of genes in the human genome came as a surprise to researchers, who were expecting we would have 70,000 to 100,000 genes. These earlier, much higher, estimates were influenced by estimates at the time that human beings make use of at least 90,000 different proteins. Why should the amoeba and the fern have such huge genomes (in millions of base pairs)? Perhaps more intriguing, the burgeoning field of proteomics (the study of the *proteome*, which is the complete set of proteins produced by an organism) is rapidly leading to some fascinating discoveries, one of which is that each protein specified in the genome likely has from 50 to 100,000 variants produced by cutting and splicing operations.[73] As of 2012 it was estimated that humans use as many as 5 million different proteins!

Species Name	Approximate Genome Size (millions of base pairs)
Fruit fly	180
Fugu (puffer fish)	400
Boa constrictor	2,100
Human	**3,100**
Locust	9,300
Onion	18,000
Fern	160,000
Amoeba	670,000

Table 11.4 – *Sample Genome Sizes*[74]

Other mechanisms for cell alteration are also coming to light. A phenomenon called "gene jumping", which has the effect of copying small segments of DNA into new locations on the DNA strand, and thereby altering protein replication and structure has recently been reported.[75] (Gene jumping helps explain the existence of introns in DNA, as well as pointing the way to methods by which millions of different proteins can be manufactured using only 20,244 genes.)

This is a rapidly advancing field. Stay tuned for new discoveries.

On average, 99.9 per cent of our genome is the same as that of other humans, which means that there are only about 3 million base pairs out of the 3 billion base pairs in our genome that vary between us. These differences give rise to many of the visible and invisible variations between humans. The (slightly) different versions of the same genome are referred to as *alleles*.

If all the human chromosomes were strung out in a long line, the DNA would be about two metres in length. This is the more remarkable if one realizes that each human has about a hundred trillion cells, each with about two metres of DNA! Clearly, as with the proteins, DNA must be very thin as well as being multi-folded or coiled up in order to fit in a cell.

In the circumstances it is easy to understand why Watson has concluded, along with "most of [his colleagues] in molecular biology" that "the essence of life is complicated chemistry and nothing more."[76]

Darwin astounded his contemporaries with the discovery that one did not need to invoke divine intervention to account for the many different species found on Earth; Crick and Watson and their many colleagues (especially, at the beginning, Franklin and Wilkins), have opened our eyes to how this is possible.

That's it for today, Davey, my friend. I'm heading out for a walk in the sunshine. You can come with me if you want.

– Not today, thank you. I've got a headache!

The Life of a Cell

> *It cannot be said too often: all life is one. That is, and I suspect will forever prove to be, the most profound true statement there is.*
> BILL BRYSON, *A Short History of Nearly Everything*, 2003

I took a break from my dialogues with Davey over Christmas and for the month of January. I had too much else on my plate, and Davey seemed not to be disturbed by the proposed delay.

February in Vancouver is a month of contrast, chill winds and occasional snow on some days, but warm sunshine, pert and colourful crocuses and clear blue skies on others. We were having a good weather day when my next meeting with Davey came due. I could not resist the temptation to have our dialogue in the open air down, by the sea. So, I dressed warmly, armed myself with a comfortable garden chair, my notes and a thermos of coffee, and set out for Jericho Beach, which, thankfully, is nearly deserted at that time of year, even on sunny days. By then I knew that Davey could find me with ease, wherever I happened to be. I often wondered if he always knew where I was and what I was doing, or only when he specifically "tuned me in". It was on my mind to ask him some time, but I had other more pressing questions about his past life.

There were times when I wondered why I was working so hard to please a disembodied voice. This likely happened most frequently while I was working on the section on microbiology. As a youngster biology had not attracted me. I had no desire either to dissect frogs or memorize a lot of Latin names for animals and plants; both activities seemed to make up a large part of the biology curriculum. Rightly or wrongly, I concluded that biology was boring while physics was "fantastic". By the time I got to graduate school, the fields had levelled quite a bit. Physics was still challenging and interesting, but so was biology – in fact there seems to have been a wholesale migration of physicists to bioscience, with Francis Crick likely the best-known transferor. I did not move fields but rather chose a non-academic career in business and government. That made my work to catch up on my ignorance of the biosciences doubly challenging. If Davey noticed, he didn't remark on it. But increasingly, as the dialogues progressed, he stopped me to pose questions and make remarks.

Some of these remarks are unforgettable, including:

"How strange that one animal should develop to so totally dominate the other animals – at least for a short time. I wonder how long this dominance will endure. I suppose it will last for a few more years yet."

"It is wonderful to have a life form to examine that is so far behind us in its development. I can hardly believe my luck."

I had been ready for the last remark when it came. With a voice pitched satisfactorily high on the tartness scale I suggested that he must therefore be in a very good position to help us poor humans to do better at understanding our circumstances.

Davey then launched into quite an elaborate apology that seemed to boil down to a reluctance to tamper with a universe he did not yet fully understand, combined with a weak promise, once again, to tell me more later.

At any rate, he said he enjoyed our last rather challenging meeting, but he was a little less sure that he wanted to learn about how the cell works. I teased him at that point.

– All I know about the development of life in your universe is what you have told me, which as you well know is almost nothing. Nevertheless it is hard for me to imagine that life in your universe would have developed so very differently from life in ours. Your materials may be very different, and your surroundings, but somehow, your universe, too, must have seen the evolution of a "heredity code script" to enable life forms to reproduce as postulated by Schrödinger,[*] and there must have been a supporting system for nurturing and reproducing that code script.

– Why do you think I am so interested in this subject? If I already knew the details of how the Envergurians and I came about, why, unless for reasons of pure curiosity, would I trouble myself to learn how humans evolved? After all you are not, you must admit, a

---

[*] See Dialogue 11 for a brief discussion of Shrödinger's role.

stunningly intelligent race. There is a lot you have failed to grasp not only about how your universe is put together but also about how you yourselves came about!

– But if you and your kind are so terribly intelligent, how is it that you, too, lack an understanding of how you came about? I know you went through a catastrophic period when you lost a lot of information about your Envergurians, but surely someone as supremely intelligent as yourself would by now have been able recover at least that part of Envergurian data that related to their ability to reproduce. And the same should be true for you, too!

– Mmmm. I think I understand how you would feel that way, but recovery of that information is much harder than you might imagine. Once I understand your story better, I shall be able to tell you more. In the meantime, please proceed to tell me about the life of the cell.

– Promises. Promises. Always promises. I'm getting tired of this!

It reminded me of the White Queen's observation to Alice in *Through the Looking Glass*: "The rule is jam to-morrow and jam yesterday – but never jam to-day."

But I was not going to let Davey get me down on a day when the sun shone and the birds basked and all seemed right with the world. So, without further comment, I launched into my prepared notes on the lfe of a cell.

A cell can be defined as the structural and functional unit of all living organisms. It is generally accepted as the smallest unit we associate with life. Viruses are smaller but are incapable of self-replication. They need to invade a living cell and commandeer its reproductive capabilities in order to reproduce. The word "cell" is derived from the Latin word *cella*, a small room. It was coined by Robert Hooke (1635-1703) – a brilliant English scientist and colleague of Newton, after viewing (and drawing) cells of cork under a microscope of his own making.

Even when examined at the most simplistic level, the activities a cell must perform look very challenging if it is to be successful both in maintaining itself and ensuring that there is at least one replacement cell when it fails. Our bodies contain some 100 trillion (10^{13}) cells, some of which are replaced very frequently (such as cells in our outer skin), and some of which last a lifetime (such as most of the neurons in our body, about 0.1 per cent of all our cells). Of course, we multicellular life forms can afford to see many of our cells die and be replaced, a luxury that is not an option for unicellular creatures. Incidentally, our human bodies play host to about ten times as many unicellular microbes as there are human cells.

Many of these guests live in our intestines. Fortunately the microbe cell is very much smaller than a human cell, so there is lots of room in the inn for these particular guests, most of which are benign, and many of which are important to our health.

Let's just look at the basic list of tasks that all the cells in the body have to be capable of:

1. Taking in nutrients

In order to operate at all, a cell needs to have on hand the materials it needs to fabricate the twenty amino acids it uses to construct thousands of different proteins, enzymes and RNA. And, of course, it also needs an energy source in order to fabricate this long list of chemicals. By and large the needed materials and fuel are absorbed through the cell walls. In multicellular life forms there are cells whose special function it is to ensure that supplies and fuel are distributed to other cells in the organism (think, for example, of haemoglobin in our blood stream, which delivers oxygen and some other gasses to cells). Unicellular life forms have no such luck. They are dependent on food and fuel arriving on their doorstep more or less by happenstance, or else, for those cells with a means of locomotion, on encountering the nutrients they need as they wander about. Otherwise they die.

2. Controlling the chemical manufacturing process

The cell needs a means of determining when it is time to manufacture more of a particular protein or RNA sequence, buttressed with a means of initiating manufacture of the needed chemical. It is also important that the cell be able to determine when there is enough (or too much) of a particular ingredient, so that manufacture of it can be slowed or halted.

3. Picking the right recipe

Having determined that more of a particular substance is needed, it is then necessary for the cell to pick the correct recipe (i.e. gene). If the cell in question is part of a multicellular being, then it also must be able to deal with a lot of information contained in introns, a lot of which will be superfluous for that particular cell type. (Recall that unicellular life forms generally don't have any introns.) Given that there are about 255,000,000 base pairs in the largest human DNA strands (chromosomes 1 and 2), finding the right gene, especially if it comes in multiparts strung out along the DNA, is no mean task.

4. Delivering newly produced material to where it is needed

The longest nerve cell in the human body is often over a metre long and reaches from the spinal column down the leg to the foot. It enables us to lift the sole of our feet off the ground and walk on our heels. I learned about this nerve the hard way, when a misguided oar stroke in an "old-timer's" rowing eight caused several disks in my spinal column to bulge and scrape some insulation off the nerve fibre, thus shorting out the activation signal to the muscle, and leaving me with something called *foot drop*, a condition where my left foot made a flapping sound as the sole of my foot flopped down as soon as my heel hit the ground. Over a year later, the foot drop was much better, but it is a slow supply line to get chemicals correctly distributed over such a long distance. There is both a fast (20–40 cm/day) and a slow (0.02–0.5 cm/day) transport system for material in the cell. At the slow end of the scale, it could take well over a year for some proteins to get from the nerve nucleus where they are manufactured, to a location down near the foot. Even using the fast transport system, transport can require two and a half days for deliveries to the foot.

5. Cell replication

When I learnt biology in school, I was taught that although most creatures that I was familiar with reproduced by sexual means, some creatures, notably bacteria, reproduced asexually. Alas, real life turns out to be rather more complicated. Some species, for example the ant, can reproduce either sexually or asexually, and all of us multi-celled life forms, even though we may reproduce by sexual means, regenerate (and replace) cells in our body through (asexual) cell division billions of times over in our normal life span.

   Most body cells last only for a matter of days or months. (Nerve cells are almost unique in their longevity.) Hence cells need a method to replicate themselves. It turns out that they also need a method called *apoptosis* to trigger their own demise. Cancer often results when the cell replication process is faulty.

6. Supervising cell function

Cells need a "command and control" set of functions to regulate cell growth. This is particularly true for cells that are part of a multi-celled life form, where development is an intricate symphony of co-operation between cells.

   If all this sounds complicated, it is because it is complicated. Indeed, there is a lot we still do not understand about these processes. For it is one thing to talk about DNA, RNA and introns, but things rapidly get more complex. There are three major varieties of RNA, messenger RNA (*m*RNA), ribosomal RNA (*r*RNA) and transfer RNA (*t*RNA), and a variety of sub-species, as well. Furthermore, as I mentioned last time, it has been, to put it bluntly, confusing and a little unsettling that human beings, amongst the most complex of life forms, have only about 22,044 genes, scarcely more than a 1000 cell roundworm (19,500 genes), and substantially fewer than corn, which has 40,000 genes. It also seems a little peculiar that we should share 88 per cent of our genomes with mice, and 99 per cent with chimpanzees. I have never mistaken a relative, friend or even an enemy for a mouse or a chimpanzee. Somehow, the math just does not seem to compute!

I took a partial step forward in understanding this conundrum when I read Richard Dawkins on the subject. Don't think of the genome as analogous to a blueprint – a detailed plan of construction – he admonishes, but rather as a dictionary, which contains words that might be used to write a number of completely different books.

> When a person is made, or a when a mouse is made, both embryologies draw upon the same dictionary of genes: the normal vocabulary of mammal embryologies. The difference between a person and a mouse comes out of the different orders with which the genes, drawn from that shared mammalian vocabulary, are deployed, the different places in the body where this happens, and its timing. All this is under the control of particular genes whose business is to turn other genes on, in complicated and exquisitely timed cascades. But such controlling genes constitute only a minority of the genes in the genome.[77]

This is the supervisory function referred to in point six above. It seems that there is some combination of supervisory genes and a variety of specialized RNA molecules nested in the introns that together contain the "architectural plans" critical to our shape and form. The ENCODE project reported discovering almost three million instances of one particular start-code sequence (called DNase 1 hypersensitive sites, or DHSs), which signal the start point on the DNA strand for coding these specialized RNA molecules. All living things draw from the same set of building blocks (i.e. proteins), which are then assembled in different ways to make different life forms much as a huge variety of different stories can be concocted from words that all appear in the same dictionary.

But it turns out that this, too, is a bit of a simplification. Even though Dawkins counsels us not to think of a genome as a recipe "unless it is properly understood", it seemed to me as I soaked in the hot shower one morning that there might be a useful analogy between gene expression and the presumed activities of typing monkeys. These thoughts were triggered by reading a highly informative article by Gil Ast.[78] After hearing what I am just about to say, you may agree with my wife that I spend too much time in the shower, but here goes!

The original riddle goes something like this. If you sat a million monkeys in front of a million computer keyboards (in the old days, they sat in front of typewriters), with each monkey punching keys at random at a fixed rate, five keys per second for example, how long would it take for the complete works of Shakespeare to emerge from all this random typing?

Of course, it makes a big difference if Shakespeare's works have to appear in perfect order, and if all the punctuation has to be correct, but the main point of the tale is that even highly improbable events will happen if you wait (or type) long enough. I have not even attempted to calculate an answer to this riddle in any of its variations, but the answer will surely be a very long time, irrespective of starting assumptions.

Now let's look at a modified monkey tale to see if it can help us to sort out how the cells in our body actually make the step from the DNA strand to production of a particular protein when more of it is needed.

Imagine that you have a recipe book that has been severely monkeyed with. The book contains more than enough recipes for any meal you might want to prepare in your lifetime, but there is a problem. Before the book was published, the text was all nicely arrayed in perfect order in the computer, but one of those darn monkeys hopped up in front of the computer while the editor was out to lunch and before he or she could send the book off to the printer. Unfortunately the monkey set to and typed a lot of lower-case random letters of the alphabet into the book. Fortunately for us, one of his favourite keys was the key that advanced the cursor to the end of the sentence, so that what emerged when he had finished, and hastily escaped as he heard the editor return from what must have been a very long lunch, was a finished recipe text that had long strings of gibberish lower-case characters inserted between sentences, but not in the middle of any actual recipe sentences.

There is, I think, a useful analogy here with the human genome. As you know, Davey, strings of useful codes for proteins called *exons* are interspersed with code strings called introns. The primary RNA transcript makes a faithful copy of the lot, but, in the next stage, the introns are "spliced out" of the transcript and discarded.

Let's go back to the recipe book again and try to imagine how the splicing out of unwanted DNA sequences might be accomplished.

Suppose that you want a recipe for muffins. In normal circumstances, you would likely turn to the index to find out what page to go to, but of course the index will not be correct any more, as there are now far more pages than there were, thanks to the monkey. If the analogy with the human genome is maintained, where introns (or monkey business) make up 98 to 99 per cent of the genome, you would actually have 50 to 100 times more pages in your book than were in the original. If you possessed the book in computer format, you could institute a search for all occurrences of the word "muffin", which were typed in bold capitals, the style for recipe headings. You might then go to the first such location and save all the information between the occurrence of this heading and the next heading. If you did so, you would have made a "primary RNA transcript" of a muffin recipe, but it would still contain 98 to 99 per cent gibberish (or intron material) typed in by that wretched monkey. It turns out that there are proteins in cells that are adept at finding the place on the DNA strand where the desired gene begins, analogously to the code "**MUFFIN**", which you used for your search. These are part of the gene expression machinery referred to by Dawkins in the quote above as "controlling genes".

So, now that you have arrived at the start of "**MUFFIN**" you are faced with about fifty pages of muffin recipe, which, if all the gibberish could be excised, would boil down to less than a page. Since all sentences start with a capital letter and end with a period, and the monkey only typed in lower-case letters of the alphabet, you can assume that every capital letter will be the start of a useful sentence (i.e. an exon sequence), and the end of the sequence will occur when you encounter the next period. (Assume for the moment that the recipe book used no proper names requiring capital letters and also no abbreviations requiring periods.)

A search conducted on this basis would be analogous to what Gil Ast refers to as the "five small nuclear RNA (snRNA) molecules . . . [which] come together with as many as 150 proteins to form a spliceosome that is responsible for recognising the sites where introns begin and end, cutting the introns out of the pre-mRNA transcript and joining the exons to form mRNA."

But most recipe books offer you options. You might prefer orange muffins one day and bran muffins the next, perhaps with or without raisins added. Humans are believed to manufacture as many as five million different proteins, but have only about 20,000 genes, not all of which code for proteins. The recipe book writer might decide to add the sentence "Add 2 ounces of grated orange peel if you want to make orange muffins instead of plain ones" at an appropriate location in the recipe text. Alternatively, in the general descriptive text between recipes, he or she might have written, "You can add raisins to any of these recipes to provide variety and pleasure to those with a sweet tooth."

Ast and his colleagues have found that analogous techniques are employed in gene transcription. They have discovered a separate regulatory system that can direct the splicing process to special sites. A sentence added within the recipe is analogous to an "optionally transcribed" piece of exon, which can be either included or excluded depending on the presence of the appropriate marker. The effect of skipping just a single exon at a specific location in the fruit fly genome, for example, will set in motion a chain of events that will result in the birth of a female rather than a male fruit fly.

An alternative situation occurs when a useful (exonic) sentence is included amongst the (intronic) "gibberish" typed in by the monkey. In our analogy, an example would be the instruction to add raisins. This technique is now known to be used frequently in nature, especially in those species possessing a high proportion of introns to exons. Indeed the apes (including humans) generate a specific "mobile genetic element" called an *Alu*, which makes it easier for useful new genetic variants to arise, be detected and be spliced out of "gibberish" intronic sequences.

According to Ast, as many as three quarters of our genes are subject to alternate splicing options. Furthermore, it appears that about one quarter of these variants of the original gene are specific to a particular species, thus providing another reason that I don't look at all like a mouse or a chimpanzee, even though we share most of our genes in common.

That still leaves unexplained how it is that a liver cell, let us say, is able to command the DNA it contains to produce the *m*RNA for the precise protein it needs at any point in time. To go back to the monkey and muffins recipe analogy, how does it initiate and send the command "**MUFFIN**", which then finds the starting point for the muffin gene? For that to happen, something has to signal, "I need more muffins around here", and cause something else to head over to the DNA strand to find the start of the **MUFFIN** recipe.

By 2003 some interesting clues to this process had become apparent.

As we've already discussed several times, researchers were very puzzled to find that humankind, despite the complexity of our brains, has fewer genes than many plants and other animals. An additional piece of the puzzle was that despite the relatively small number of genes we possess, we have more introns than any other animal. In fact it looks as though, unlike the genes themselves, the number of introns in the gene of a given life form roughly scales with its complexity. Furthermore, our gene replication system goes to just as much trouble to copy many introns faithfully as it does exons, which, when properly conjoined, produce the proteins that are the building blocks of life.

All these clues led to a very strong suspicion that just maybe many introns perform a useful function, or perhaps several useful functions, suspicions that were amply confirmed in 2012 by the ENCODE project, as already discussed.

We know from what we've talked about today that some segments of introns can be woven into *m*RNA to provide variants of the basic protein specified by the gene. But it now appears that although protein coding requires exons, the introns can and do fabricate segments of RNA that have functions other than the production of proteins. One estimate is that "easily half" of all the different RNA sequences produced from a mouse gene are for purposes other than to produce proteins.[79] These RNA segments can and often do directly alter the behaviour of a cell. In particular, it has been found that some RNA segments form into what have been called *riboswitches*. These segments have one end that acts as a sensor, which on collision with a particular target causes the other end of the RNA segment to unfold and produce a protein in the standard way. One can imagine a **MUFFIN** riboswitch that did nothing as long as it was colliding with **MUFFINS** in the cell, but if **MUFFINS** started to become scarce, it would have an increasing likelihood of colliding with something other than a **MUFFIN**, thus increasing the likelihood of triggering the genesis of another batch of muffins.

It turns out that even a relatively small dictionary (number of genes in the human genome) can be used to generate far more words than there are in the dictionary (DNA) by splicing in new syllables to existing words, and leaving out syllables in others. The human genome seems to be particularly adept at this process. Like other animals, we have some unique features that have given us a particular set of competitive advantages. We are not very fleet of foot, nor do we possess outstanding strength, a particularly good sense of smell or especially acute eyesight. It is at least possible that our greatest competitive advantage lies in our intron sequences, sequences that have permitted us to develop new ways of producing exotic proteins as well as some special RNA sequences to better control protein production. This kind of capability has almost certainly been especially useful for the development of our brains, where so many different proteins are required.

Perhaps because our brains are perceived by us to be so central to our life and our survival, we have a tendency to forget how many life-affecting decisions are made autonomously. We also often forget how many unicellular animals thrive without having any brains at all, and, indeed, how our own cells lead a quasi-autonomous existence that most of the time helps keep us alive, but which can work to our detriment.

Consider, for example, the life of a thecate amoeoba, as described by Brian Ford:

> What can simpler forms of unicellular life do, more than sit in a pond and vegetate? To me, a fundamental principle of life is that the behaviour of complex life-forms is a manifestation of what single cells can perform. A thecate amœcba can select grains of minerals from its watery environment, pick them up, and fit them together to produce a delicately constructed flask-like shell within which it lives. This is a task of great complexity, and the building of walls by humans has resonances of the same remarkable ability.[80]

Microorganisms can navigate, too. Many rod-shaped bacteria, which inhabit silt, develop granules of haematite, with which they sense the direction of Earth's magnetic field. Some amœcbæ can form a capsule when the environment dries up, by means of which they can travel through the atmosphere, carried on the breeze only to re-emerge when a suitable environment is encountered. An analogy with human space travel is discernible. We may have been taught that every aspect of human life is controlled, directly or indirectly, by the workings of the brain, but this view is certainly flawed. As we talk, granulocytes in my throat are very likely identifying a *Streptococcus* that I recently inhaled. They identify the newcomer as a potentially dangerous intruder and signal to each other to congregate around the proliferating pathogens and attack them. The bacteria are ingested, digested and – all being well – soon destroyed, all without any knowledge of or intervention by my brain. Mechanisms like this are at work throughout our lives; every minute, some such event takes place.

The key fact is that these responses of cells lie in the purview of the cells themselves. They reach their own decisions, make their own "judgments", initiate action as a team and regulate their own motility without the intervention of the human (or any other) brain. We do not control such cells consciously, subconsciously, unconsciously or in any other fashion. Single cells have a proclivity to conduct their own affairs for the good of the whole body, but they are not under its control. Countless complex mechanisms of this sort maintain our lives. The regularization of blood flow within the capillary bed, the timing and extent of proliferation of cells within organs of high cell turnover – like the liver – and the apoptotic sequence,* are all aspects of the complex choreography conducted by the cells themselves and controlled by their own inherited chemistry as spelled out in their DNA.

It is clear that while our brains enormously enlarge our range of activities, single-celled creatures lacking any brain at all (in the sense we think of brains) survive and, indeed, carry out some quite sophisticated tasks in order to do so. And we multi-celled creatures who have evolved from unicellular life forms, continue to have many of our "activities" controlled at a cellular level.

Within the cell, life is ordered by thousands of proteins, each fabricated according to instructions encoded in the cell's DNA from a very small subset of candidate organic molecules, essentially twenty amino acids.

– Strange isn't it, that your brains actually control so little of your life. You seem to be made up of trillions of cells that lead a more or less independent existence, and yet we are still able to converse.

What happens if some of your cells start to act up? What do you do then?

– Well, sometimes you die, other times you just get sick. Cancer and the auto-immune diseases are examples of afflictions we can acquire without the need for any foreign bacteria or viruses to attack us.

– That is very odd. It seems like a huge design fault. I must follow up with the other humans I am talking to, to see what they have to say.

– By all means, and please be sure to let me know what you learn.

---

* *Apoptosis* is a process of deliberate life relinquishment by an unwanted cell in a multicellular organism. It is a part of the ordered cell replacement process of multicellular life forms.

DIALOGUE 13

Very Small Animals

*[I]t is the bacteria, including archaeans, who display the fullest
spread of chemical skills. Bacteria taken as a group are the master
chemists of this planet. Even the chemistry of our own cells is
largely borrowed from bacterial guest workers, and it represents a
fraction of what bacteria are capable of. Chemically, we are more
similar to some bacteria than some bacteria are to other bacteria. At
least as a chemist would see it, if you wiped out all life except
bacteria, you'd still be left with the greater part of life's range.*
RICHARD DAWKINS, *The Ancestor's Tale*, 2005

Spring was in the air and the sun was shining brightly the day I next met
Davey. It was mid-February. Daffodils in all their various shades of yellow,
from near blonde through lemon to gold and amber, were blooming (a few
weeks early) in profusion. I drove out to the Reifel Bird Sanctuary in the
Fraser River delta and found a secluded bench in the sunshine, with a
westerly view out toward the snow geese in the salt marshes.

As I had already told Davey, when I was a young man I studiously
avoided having anything to do with biology. I felt faint at the sight of
blood, the thought of dissecting a frog's leg, a common laboratory
undertaking for students at that time, disgusted me, while the very idea of
having to commit to memory long lists of families, orders, classes and
phyla positively repelled me.

As the last two dialogues so tellingly attest, I was wrong to be put off by
the subject as it then was. It has turned out that the most interesting and
rewarding scientific advances in my lifetime have occurred in biology,
perhaps especially in the interdisciplinary fields of biochemistry and
biophysics.

It is easy to lose sight (metaphorically speaking) of what we cannot physically see. Thus most of us are little interested in (though perhaps vaguely fearful of) bacteria, those unseen, unicellular life forms we have all heard about but which rarely seem to impinge on our daily lives. Yet these tiny creatures are our relatives – albeit distant relatives – and knowledge about how they live and operate is important to gaining an understanding of how we ourselves survive in an otherwise hostile environment.

It was with words like these that I introduced the subject of our thirteenth dialogue to Davey, after first having settled myself on the aforementioned, secluded, sunlit bench.

– You need not be so apologetic. As you know I am certainly not human, but I am very interested in the tiny life forms you propose to explain to me. I expected that you were going to tell me that you share a common ancestor with these beings. It is at least possible that a similar development took place in our universe, even though, sadly, we have no way of finding out directly.

- Yes, yes, I know. You used to know everything, but you have since gone through a massive loss of data, so now you know almost nothing, and what little you do know you can't tell me just yet for fear that you will cause unspecified damage to humanity if you do.

I was growing mighty weary of his lack of communication when it seemed to me that he could have been so helpful, and had decided to keep reminding him of this in the hope that I could convince him to tell me more about himself and his universe. Davey said nothing further on this occasion, so I continued with my dissertation on very small animals, with only a raised eyebrow to convey my dissatisfaction – which of course he could not see in any case!

– On October 8, 1676 Anthony van Leeuwenhoek, a self-trained Dutch scientist, wrote a seventeen-page letter (in Dutch) to the Royal Society in London. In it, he detailed some of his discoveries, enabled by his careful attention to the manufacture of his own microscopes. An English translation of part of his letter reads as follows:

> The sixth of August [1676] . . . This was to me, among all the marvels that I have discovered in Nature, the most marvellous of all; and I must say that for my part, that no greater pleasure has yet come to my eye than these spectacles of so many thousands of living creatures in a single drop of water moving among one another, each individual creature with its particular movement. And if I said that there were one hundred thousand in one droplet . . . I should not err. Others viewing this would multiply the number by fully ten times, but I state the least. My method of seeing the

very smallest of animalcules I do not impart to others; nor how to see very many animalcules at one time. This I keep to myself alone.[81]

Note the reference to his unwillingness to let anyone else know how he gained his knowledge, in this case the secret of how he made and used his microscope. As is so often the case, it is through the development of new instruments that scientific knowledge makes its greatest advances. Scientists are therefore often parsimonious in their descriptions of their instruments, at least until they move on to a new instrument, thus (they fervently hope) conserving new discoveries yet to be made with their instruments for themselves.

A few years later, in September 1683, van Leeuwenhoek wrote again. This time his letter really attracted attention. He had turned his focus and his microscope to the examination of tooth scrapings and spittle both from his own mouth and from the mouths of others. He actually made sketches of six categories of bacteria he had observed. He went on to describe them in colourful language. For example, "the biggest sort . . . had a very strong and nimble motion, and they shot through the water or spittle as a Pike does through water." Another group was likened to "a big swarm of gnats or flies, flying in and out among one another."[82]

The very idea that we are all infested with bacteria came as a great shock at the time, and is not fully understood by most laymen even to this day.

In the 1860s Louis Pasteur of France concluded that some of these tiny organisms were the cause of human disease. He discovered a way to kill most of the harmful germs in milk by a process now called pasteurization. In the same era, attention focussed on water purity, sanitary systems and hospital sanitation as ways to reduce the transmission of illness.

It turns out that quite apart from their well-known medical significance, bacteria provide many of the clues we need to understand the origin and evolution of life. In terms of intelligence, these life forms come right at the bottom of the scale. Since they possess only a single cell, that cell cannot be a neuron (which could not survive on its own), so they don't even possess a single brain cell, let alone the approximately 100 billion neurons possessed by humans. Yet despite what may seem to us to be such a serious shortcoming, this group of unicellular life forms taken as a whole, is by far the most successful. It has been estimated that single-celled life comprises at least seven times the biomass of the more complex multicellular life with which we are most familiar. Furthermore, some species of bacteria are known to have been in existence for as long as 3.8 billion years, in stark contrast to *Homo sapiens*, which is usually reckoned to have been around for a puny 150,000 to 200,000 years.

And there is yet another important reason for an interest in bacteria. For buried in their (and our) DNA is a series of clocks and linkages that has allowed specialists to work out the interrelationships between the many thousands of unicellular species so far identified, as well as their likely relationships with us, the multi-celled creatures and plants. The revelations are fascinating.

Earth had no free oxygen in the atmosphere for its first 3 billion years of existence. It is only in the last 1.5 billion years or so that there has been oxygen. The oxygen was provided courtesy of single-celled animals called *cyanobacteria*, which give off free oxygen in the process of using energy from the sun to make carbohydrates from water and carbon dioxide. For the first few hundred million years that they provided this service, the oxygen was all taken up by the iron dissolved in the ocean, causing it to be precipitated out (i.e. come out of solution) as ferric oxides, oxides that are now our major source of iron ore. Once the dissolved iron was all gone from the sea water, the oxygen bubbled its way into the atmosphere, making it possible for a new class of unicellular beasties, called *eukaryotes*, to come into being. All eukaryotes require oxygen in order to produce the energy they require to live. Almost all of the animate life forms that we non-specialists now know and see are multi-celled eukaryotes, including, of course, ourselves and the trees, plants, birds, fish, bugs and reptiles around us.

Cyanobacteria belong to the domain of *Archaea*, one of three fundamental domains that together encompass all life forms. The other two are the *Eukarya* and the *Eubacteria*. Eubacteria and Archaea are almost all single-celled. There are likely to be between four hundred thousand and 4 million different species of Eubacteria and Archaea, though as I speak, only five thousand species have been identified.

Just think of it! There are trillions of tiny unicellular life forms all around us in almost every imaginable nook and cranny of our planet, as well as inside us. They are the epitome of sophisticated self-replicating robots, yet they have no sensors to carry a message to their brains about the outside world. Indeed, they have no brains at all and live strictly according to the rules laid out in their DNA! For most of human history we have been sublimely unaware of these beings, yet collectively they are essential to our existence, as well as being the source of some nasty diseases.

Almost all cells are self-replicating robots in the sense that they: 1) take in nutrients; 2) convert the nutrients to energy; 3) carry out specialized functions, including the synthesis of the many different proteins they need to stay alive; and 4) reproduce by cell division. [In Figures 13.1 and 13.2 the two basic types of cell, prokaryotes and eukaryotes, are shown. Unfortunately, these diagrams, like all the others, could not be seen by Davey.] The cells of Archaea and Eubacteria are *prokaryotic*.

The term *prokaryote* literally means "before there was a nucleus." These cells are believed to resemble the earliest life forms in that their DNA is lodged in the main cell body, whereas the *eukaryotes* have a nucleus surrounded by a protective membrane in which the cell DNA is housed and kept separate from most of the cellular chemistry required to maintain cell viability.

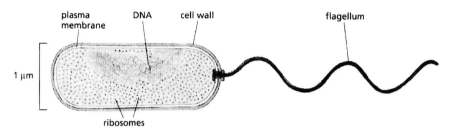

Figure 13.1 – *Diagram of a typical prokaryotic cell. In this case, the drawing is of a bacterium (*Vibrio cholera*) that has a rotating flagellum at one end. The flagellum acts as a propeller, moving the cell forward..*

Prokaryote DNA also looks very different from that of the eukaryotes. It is not grouped into chromosomes as in eukaryotes, but rather consists of a single loop of DNA. During cell division, replication begins at a single origin and proceeds in both directions around the loop. The rate of replication is approximately one million base pairs per minute. It typically requires forty minutes to complete the process, but another round of replication may begin before the previous one has finished, thus allowing cells to divide every twenty minutes in ideal circumstances.[83]

Prokaryotic cells are typically between one and five microns (millionths of a metre) in diameter, while eukaryotic cells typically have a diameter of ten to one hundred microns. A wide diversity of cell shapes can be observed, including the rod shape. The three most common shapes are *cocci* (round), *bacilli* (rod shaped) and *helical* (spiral).

It is tempting to label the prokaryotes as "simple and old", and the eukaryotes as "sophisticated and modern", but such generalizations don't hold up. For one thing, there is no such thing as a simple cell. Prokaryotic cells contain between three and five thousand genes, coding for a similar number of different chemicals needed to keep them functioning and able to "divide and multiply". We should not be surprised by the complexity of the simplest cell, which, as already mentioned, is effectively a self-replicating autonomous robot.[84]

And although it is true that prokaryotic cells existed on Earth long before the appearance of the first eukaryote (perhaps as much as 2.3 billion years later), the prokaryotes we observe under a microscope today are every bit as modern as we are. Based on the accumulated evidence derived from an assessment of similarities and differences in our DNA, all of us are descended from a common ancestral cell, though not always, as we shall see, by well-understood Darwinian evolutionary processes.

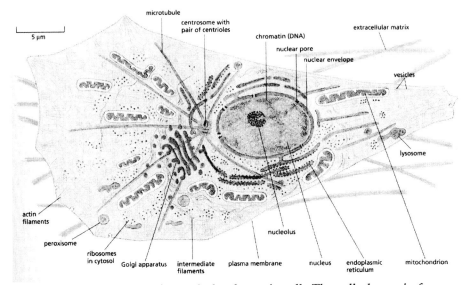

Figure 13.2 – *Diagram of a typical eukaryotic cell. The cell shown is from an animal, but plant and fungus cells are similar, as are the single-celled eukaryotes, such as yeasts and protozoa. Plant cells are surrounded by a tough cellulose outer wall and also contain chloroplasts, which convert solar energy to chemical energy. All eukaryotes need access to oxygen to fuel the mitochondria, which provide the energy needed to keep the cell functioning. (See Appendix 3 for some fascinating information about mitochondria.) In general, eukaryotic cells are about ten times larger than prokaryotic cells in linear dimensions, and about a thousand times larger in volume.*

Consider for a moment the continuum amongst animals ranging from single-celled self-replicating creatures with no brains at all to multi-celled self-replicating creatures with very large brains. Somewhere on the continuum we start to encounter animals capable of visualizing situations and consciously thinking through alternative actions. It is known, for example, that dogs are capable of such thought processes, as are crows[85]. At some level, too, there is conscious communication between animals of the same (and sometimes different) species. Only humans, it seems, have succeeded in developing complex verbal and written communication skills, the latter only in comparatively recent times. At least in my time, no reputable scientist disputes that we are the most intelligent of species known, but most believe that it is the invention of a sophisticated language and the development of writing skills in the last twelve thousand years or so which have seemed to set mankind so far above other species. Were it not for these remarkable skills, we would not likely live very differently from some of our ape brethren.

Eukaryotic cells first appeared on Earth 1.5 to two billion years ago. That the DNA in each cell is enclosed in a protective membrane gives these cells some important advantages. One major advantage is that much more complicated DNA code sequences can be transcribed with acceptable error rates. The arrival of eukaryotic cells enabled the blossoming of life forms consisting of many different cells, all derived from the same chromosomes, but having specialized functions – such as nerve cells, blood cells, kidney cells, etc. Multicellular life forms were able to grow considerably larger than their single-celled brethren and had the additional advantage of carrying cells with specialized functions.

Since Darwin's time most scientists, and many of the rest of us as well, have believed that all known life forms were descended from an original live cell. The evidence for this had been persuasive but not conclusive. The ability to perform detailed analysis of DNA has changed all that. The DNA evidence is conclusive. [An important part of the story of why this is so is outlined in Appendix 2: Notes on Evolutionary Biology.*]

Subsequent to the discovery of DNA in 1953, and the later evolution of tools for DNA analysis, scientists have been able to discern evolutionary developments with much greater accuracy than was previously possible, especially insofar as the evolution of single-celled life forms is involved. For it has turned out that not only is DNA analysis the most powerful tool available for determining the lineage and relatedness of plants and animals we can see and analyze, it is the only reliable tool we have available to determine speciation amongst life forms we cannot see with the naked eye, such as bacteria. By the late 1960s, comparative analysis of the DNA of different life forms began, and what a wonderful set of revelations it is yielding! The centuries'-old kingdoms of animals and plants have become mere subplots of a much larger schema. In addition, we have had to revise our Darwinian notion that all life evolved gradually by progressive adaptive changes from the first cellular life form.

By 1866 Ernst Haeckel had produced a phylogenetic tree of three branches, plantae, animalia and protista. The protists shown in the figure

---

* Note that while Davey insisted that I take him through some of the details of why the evidence for a common ancestor of all known life forms is so conclusive, the average reader may not be as demanding of details, hence the consignment of these notes to an appendix.

were a sort of "catch-all" category of life forms that were not animals, plants or fungi. Most are single celled, but the huge domains of Eubacteria and Archaea were left out. We simply did not yet have the tools available to categorize such tiny things.

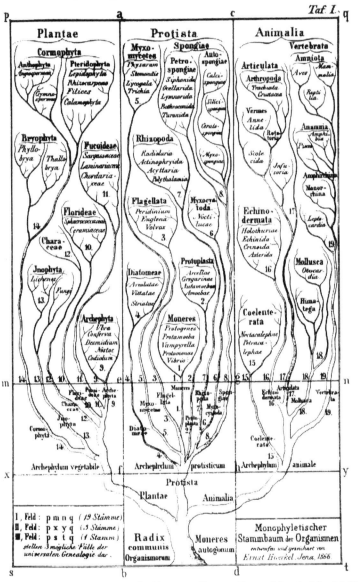

Figure 13.3 – *The Tree of Life according to Ernst Haeckel in 1866*.[86]

This nineteenth-century tree can be compared with a more modern tree. The modern much-revised tree still shows only three main branches, but the protists have gone, moved back into the domain of the eucharya where genetic analysis shows they belong. Meanwhile a whole new domain of single-celled creatures called Archaea has been discovered.[*]

Phylogenetic Tree of Life

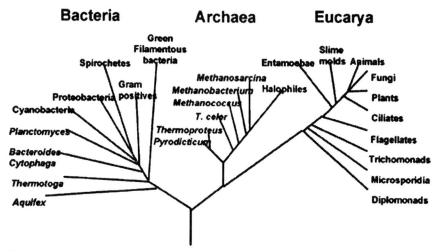

Figure 13.4 – *A Modern Phylogenetic Tree of Life as conceived by Carl Woese, incorporating knowledge gained from DNA analysis. Note that animals (including humans) are relegated to a minor position in the "Eucarya" branch (or domain). Most of the life forms listed possess only a single cell.*

As Ford Doolittle has pointed out:

> [M]icrobes were the only inhabitants of the earth for the first half to two thirds of the planet's history; the absence of a clear phylogeny (family tree) for micro-organisms left scientists unsure about the sequence in which some of the most radical innovations in cellular structure and function occurred. For example, between the birth of the first cell and the appearance of multicellular fungi, plants and animals, cells grew bigger and more complex, gained a nucleus and a cytoskeleton (internal scaffolding), and found a way to eat other cells.[87]

We have learned that microbial DNA can change, not only through the now-familiar Darwinian route by which genetic changes that confer survival benefits are passed on to progeny, but also by means of a gene-swapping mechanism called horizontal gene transfer, by which single-celled creatures are able to acquire whole new genes, rather than just

---

[*] See Appendix 2 for more background on Archaea.

genetic variants, or alleles.* In the very early stages of life on Earth, it is believed that there was extensive horizontal gene transfer taking place amongst all existing (single-celled) animals. Thus it now appears that early species development would not have relied very much on the slow evolutionary process whereby individual genes undergo gradual small changes in succeeding generations.

Horizontal gene transfer still occurs amongst single-celled life forms and is a major mechanism by which bacteria can gain immunity from drugs aimed to control them simply by swapping in a gene from another bacterium that already has such immunity. Comparative studies of DNA across all domains of life have led to the inescapable conclusion that even lowly bacteria are cousins of ours, and that all life forms must have emerged from a first living cell.

However, as we shall see in a couple of weeks, we still have a lot to learn about how that first living cell might have come into being.

– Don't tell me you are stopping there, just as things start to get really interesting! It comes as no surprise to me that all life on Earth is almost certainly derived from a common ancestor cell. The two really interesting questions are firstly, how did the first living cell come about, and, secondly, how will life continue to evolve on your planet from here on out into the future? The known history of our universe may not be relevant, but it does give one pause for thought . . .

– You don't say!

My response was larded with as much sarcasm as I thought the three words would bear.

– Are you ready at last to tell me more about your universe? Or perhaps you have another headache, or your colleagues have sadly forbidden you to say more?

– Soon I will tell you more, but not just yet. I need to know more about you and your fellow human beings first. And by the way, what makes you think I have colleagues?

I sensed that he had departed without waiting for an answer. He was somewhere back in his own universe already! My frustration was tempered a little by the realization that almost every time we spoke, he left me with a tantalizing piece of new information. This last comment of his really set me

---

* See Appendix 2.

back on my heels. Was it possible that Davey was the lone inhabitant of his universe? He hadn't said as much, but he had certainly raised the possibility. If so, what implications might that have for the future of humanity? I was no longer quite so sure that I really wanted to hear about the history of life in Davey's universe.

Adaptation by Gene Selection: A New Research Frontier

> *I love science, and it pains me to think that so many are terrified of the subject or feel that choosing science means you cannot also choose compassion, or the arts, or be awed by nature. Science is not meant to cure us of mystery, but to reinvent and reinvigorate it.*
> ROBERT SAPOLSKY, *The Trouble with Testosterone*, 1997

– I've got a bone to pick with you!

It was unmistakably Davey's voice, but we had no meeting planned for that leisurely hour in the morning, and it was only two days since our last dialogue. I was sitting in the sunshine with a rapidly cooling cup of coffee beside me, catching up on the confusing financial news in the paper.

I put down the paper and took a deliberately slow sip of coffee while I wondered what could have prompted this unscheduled outburst.

– What kind of a bone do you have in mind?

I spoke as evenly as I could, while my mind raced in a vain attempt to understand what might have driven him to burst in on me so suddenly.

– What do you mean?

– Just what I say. Do you want to pick a ham bone? A skull bone? Or perhaps an Envergurian bone, if they had bones? I don't think you ever told me whether they have bones or not.

– Don't do this to me, Peter! I have heard over a hundred people say, "I have a bone to pick with you", and the subject of the type of bone has never come up. Why is it suddenly so important to you?

– Maybe it's just because you are so secretive about yourself and your damned Gurians. You haven't even told me what you eat for breakfast!

– You never asked. Actually, I don't eat breakfast. But I do need to talk to you.

I have been relying on you to give me important insights into how you humans came about and how you live and think, but I believe you have been leading me astray, and I don't understand why you would do that.

– Perhaps you had better tell me a little bit more about your concerns.

– Well, you told me a lot about Darwin some weeks ago and convinced me that he really was the one who had finally solved the difficult problem of understanding how it was that humans, as well as millions of other species, came about. Despite a long-held belief that living beings underwent changes as a result of their life experiences, and that these changes were then inherited by their offspring, you convinced me that this was not the case at all. Instead, you told me that there are lots of small random errors made when parents pass on the information needed to produce the children, and a small percentage of these changes actually enhance the odds that the children will survive long enough to have children of their own. In this manner, a number of new species have gradually evolved through a process of survival of the fittest.

The idea that species would undergo inheritable changes as a result of life experiences rather than random changes in the genetic code would seem to make a lot more sense if you stand back and think about it. Certainly Jean-Baptiste Lamarck famously thought so, as did Darwin's grandfather, as well as the infamous Soviet-era agronomist Trofim Lysenko,[88] and many others.

– And you now think Darwin and I were wrong and they were right?

– No, not wrong exactly. But it seems that Darwinian evolution is not the whole story of how life has developed in your universe.

– Indeed not. But it really is most of the story.

As we discussed very briefly only two days ago, there is also a phenomenon called horizontal gene transfer by which entire genes can be transferred from one unicellular creature to another. It is the mechanism that, in a number of cases, has allowed some bacteria to acquire immunity from drugs designed to help cure humans of those diseases. However, horizontal gene transfer does not seem to work for humans or any other form of multicellular life.*

– Of course that's interesting, and I do remember that you have already mentioned this process. I may want you to tell me more later, but in the meantime I want to know why you haven't ever even mentioned epigenetics, which, as I have been told by others is the study of changes in gene expression; as opposed to genetics, which focusses on changes in the gene itself. I have just been listening in at a conference on the subject and have learnt that the external environment can have a huge influence on the development of many animals, certainly including you humans. This influence seems to arise through chemical mechanisms that have the effect of switching selected genes on or off. Apparently, whether or not a gene is enabled or expressed can lead to some phenomenal changes in the behaviour and/or appearance of the animal affected. This is a surprising development in light of what you have previously told me.

  Why would you not have told me about epigenetics before now?

– Don't worry. There is no sinister design at work. I simply had not yet decided when I should tell you about this phenomenon. I wondered about telling you when we were discussing evolution and genetics, which would be about now, or perhaps later when we are discussing the brain and how it works, since some of the most important epigenetic effects insofar as humans are concerned seem to be on the developing brain in early childhood.

  As you may already know, there is a mechanism by which the initial stem cells in the human egg give rise to daughter cells that develop specializations, such as becoming liver cells, blood cells or neurons. Once the change occurs, it is passed on to subsequent generations of those cells in the body. Of course these subsequent generations are still a part of the same person. The changes are not generally passed on to subsequent generations of humans, though there may be exceptions. The field of study is so new, that this is but one of many mysteries as yet unresolved.

  A vital part of the cell specialization process involves the deactivation of genes in the DNA that are not used in the manufacture of the cell in question. As previously described, one of the most common ways that the gene is switched off is by attaching a small blob of carbon and hydrogen called a methyl radical at an appropriate location on the gene. Daughter cells of the original cell will reproduce this methylated or switched-off gene

---

* See Appendix 2 for more information on horizontal gene transfer.

in subsequent generations. As I am sure you can appreciate, the methylation function is essential for all multicellular life forms, where only some of the total number of genes are needed for the reproduction and maintenance of any given cell.

It should not come as a surprise then to discover that over the long course of evolution, gene methylation and related phenomena have been adapted for uses other than cell specialization. Perhaps the most important of these is to adapt cell behaviour or characteristics to particular circumstances the life form may encounter, particularly in its early development phase. Some of these adaptations came as quite a surprise to those studying the phenomenon of epigenetics.

For example, consider the locust (*Shistocerca gregaria*), an insect that can exist in two different forms. It may be either "short-winged, uniformly coloured and solitary", or "long-winged, brightly coloured and gregarious".[89] Which form the locust takes is determined by which of its genes are switched off. In its gregarious form, the locust is commonly thought of as a plague, since swarms of millions of itinerant locusts devour fields of grain in short order and then move on. The metamorphosis of the locust from its solitary to its swarming migratory form is triggered by crowding. Several processes are involved. For example, repeatedly touching the outer surface of the femur with a fine paintbrush (to simulate close proximity to another locust) can produce a transformation from the solitary to the gregarious form in about four hours, while the smell of other locusts can induce dark colouration in solitary young locusts. The progeny of these gregarious locusts may also be born as gregarious migrating locusts for several succeeding generations if the mother locust covers her eggs with a foam that apparently ensures the appropriate genes remain switched off. It was not until 1921 that a Russian biologist named Boris Usarov discovered that the two different forms of locusts were in fact two different versions of the same insect. Up to that time, the origin of the migratory form of the insect was a mystery.

So, gene methylation shows up in a number of different circumstances. For another example, the temperature of the embryo influences the sex of a number of species, including the snapping turtle, the American alligator and the red-eared turtle.[90] If the embryo of a snapping turtle is below 22°C or above 28°C at a critical time in its development, it will almost certainly be female. Between 25° and 27°C the odds are better than 50 per cent that the embryo will be male. On the other hand, American alligator embryos are almost certain to be male if their temperature is between 31.5° and 34°C. In a very minor footnote to the history of science, apparently Aristotle wrongly believed that warm human semen was more likely to result in male heirs, and so advised elderly men "to mate in the summertime if they wanted male heirs".[91]

Temperature and crowding are not the only external variables that can influence embryo development. Authors Gilbert and Epel list the following additional factors: nutrition, pressure and gravity, light, and stress (including the presence of predators).[92]

Thus gene methylation and related phenomena, which initially evolved in a manner that permits the manufacture of many different cell types from identical DNA strands (thus making possible the development of multicellular life forms), has further evolved to enable some life forms to better adapt to their surrounding ecosystem without any alteration in the genome itself. In other words, the ability to switch genes on and off – a necessary function for the evolution of multicellular life forms – has been adapted through evolution to permit species to alter form or behaviour to enhance their success in varying environments. That is to say, epigenetics has enabled short-term adaptive genetics.

The cardinal function of short-term adaptive genetics is that it permits a rapid adaptation of life forms to the particular circumstances they encounter. Locusts, for example, can cope much better with overcrowded conditions than they otherwise would by adapting their behaviour (swarming) when overcrowding occurs.

You may recall that several weeks ago we discussed how very different some human societies have been from each other.[*] For example, we touched on the extreme difference between Dobu Island society, which Ruth Benedict characterized as one that "put a premium on ill will and treachery" and Zuñi society, which thrived on mutual tolerance and trust.[93] One could well imagine that babies born into Dobu society might need to be aggressive, selfish and nasty to survive and prosper, while the opposite might be true for those born into Zuñi society.

As I shall explain in a few week's time when we discuss brain development, it turns out that "bad mothering" tends to result in offspring who thrive in Dobu-like circumstances, whereas careful loving mothering is likely to result in children who would thrive in Zuñi-like societies.

There are several different ways that gene activity can be altered to bring about the necessary changes in the embryo or, in some cases, in the adult animal, and there is still much to be learned about the details of these mechanisms. DNA methylation is one of these ways.

Another involves the attachment of either a methyl or an acetyl radical to large molecules called *histones*, which act as reels on which the DNA is coiled. Histones that are heavily methylated tend to result in reduced

---

[*] See Dialogue 10.

transcription of the genes wound onto the histone, while histones that are heavily *acetylated* render the associated genes much more easily replicated.[94] (The acetyl radical is composed of carbon, hydrogen and oxygen.) When you recall that more than half of human genes are specific to cells in our brains, it is clear that mechanisms are needed to cloak large groups of brain-related genes so that they do not turn on in blood, lung or kidney cells, for example, and vice versa.

A number of other gene-switching mechanisms are either known or suspected. Epigenetics is a relatively new field, the subject of intense investigation – and for good reason.

It has been found, for example, that all humans are born with roughly the same number of sweat glands, but that initially none of these glands function. The glands are activated gradually during the first three years after birth, and the activation rate is dependent on the temperature. For this reason, those born in hot climates are much more likely to remain healthy in hot climates when they are adults than the rest of us, since they have more functioning sweat glands to help keep them cool.[95]

The likely fetal origin of adult diseases is another area of very active research relative to epigenetics. For example, it seems probable that malnourished foetuses are programmed to conserve energy and store fat through the activation or not of specific genes. This obviously promotes species survival if they subsequently live in an environment where food is in short supply. But if they find themselves in an environment where food is plentiful, then their energy-efficient metabolism is likely to lead them to become fat, and, in addition, to suffer from heart and kidney problems. This research has since expanded to link "prenatal influences with other types of adult illness, including lung disease, cognitive development abnormalities, breast cancer, prostate cancer and leukemia."[96]

The human infant is unusual among animals in that an abnormal amount of development takes place after birth. This is particularly true of mental development, because there is a limit on the size of head (and hence of brain) capable of navigating the mother's birth canal. There is thus a growing sense and a gathering body of evidence that suggests that adult physical and mental health are particularly strongly influenced by the physical and chemical environment of the foetus and the young child.

In particular, human mental development takes place according to an intricate schedule, with sensing pathways, such as vision, hearing and touch, starting before birth, stimulated by both external and internal events. Neural pathways that relate to our capacity to cope with what our sensors tell us about the outside world follow in short order, and are particularly active soon after birth. Their timely development is heavily dependent on the sensory neural pathways that precede them. Language development follows once the coping pathway development is well launched, and that is followed by development of the higher cognitive functions.

Disruptions to this schedule can have serious consequences for the mental developments scheduled to come later. Perhaps the most dramatic illustration of the sometimes catastrophic consequences of delay is the known fact that if kittens, all of whom are born blind, are blindfolded during the first week that their eyes open, they will remain blind for life, since essential elements of the visual processing area in their brains will not have received the appropriate stimulus on schedule.

Research is showing that the later-in-life consequences of inappropriate care of young children is subject to analogous effects.[97] For example, it has been established that children who are most at risk of not learning to regulate aggressive social behaviour during early childhood "have mothers with a history of anti-social behaviour during their own school years, mothers who start child-bearing in their adolescence and who smoke during pregnancy, and parents who have low incomes and difficulty living together."[98] Similarly, children with poor verbal skills at age three are very likely to trail in language and literary skills later in the school system.[99] One important mechanism by which this chain of events is implemented appears to be through the methylation of genes that control the release of *cortisol* in the brain. Cortisol is an important hormone that (amongst other functions) affects human response to stress. The methylated genes can generally be expected to affect behaviour for the lifetime of the gene's owner.

– Well. I'm glad I asked. I think I now understand why so many of the speakers at the conference I listened in on kept harping on the importance of early-childhood care.

– Quite so. The linkage between the quality of early childcare and the quality of later life, insofar as mental health, addiction, obesity, diabetes, high blood pressure and criminal behaviour are concerned, is revealed in study after study.[100] This type of research is behind pressure for the creation of greatly enhanced early childcare systems in many developed countries.

Indeed, it would seem that any serious attempt to bring down crime rates and reduce Medicare costs must include programs to ensure that our youngest children are free from child abuse and neglect, and have every opportunity possible to undergo healthy and timely mental and physical development.

– If it is as obvious as that, why doesn't every country have such programs?

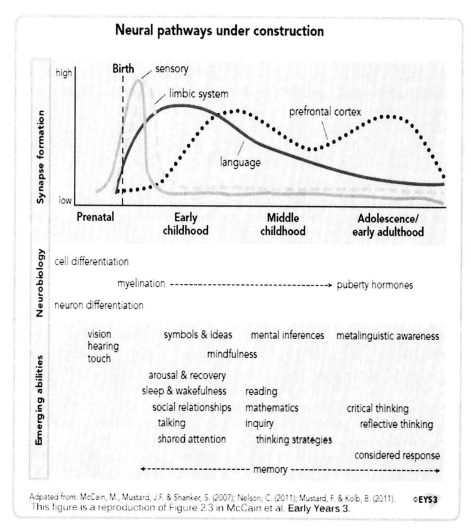

Figure 14.1 – *Early Brain Development. Synapse formation showing the time sequence of brain function development in humans.*

– A good question, Davey. Sadly it takes a long time for such information to become sufficiently embedded in the minds of the people and of their political leaders for appropriate action to be taken. In Canada, some provinces are moving ahead at a measured pace, while others have not yet seen the importance of such actions.

– It sounds to me as though you need more women involved in politics. Then you might get more progress.

DIALOGUE 15

Emergence

*Why should a group of simple, stable compounds of carbon,
hydrogen, oxygen and nitrogen struggle for billions of years to
organize themselves into a professor of chemistry?*
RICHARD PIRSIG (ATTRIBUTED)

*The universe could so easily have remained lifeless and simple –
just physics and chemistry, just the scattered dust of the cosmic
explosion that gave birth to time and space. The fact that it did
not – the fact that life evolved out of literally nothing – is a fact
so staggering that I would be mad to attempt words to do it
justice. And even that is not the end of the matter. Not only did
evolution happen: it eventually led to the beings capable of
comprehending the process, and even of comprehending the
process by which they comprehend it.*
RICHARD DAWKINS, *The Ancestor's Tale*, 2005

– I'm rarin' to go! This is a really interesting part. Just how did life
get started on the planet Earth?

– Don't get your hopes up too high. You once told me you were not keen to
learn about carbon chemistry. Well, this is where you get a brief
introduction to what you said you did not want to know! Furthermore, we
humans have not yet quite figured out how life first began, even though we
have succeeded rather well in filling in how life developed once started –
but as you know, most of the filling in has been done over the last thirty
years or so, so even this knowledge is quite recent.

– Yes, yes, yes. But let's get on with it!

– Too bad you can't see me. It is a lovely, sunny, warm spring day, and I
am sitting outside in the garden. If I sound a bit less enthusiastic than you,

it may be because I turned down a good game of tennis in order to have this particular dialogue with you!

In fact I was feeling a bit resentful that day at this intrusion on my retirement. Furthermore it was mid-March already, in fact quite precisely mid-March, the anniversary of Julius Caesar's assassination in Rome in 44 BC, and for centuries, because of this, a cause for cautionary superstition, as in "Beware the Ides of March" (whether or not the rest of us have ambitions to become emperors or tyrants)!

– Ah, yes, tennis. I have never seen it played, only heard the sounds of the ball being hit, and the players cursing themselves, or sometimes even cursing their opponents. I thought that was a sign of what someone called "bad sportsmanship". It sounds like a strange game. And the system of scoring seems to defy all logic – but I suppose it must have its uses. Probably goes back to some bundle of instincts you have all inherited. But never mind, let's get started.

So, start I did.

– It is currently believed that the first simple life forms appeared on Earth about 3.8 billion years ago, or about 700 million years after the planet was formed, although this assertion has been challenged.[101] Photosynthesis by blue-green algae is thought to have started at about this time and resulted in the formation of *stromatolites*, arrays of unicellular life, admixtures of both living and dead creatures, which build up somewhat in the manner of coral. Living stromatolites have been found in Australia, so it is still possible to observe the life cycle of these most ancient of known life forms. As I mentioned in our dialogue about bacteria, [Dialogue 13] it is believed that stromatolites and other early life forms gave off enough oxygen to provide a substantial atmospheric concentration of the gas, thus enabling the later evolution of more complex, oxygen-consuming species. However it was another 3.2 billion years or so before there were signs of complex invertebrate life captured in fossils which are dated to the beginning of the Cambrian period starting about 542 million years ago. The period got its name from Cambria, the classical name for Wales, where Cambrian period fossils were first observed.

What actually happened in the evolution of life prior to stromatolites, and when, is still a mystery, but there are a lot of theories out there. In 1953 a graduate student named Stanley Miller, working under Harold Urey at the University of Chicago, attempted to reproduce in the laboratory conditions similar to those in which life may have begun at an early phase in Earth's history when water, ammonia and methane were believed to have been plentiful. This research surprised everyone by showing the ease with which a variety of amino acids, the basic building blocks of life, could be produced. However, as it turned out, from that point it is still a big step from the availability of the relatively simple amino acids to the formation of the very complex protein molecules essential to life. Such proteins may easily comprise over ten thousand amino acids, each one in an order and orientation critically important to its function.* Fabrication of such proteins seems to require an appropriate template, such as that provided by RNA or the more complex DNA molecule. But of course RNA did not exist prior to the appearance of life, since a living entity is required to produce it.

Current thinking is that common clay could have provided the necessary template for the first, relatively simple proteins, albeit as a result of one or two chance occurrences in many many billions of interactions. From such very simple beginnings, over a period of about a billion years, a full-blown reproductive system for very simple life forms could have evolved, the very small odds of any single encounter advancing the development of life being offset by the very, very large number of such encounters over the time period.[102]

Thus, although we can now trace our evolution back to a common ancestor of all life forms, who or which existed about 3 billion years ago (plus or minus about 200 million years), we simply do not yet know how the very first living cell came into being.

For one thing, the evidence of the necessarily very simple early life forms has almost certainly been completely destroyed. As the number of different life forms increased, so would have competition for survival. Competition, especially in the early years when life forms were at their simplest, almost inevitably gave the initial advantage to a more complex being better adapted than its predecessor to its surroundings, although, as you will soon find out, greater complexity is, in itself, no sure-fire recipe for survival.

For another, we don't yet know how simple the simplest possible life form could be. By examining the common genes amongst bacteria that are known to have a very small gene count, it has been estimated that the genome of a self-sufficient self-replicating bacterium could be as small as 315,000 base pairs, comprising about 250 genes.[103] But this estimate was arrived at by what is called a top-down approach – that is, by looking at what we can observe in nature – rather than from a bottom-up process of counting, one by one, which genes are absolutely required. Genetic analysis has somewhat surprisingly shown that the ancestors of many bacteria currently being studied previously had larger genomes. The route to

---

* As already noted, titin, one of the largest-known proteins, comprises approximately 27,000 amino acids.

survival for these particular bacteria was to make their genome less complex. What if you start with no genome at all in a soup of chemical compounds, and one "compound" emerges that is self-sustaining in its particular environment, as well as being self-replicating? What might it look like? We just don't seem to know at this time.

Most experts believe that there was a long period when Earth was host to an ever richer variety of chemicals, some of which reacted with each other to form increasingly more complex molecules.

I have already mentioned one theory of how life began using a template of clay to form the first fragile strip of RNA. At this stage of our understanding, there are many theories to be tested. Some hold that the first self-replicating cell was born in oceanic hot vents rather than in a bed of clay. This idea has led to an interesting space research proposal. It is almost certain that one of Jupiter's moons, Europa, has hot vents under its sea ice, as passing sister planets exert strongly varying gravitational pulls on it, providing the kind of stresses and frictional heating that are likely to give rise to such vents. Plans are afoot to launch a satellite sophisticated enough to land on the ice-covered Europan ocean, drill through the ice and then launch a submersible to the ocean bottom to search for life. This is likely our best opportunity in the nearer term to discover life elsewhere than on Earth.

I suppose that for molecular phylogenists who study genomes and inheritance pathways, it comes as no surprise at all to learn, as Richard Dawkins makes clear in *The Ancestor's Tale*, that the apple tree in my backyard and I both had a common ancestor somewhat over a billion years ago (it is perhaps my 50 millionth cousin!), or that the robin on the lawn and the girl next door share a common ancestor who lived only about 510 million years ago. But for the rest of us, it gives pause for thought.

Robert Hazen, whose video lecture series on the origins of life[104] have provided me with a useful background on the subject of emergence, quotes the famous German philosopher, Immanuel Kant (1724–1804), as follows:

> *It is absurd for men . . . to hope that another Newton will arise in the future, who shall make comprehensible by us the production of a blade of grass according to natural laws which no design has ordered. We must abundantly deny this insight to men.*

But it is just this Kantian "absurdity" that an army of scientists, primed by our newfound (since 1953) understanding of the genetic code, is abundantly confirming. As I suggested in our previous dialogue, as far as I am aware no serious scientist questions any longer that all of life as we currently know it derives from a common ancestor. The evidence is just too compelling.

The very idea that all life forms on Earth are related was repugnant to earlier generations. Indeed, there was little scientific evidence to support such a contention prior to the publication of *On the Origin of Species* by Charles Darwin in 1859. But serious study of the origins of life did not receive much funding support until the late 1990s, after the work of molecular phylogenists was starting to establish a common strand to all life going back billions of years.

As Dawkins succinctly wrote concerning the likelihood that all life forms on Earth share a common ancestor[*].

> *We can be very sure there really is a single concestor of all surviving life forms on this planet. The evidence is that all that have ever been examined share (exactly in most cases, almost exactly in the rest) the same genetic code; and the genetic code is too detailed, in arbitrary aspects of its complexity, to have been invented twice. Although not every species has been examined, we already have enough coverage to be pretty certain that no surprises – alas – await us. If we now were to discover a life form sufficiently alien to have a completely different genetic code, it would be the most exciting biological discovery in my adult lifetime, whether it lives on this planet or another. As things stand, it appears that all known life forms can be traced to a single ancestor which lived more than three billion years ago. If there were other, independent origins of life, they have left no descendents that we have discovered. And if new ones arose now they would swiftly be eaten, probably by bacteria.*[105]

Thus, the mystery of how the first self-reproducing cell came about still remains. There is a huge difference in complexity between the organic chemicals that have been generated by simulating conditions on Earth's surface, as well as deep under the sea, 3 to 4 billion years ago, and that, at best, contain only a few base pairs, and the simplest life forms we have observed, which have thousands of base pairs.

– This is all very disappointing.

Davey's voice was agitated.

– You people just don't know how life began! It is all very well to talk about low probability events, but unless you have a pretty clear idea of how this unlikely creation event happened, you might as well say

---

[*] Dawkins defines the word "concestor" to mean the last common ancestor.

that some god created life.

– What's wrong with you, Davey? You don't seem to be any more patient than we humans, even though you claim to be superior. I have made no secret of the fact that we still don't understand how life began, but as little as sixty years ago we didn't even understand how the genetic code was constructed. The very idea that I may be related to an apple tree was pure speculation. Now we know beyond reasonable doubt that it is so. Do you know how you were created?

– As a matter of fact, I do.

– Fantastic! That is really interesting! Do you think that knowledge will help us to understand the creation of life on Earth?

– Not really – and no, I cannot explain why just yet, but please believe me that I shall explain why before our dialogues end.

A disappointing start to our dialogue! No explanation yet from him, just more promises. And, in addition, a statement that understanding how Davey's ancestors came about will not help me to understand how we came about! If Davey already knew about his origins, why on Earth would he be so interested to know how we came about? The mystery deepened. I sighed loudly.

– Oh stop it! I really am sorry to disappoint you. But please carry on with our discussion. I, too, am disappointed, but I still do want to hear your best assessment of what is known about your origins.

I sighed again, stared unseeing at my notes, and started in to describe their contents. I already knew what I needed to say to finish the dialogue.

– There are three questions we need to answer with respect to emergence. First, you should hear me out on theories about where and how life originated on Earth. You cut me off just now before I had finished. Second, it is important to know a lot more than we do about the probability that life will evolve spontaneously. It will help us to determine the likelihood of life elsewhere in our universe. (Of course, you, Davey, might save us a lot of trouble by letting us know what you have learned about our universe!) Finally, it would be very interesting to know why, of all the possible chemical systems that life on Earth could have used, the particular set of amino acids and proteins that are used in all life on Earth was the one that prevailed.

So, if it is all right with you, I shall now do my best to address these three questions.

1. Where and how did life originate on Earth?

Three sources are under active investigation.

First, there are variants on Charles Darwin's speculation that life may have started " . . . in some warm little pond, with all sorts of ammonia and phosphoric salts, light, heat, electricity, etc., present, that a protein compound was chemically formed ready to undergo still more complex changes . . ."[106] The Miller-Urey[107] experiments of 1953, which I referred to earlier, were the first serious experiments aimed at testing such an hypothesis. A host of organic compounds were produced when methane, ammonia and hydrogen, all of which might have been present over Earth's oceans four billion years ago (though this is disputed), were mixed and subjected to heat and simulated lightning. Some of the compounds produced from this and successor experiments yielded important components of DNA and RNA. It appears that most of the twenty amino acids coded for in DNA and RNA were present, as well as some amino acids that do not occur naturally. However, it has proven extremely difficult to move to the next step where strands of RNA or protein are fabricated from these amino acids. One problem is that the postulated sources of energy, primarily lightning and ultra-violet radiation, even as they enable the production of small organic molecules, tend to break down the much larger chains of organic molecules (called *polymers*) needed to get life started and sustained.

A second speculative locus postulated for the origin of life is in or near hot vents, located deep in the ocean, well away from the large jolts of energy simulated in the Miller-Urey experiments, but fuelled by high temperatures (up to 250°C) and reactive chemicals, especially compounds of sulphur such as hydrogen sulphide. A "hot" contender as the primary builder of the first living cell is what Hazen calls the "Grand Hypothesis of Günther Wächterschäuser."[108] The hypothesis relies on the fact that the iron sulphide mineral pyrrhotite (FeS) is commonly found as a deposit at the mouth of hydrothermal vents. This mineral is highly reactive and can initiate a series of cascading reactions making use of carbon dioxide and other simple molecules present in the water near the vents. At this time the hypothesis seems fairly close to providing a credible route to a rudimentary life form, but so far no one has come close to success.

Progress in research in hot-vent chemistry is greatly hindered by the difficulty and danger of doing experiments at one thousand or more atmospheres of pressure and temperatures ranging from 200–400°C. There are some interesting changes to some molecules at such temperatures and pressures. One important example is that the dielectric constant of water changes from its normal very high value of about eighty down to about twenty. (The dielectric constant is a measure of the homogeneity of charge on a molecule, which in turn strongly affects the degree to which a molecule is attractive to other molecules.) The normal high dielectric constant of water results in it being a very good solvent of many chemicals, but a very poor solvent of most organic chemicals, and hence a poor milieu in which various organic chemicals might mix and match to assemble the first life form. However water with a dielectric constant of only twenty will have a very different chemical behaviour from water at normal pressures. At this time, very little is known about organic chemistry at one thousand atmospheres of pressure, but this situation is starting to change.

Now that we know how proteins are produced in living cells, researchers are focussing on how RNA and DNA structures evolved to make such production possible. The general consensus is that the first forms of life would have used RNA rather than DNA for reproductive purposes. But even RNA can be complex although it is much simpler than DNA. It is also much more prone to errors in reproduction, a fact which could be advantageous up to a certain point, as the rate of generation of potentially successful variant RNA forms would be quite high. A number of frameworks on which an RNA molecule could conceivably self-assemble have been proposed. As already mentioned, one candidate for such a framework is clay, which has an interesting layered crystalline structure, with room between layers for hydrocarbon molecules to adhere and form linkages with each other. Other candidates include a variety of minerals with flat crystalline facets that could act both as an assembly frame for RNA molecules and provide energy through chemical interactions.

Those who favour hot vents as the most likely location for the commencement of life tend to favour mineral platforms, while those who believe that life commenced in a primordial soup tend to favour clay. Neither hypothesis may be correct. No explanation has yet proved totally satisfactory.

The third hypothesis concerning the origin of life on Earth is that it came to us from space, not in the form of Martians or Vegans or other imagined representatives of advanced civilizations landing in spacecraft, but rather in the form of organic molecules that have hitched a ride on a meteorite. Over 140 different compounds, many of which are organic, have been detected in giant molecular clouds in deep space, and we know that some quite simple organic compounds can and do survive a hot entry through Earth's atmosphere. However, most observers think it unlikely that something as complex as an RNA molecule or a protein could survive the trip.

Even if one rejects the hypothesis that life arrived on Earth from outer space, there remains the issue of whether there is life on other planets close enough to us that we might observe it, and even more interesting, whether such life forms are likely to employ a carbon chemistry similar (but likely not identical) to our own, or whether life forms beyond Earth have evolved down very different paths.

It is believed that Mars at one time had a climate more amenable to the development of life than Earth, hence the strong interest in scientific circles in searching for life on Mars. As already mentioned, those particularly interested in hot vents are also keen to look for life on Jupiter's moon Europa, which is bathed in an ice-covered ocean and likely has hot vents.

2. Just how likely is it that life will evolve spontaneously and under what conditions?

The answer to that question will provide some needed facts relevant to our endless speculation about extra-terrestrial life forms. The laws of probability strongly suggest that, given the vast number of planets in the universe, life does exist elsewhere in our universe. But does it exist in a million other locations? Billions or quadrillions of locations? Or is it possible that life is unique to Earth? And what about intelligent life? What, for example, are the odds that other planets will have evolved multicellular critters that can think?

Thinking related to this important question does not seem to have evolved much past the famous Drake Equation we talked about in our fourth dialogue.

3. How is it that our particular form of carbon chemistry life was the one that evolved on Earth, and what other possible chemical life forms are there?

Many physical phenomena have a delightfully simple conceptual framework, but in their implementation they exhibit a formidable complexity. We have seen, for example, how Einstein's two theories of relativity (Special and General) stem from a basic assumption that the speed of light is constant when viewed and measured by an observer, irrespective of the velocity or rate of acceleration of the light source relative to the observer, yet the details of General Relativity in particular, are beyond the mathematical skills of most of us.

So it is with the emergence and the evolution of life.

At the centre of any discussion of life on Earth is the carbon atom, which has six positively charged protons in its nucleus and six negatively charged electrons in orbit about the nucleus, together giving the atom a neutral charge, and an atomic number of six. You may already have heard from others that the orbiting electrons of the elements form into shells. The inner shell contains only two electrons before it becomes "full up". The next shell will take a maximum of eight electrons, so the neutral carbon atom, will have a full inner shell of two electrons, and the remaining four electrons will orbit in the next shell, leaving four vacancies before the shell is full up (with eight electrons). The most common form of chemical bonding occurs because elements are in a lower energy state when the orbital shell is full, and an element such as carbon can effectively fill up its outer orbit by "pairing" its four outer orbit electrons with four electrons from the outer orbits of other atoms. In this way, atomic charges can remain neutral, while the sharing of electrons allows participating atoms to have apparently full outer orbit occupancy. Thus hydrogen, which has one proton in its nucleus and one electron in orbit, can pair up with one of the four carbon outer orbit electrons. If four hydrogen atoms pair their electrons with a single carbon atom, all will achieve stable outer orbits. This combination of four hydrogen atoms with one carbon atom is called methane and is represented chemically as CH_4. It is often shown pictorially as:

Methane
H
H-C-H
H

where the single lines drawn between the hydrogen (H) and the carbon (C) atoms denote the sharing of a single electron pair. If two electron pairs are shared as often happens when atoms bond with oxygen for example, a double line is drawn, and so on.

Ethane
H H
H-C-C-H
H H

Carbon can also share a single electron pair with another carbon atom, with hydrogen providing the remaining three available bonds. In such a case, the resulting compound is ethane (C_2H_6), which is depicted on the left.

Methane and ethane are but simple compounds commonly encountered in the discipline of, which can be described as the study of chemical compounds of carbon other than a few very basic and simple compounds such as oxides, carbides and carbonates. Studies of the chemistry of life are an important part of organic chemistry, but do not comprise the whole field by any means, as there are many carbon compounds that are organic but play no role in the chemistry of life.

Many of the basic organic compounds employed by living cells are combinations of the elements hydrogen (H) (with one gap in its outer ring), carbon (C) (with four gaps in its outer ring), nitrogen (N) (with three gaps in its outer ring) and oxygen (O) (with two gaps in its outer ring). These elements can bond in a variety of different ways to form simple compounds, such as the following:

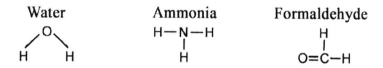

Water

Ammonia

Formaldehyde

These simple compounds (*monomers*) can, in their turn link up into more complex molecules, such as amino acids, sugars and cellulose. Furthermore, these groups commonly form into rings and long daisy chains of repetitive structures (*polymers*, or "many monomers"), often many hundreds and sometimes thousands of atoms in length, all of which assume characteristic geometries; shapes that play a major role in defining the nature of reactions between different organic compounds, much as differently shaped keys will only fit with certain locks.

Adenine ($C_5H_5N_5$), which is notably important because it is one of the four key codons in RNA and DNA, is essentially a combination of five molecules of hydrogen cyanide (HCN). Cellulose, often written $(C_6H_{10}O_5)_n$, is a polymer chain of indeterminate length (n), often several thousand monomers in length. Long strands of cellulose often link laterally with hydrogen bonds to form sheets of considerable strength, as we all know from our daily encounters with wooden products of many kinds that capitalize on the strength of the cellulose bonds in wood.

Adenine

Cellulose – Two Molecules in the Chain

– I suppose you are telling me all this just to annoy me!

Davey's voice had a peculiar low whine to it.

– You know very well that I cannot see any of your diagrams, so why do you even talk about them?

– Quite right. I knew you would start whining soon. How do you think I feel? Not only do you not show me pictures of your universe, you won't even talk about it! Anyway, why can't you see what we are doing down here?

– That much I can answer. It is easy really. I have no means of detecting what your scientists call electromagnetic fields. I gather these fields are all around you. In fact, it is likely fortunate that I cannot detect electromagnetic waves, as it may be that I would simply have been overwhelmed by them, and hence unable to detect other phenomena, such as sound waves.

– Very interesting! So, you can hear the dog bark, but you cannot see it. You hear cars collide, but fail to see the damage. You hear the ocean waves crashing on the barren rocks, but cannot tell a barren rock from a pebble! You cannot even see the difference between me and a mountain. How do you manage to understand life in our universe at all?

– Well, it certainly isn't easy, but it may not be as difficult as you think. I can analyze and catalogue every sound I hear. I then look for sounds that occur together, or one after the other, and that are co-located. Gradually I have come to understand what is going on. Your scientists have gone through a similar exercise in trying to understand the origins of your universe. There are only the faintest of clues remaining from that time over 13 billion years ago, yet careful examination of the clues has allowed you to understand quite a lot. You just have to be alert and ready for new clues all the time.

But perhaps we should get back to discussing this painful lesson in organic chemistry. I cannot really see how the molecules fit together, and they don't seem to make any noise that I can hear, so I shall just have to think up a useful analogy from my universe.

– Yes, I think you will.

I was not in my most sympathetic mood and I wanted to finish up what I had to say.

– Of course, the picture rapidly gets more complex. Other elements come into play, though in much smaller quantity than carbon, nitrogen, oxygen and hydrogen. Phosphorus is a key element in the backbone of both DNA and RNA to which the individual nucleotides are attached; sodium and potassium ions are important to intercellular communication, sulphur can replace oxygen atoms in some compounds and so on.

Note that the two sugars (riboses) for which I have diagrams that you cannot see – sorry – contain identical components but are assembled differently. Sugars and many other compounds exhibit a handedness or *chirality*, just as our right and left hands are generally the same except that one is the mirror image of the other. Organic compounds that have chirality are often prefixed with either an "L" or a "D" to denote whether its chirality is left (L apparently from the Slavic word for left, *levo*) or right (D from the Greek word for right, *dextro*). Life on Earth seems long ago to have made a choice between versions of chiral compounds and stuck with that choice wherever living cells appear. Specifically, where sugars are needed, it is always the right-handed version that is used, and wherever amino acids are needed, it is always the left-handed version that is used.

– That's interesting. How is the choice made? For example, what prevents a right-handed amino acid from being incorporated into a living cell?

– What prevents a key with a pattern on the bottom of the key instead of the top from working? It doesn't fit. Chemistry is a bit like fitting keys into locks. If the shape isn't right, or the distribution of charges on the molecules is not right, the joining of two molecules won't occur. So a left-handed sugar or a right-handed amino acid just won't form the necessary chemical bonds in living cells.

You may already be familiar with the important role that the amino acid dopamine plays in the brain. It is often referred to as *L-dopa*. *D-dopa* exists, but is not used by life on Earth. If life is ever discovered at the bottom of the ocean on Europa, it will be interesting to see if life there follows the same chirality rules as found here.

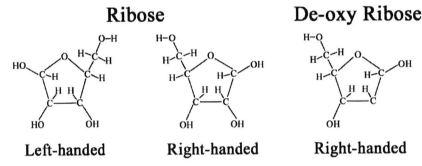

Many believe that the choice of left or right chirality occurred by chance, although Hazen reports on experiments that suggest proto-life forms assembling on the surface of some minerals could have preferentially included amino acids of one chirality over the other, with the preference depending on the location of the mineral in question. Proteins containing amino acids of mixed chirality have very different shapes from those with a fixed chirality, and are thus likely to have been non-starters for building life forms from the beginning. Imagine the added complexity of a genetic code that coded for the chirality of individual amino acids, as well as the composition of the amino acid itself!

Some experiments have been reported that seem to imply there may have been a precursor to RNA with a backbone different from the phosphate groups found in DNA and RNA, but no one has yet come up with a plausible alternative. The general consensus amongst researchers in the field seems to be that the probability of the emergence of life based on a focal atom other than carbon is very small indeed, but it is at least intriguing to speculate on life forms in other locations in our universe that are not RNA and DNA based, or that use the opposite chirality for sugars or amino acids from those used on Earth. We can only hope that some of the planetary probes planned for launch in the next few decades will return with some enlightening answers. Who knows, perhaps the NASA rover *Curiosity*, which landed on Mars in August 2012, will provide an answer for us. What do you think, Davey?

– I think it's a long shot!

– I wonder if you really know, or if you are just guessing.

In all this very brief discussion of the emergence of life, I have avoided speculation about the emergence of the all-important cell membranes, which partition the cell from a potentially hostile external environment* and, in the case of eukaryotic cells, provide additional isolation for the cell's DNA in the nucleus. This turns out to be less of a conundrum than you or I might have supposed. The lipids used in the fabrication of cell walls are longish molecules that attract water molecules at one end (they are hydrophilic) and repel water molecules at their other end (they are hydrophobic). As soon as there is a large enough concentration of lipids in a water solution, the hydrophobic ends pair up and attach to each other, thus avoiding contact with water molecules, with the two remaining (hydrophilic) ends pointing out toward the water solution. Groups of molecular pairings huddle together in sheets and find optimum stability when they have formed into closed bodies, or cells, with the free hydrophilic ends embracing water molecules both inside and outside the cell. The principal challenge in explaining the evolution of cell walls is to explain how a sufficient concentration of lipids for formation of cells could have arisen, but lipids, at least in low concentrations, are thought to have been present in the oceans from very early in the lifetime of planet Earth.

You will no doubt be pleased to know, Davey, that there are many layers of complexity that could be added to this basic explanation of the chemistry of life, but the essential message is that a small number of the many very simple chemicals available on Earth have combined into complex systems, many of which defy description. An analogy with the binary bits – the basic building elements of the computer – is not far-fetched. Out of very simple building blocks, some wondrous complex structures can be formed.

Davey let out a large and rather inhuman sigh. I took it that he was simply acting, and had likely not found my very simple explanation at all difficult to understand. On the other hand, I thought to myself, it would be very inhibiting not being able to see the diagrams of the organic molecules

---

* See Figures 13.1 and 13.2.

that are the building blocks for life on Earth.

– Yes, I am a little surprised at the extreme length and complexity of most of the molecules that are essential to life in your universe. You are clearly a very long way from understanding how the fairly simple organic molecules you can induce by such tricks as mixing some gases in the presence of heat and lightning can be combined to form RNA or RNA-like molecules, such that life forms can come into existence. That same heat and lightning so essential to creating the simple organic molecules apparently prevents the formation of the longer molecules needed to maintain life. This is a very vexing problem. Have you any ideas as to how such a miracle might have occurred?

– Not really. But I was very intrigued by the fact that some chemical reactions proceed very differently under very high pressure – the sort of pressure encountered at the hot vents under the sea. It would not surprise me if it turned out that these vents played an important role in the emergence of life on Earth.

– I suppose it must indeed be true, as you assert, that all life forms on Earth are related, though I can readily understand why some humans would find this thought repugnant and unbelievable. You are a funny species, you know!

– You have to remember that the beliefs many humans hold dear today are very different from the beliefs of their forebears of only a few thousand years ago. Those beliefs are sure to continue to change in the future, especially now that we have so much more evidence of how we really came into being.

– I agree. A few thousand years is a short time frame. But it is also true that a lot can happen in a thousand years. From what I have learned by listening to your wise men, humanity may be lucky if it hasn't destroyed itself within the next few hundred years. Based on experience elsewhere, if I were you I would worry about that.

– I suppose you are talking about the old saw that "a little knowledge is a dangerous thing", and combining it with fables about loss of innocence and the damnation that ensues when we pluck information from the tree of knowledge. Adam and Eve and all that.

You are right to pick up on this question. It is a difficult one for us humans to contemplate.

Over the last two or three decades, humankind has learned how to alter specific genes in both unicellular and multicellular life forms to change such properties as size and vulnerability to bacterial attack. Scientists are rapidly learning how to design and produce a complete genome using biological "circuitry" which mimics the roles of DNA and RNA. Such "humanmade" DNA could be inserted into living cells to promote the production of life-saving drugs.[109] Anti-malarial and anti-HIV drugs have been quoted as promising examples. Ultimately, perhaps, such techniques will permit the fabrication of a completely humanmade self-replicating cell line.

The implied assumption by humans of a role previously attributed uniquely to God will be an enormous milestone in the development of civilization. The achievement of such a milestone is, as you suggest, infused with both danger and promise.

Given your remark about "experience elsewhere", I can't help but wonder, Davey, if you in your universe have experienced such a transformational step in knowledge about life, and if so, whether there are useful lessons for us deriving from your experience. It is not a problem I expect to encounter in any meaningful way in my lifetime, but I do expect that my grandchildren may experience both the perils and the promises of artificial life in spades.

– Yes, errr, yes.

Davey was hesitant. I decided not to break the ensuing silence, as it seemed he needed time to think before saying more. At last, he continued.

– It all happened a very long time ago. In our case, the difficulties seem to have arisen much more from gaining an understanding of how to modify what you call instincts than it did from understanding how life first came about. Looking back, it was a terrible period in our history. This was the time during which many of our historical records were destroyed, so I cannot give you a full account. I am doing my best to recover some of that history. As you know, a part of my research involves looking for events in other universes that might have relevance to what happened at home. Hence my conversations with you and others on Earth. These talks have been helpful, and at times I even find it a bit emotional, although I am not supposed to have any emotions left.

As you already know, I would rather, if you can bear it, hear out your story before attempting to tell you mine. The more I know about you and your fellow humans, the more likely it is that what I tell you will be helpful.

I shall leave you now and look forward to our dialogue next week. I hope you will use at least some of the intervening time to play tennis, and perhaps to go hiking up Mount Seymour one more time. I envy you your apparent capacity for enjoyment.

– But, Davey . . .

But Davey had gone.

PART 5

A WONDERFUL AND

IMPERFECT

INSTRUMENT

My next meeting with Davey did not take place as planned. First of all, I got a bout of flu just before the appointed day, and it took me several weeks to recover fully. Then, when I did finally get back to thinking about conversations with Davey, I found I needed to do much more homework on the workings of the brain than I had planned on.

A second factor also came into play. I had debated for some time whether or not I should tell my wife about my dialogues with Davey. In general, our marriage was not one where we held secrets from each other, but we did have areas in our lives where we operated quite independently, so not much communication was expected. This was a product of the fact that during our early years of marriage I was involved in highly classified research and development related to the detection and tracking of submarines, so I was unable to share with her any details concerning my work.

I had wondered almost from the start whether or not I should tell Margaret about my conversations with Davey. Initially, I was dubious about their reality myself, and so I decided to delay addressing the issue on the grounds that it all sounded pretty unbelievable, and was in any case quite harmless to and independent of our marriage. As time wore on and Davey became a much more real figure to me, it became apparent that I should let her in on my secret.

When I finally broached the subject of Davey with her, not long after I recovered from the flu, she seemed very calm and understanding – far too much so in my opinion! I concluded that she must be humouring me, while wondering if the flu had attacked my brain in some obscure manner. I realized almost immediately that I might have thought the same way had she broached such a subject with me.

When I raised this possibility with her, she denied it, but in such a way that I was unconvinced. It seemed that she perhaps wanted to seek professional advice before either agreeing or disagreeing with me over the existence of Davey! I told her that I would suggest to Davey that she should be allowed to attend our sessions. She agreed and asked that the meeting should take place soon.

I had learnt some time ago that the most effective way of initiating a conversation with Davey was simply to say out loud in a normal speaking voice, "Davey, I am ready to talk", or similar words to that effect. I feared that Davey would not want to speak with my wife, and was rather on edge when I called on him, alone in my home office early in May, the day after Margaret had agreed to be introduced to Davey.

It turned out that I need not have worried. He seemed to welcome the idea that she would join me and suggested that I summon her immediately.

Margaret was very suspicious at first (who could blame her?), but soon started to pepper Davey with a number of questions I had wanted to ask but had not dared to. I agonized that all her questions would drive him away, but again I was wrong. His answers were always courteous, though often not very helpful.

No, he did not have a wife. Nor were there children on his planet.

Yes, he was very sorry we could not actually see him, but it had been a major breakthrough to discover a way both of detecting sounds in another universe and then of making some intelligible sounds in response.

What did she mean when she asked whether or not he could influence events in another universe? Well, he could not really answer that.

No, he wasn't hungry, thank you very much, nor would he like a drink. And thank you, he did have a place to stay.

How he wished a million times that he could see us, but in the circumstances, the fact that the room was rather untidy and the curtains needed replacing did not really matter to him.

And he quite understood that I had not been well enough or ready to continue the dialogue for the last six weeks.

The six weeks had gone by rather quickly. I did not feel entirely ready and confident of what I had to say when we resumed our dialogue, this time with Margaret present, on a clear sunny morning, just after I had mown the lawn and was feeling both warm and a little tired.

Margaret and I were comfortably settled in our chairs in the family room. We might have sat outside in the sunshine, but the prospect of a neighbour or even a relative dropping by and seeing us having a conversion with nobody was more than either of us could face.

– Hang onto your chair, Davey. The next three dialogues are a barn burner! Absolutely fascinating.

– By now I do know what a chair is – it is an artificial device that allows you humans to rest your behinds while the rest of your body remains more or less upright. I have never been quite sure what the advantage of such a halfway position is.

– Perhaps after you understand us a bit better I shall be able to tell you.

– I see.

But I wondered if he really did understand. Nor could he see the smile of satisfaction on my face at having another opportunity, which didn't come often, to treat him with the same casual superiority and school-masterly detachment as he had used on me for so long. I heard Margaret clear her throat in the same way she does every other time when she thinks I have not been quite as polite as I should have been, but Davey made not a sound. I continued.

– Like most human beings, I have been very slow to understand the extent to which my actions and those of others are driven by instincts embedded in our brains. These instincts can be denied and ignored to a considerable degree, but like a prevailing westerly wind that continuously influences the heading on a sailboat, they have a continuous presence that influences our thoughts and our behaviour. Many would deny the strength and ubiquity of this phenomenon. I hope that the glimpses into the workings of our brains that I have prepared will convince you that such a denial is not just unwise. It can be dangerous.

Our brains constantly fool and make fools of us.

When I was in my middle twenties, a drug called LSD was popularly supposed to open up our minds to the visions and ideas an undrugged mind suppressed. As with many others, I found this idea fascinating and appealing. I wondered what insights and life experience I might be missing if I did not try LSD. Perhaps fortunately for me, while staying at the home of an old friend, I consulted his father, a distinguished surgeon, on the matter. He was horrified by my question and urged me in the strongest terms not to experiment with my brain. While not totally convinced by his argument, it gave me pause. Some months later I learned via a newspaper headline that an acquaintance from school who had been a regular user of LSD, I suspect for reasons similar to those that had tempted me, had committed suicide after a session using the drug. The two events together caused me to lose my appetite for experiments with mind-altering drugs. In subsequent years, as I have learned a bit more about that wonderful chemical factory that is the human brain, I have had occasion to be thankful that, in this case at least, I learned the easy way instead of the hard way. Tampering with the intricate balance of chemicals in our brains is a risky undertaking.

Today, the perils associated with mind-altering drugs have been exaggerated to the point that many warnings have lost credibility with important segments of the population, especially young people. I wonder, Davey, if in your universe you have ever had a similar situation? Once a drug is prohibited – be it alcohol, cigarettes or mind-altering drugs – there seems to be an immediate urge by some to try the banned substance. Much better, I think, to lay out the facts in an unbiased manner and to tax heavily those substances that are clearly unhealthy.

Davey replied that he found this comment about human behaviour very interesting although, for reasons he was not yet prepared to explain, drug-taking was not a problem where he came from.

So, after a pause to digest yet another deferral from Davey, I continued.

– I do not intend in this dialogue to delve much into brain chemistry, but I do want to share with you some fascinating discoveries about the human brain. It seems to me that they shed a very interesting light on how and why we behave as we do. By the time this part of the dialogue is ended, I hope you will better understand what complex and delicate organisms human brains are, and why they are not the tidy processors of logic I once thought they were.

So, let's start at the beginning with an overview of how we came by our present, still only partial, understanding of how our brains really work. For reasons that will become clear, the research has demanded great ingenuity from the scientists, but they have been rewarded with a string of important surprises.

At this point, Margaret rose to her feet and addressed Davey.

– Davey, I think I should get on with some gardening while the sun shines. We get quite a lot of rain here, as you may know, so one has to take the opportunities to get outside jobs done when the weather is good.

It has been a great pleasure meeting you. I hope you will allow me to join you and Peter again when you are ready to tell us more about yourself. We are honoured that you have chosen us as one of the families you want to dialogue with, and I must say I am a bit relieved that you really do seem to exist. I don't know what I would have done had I concluded that you were a figment of Peter's imagination.

Before I go, is there anything I can bring you? A cup of tea perhaps, and some cinnamon toast?

– No thank you. I don't drink tea. The pleasure of our meeting was all mine! I look forward to our next meeting already, perhaps in a few weeks time.

Harrumph! I thought as Margaret left my office. I wonder where Davey learnt to be so polite and solicitous with women.

DIALOGUE 16

The Deconstructed Brain

*Men ought to know that from the brain, and from the brain only,
arise our pleasures, joys, laughter and jests, as well as our sorrows,
pains, griefs and tears. Through it, in particular, we think, see,
hear, and distinguish the ugly from the beautiful, the bad from the
good, the pleasant from the unpleasant . . . It is the same thing
which makes us mad or delirious, inspires us with dread and fear,
whether by night or by day, brings sleeplessness, inopportune
mistakes, aimless anxieties, absent-mindedness, and acts that are
contrary to habit. These things that we suffer all come from the
brain . . .*
HIPPOCRATES, *On the Sacred Disease*, c. 400 BC

– Almost every medical student is aware that Aristotle thought that
"common sense" (by which it is thought he meant the seat of
consciousness) resided in the human heart, and that the brain functioned as
some sort of cooling device for blood coming from the heart. This quaint
idea is a little more comprehensible when one understands that our brains,
when working hard, consume a lot of energy. Thus the brain is served by
large blood vessels bringing in fuel and oxygen and removing spent
chemicals. About 20 per cent of our total blood supply and 20 per cent of
energy-giving glucose is directed to our brains. While others before
Aristotle, notably some Egyptians, had concluded that "common sense"
resided in the cranium at least as early as the seventeenth century BC, it
was not until the 1600s AD that we encounter reliable descriptions of the
anatomy of the brain. By the early 1800s researchers were tracing nerve
fibres from various parts of the body back to the brain, and recognizing
their sensory and control functions. By the mid-1800s the relationship
between personality and the brain was convincingly demonstrated when a
quiet and industrious railway construction foreman named Phineas Gage

208

was wounded in an industrial accident during which an iron rod passed right through his head, severely damaging the frontal lobes of his brain. Miraculously Gage survived, but his pleasant personality did not. He became surly and combative. This period saw a blossoming of our understanding of what parts of the brain were responsible for some specific functions, so that, much as geographers and explorers named mountains and regions after themselves and others, so, as the geography of the brain slowly became revealed, we encounter locations such as *Broca's area* and *Wernicke's area*, to name two of the most famous, which happen to be associated with our ability to speak and to understand the spoken word.

By and large early research proceeded by treating the brain as an inaccessible black box, impossible to understand through physical examination. The best strategy available at the time was to observe a lot of people suffering from a variety of different mental afflictions, and to correlate these afflictions with brain damage to specific areas of the brain as determined from known injuries and/or by examination of the brain *post mortem*.

In the 1930s brain research took a giant step forward when the American-born Rhodes scholar, Wilder Penfield[110], founding director of the Montreal Neurological Institute learned how to probe the brains of patients who were wide awake, and to have them report on what effect the probing was having on their mental state. It seems strange to most of us when we first learn of this process that a patient can be wide awake and free from pain in such circumstances. But brain probing is possible because there are no pain sensors inside the skull, although the patient is given a local anaesthetic to numb the pain sensors on the outside of the skull. Of course, doctors do not have free rein to cut a hole in our cranium and probe about. The probing almost always takes place before and during operations to remove cancerous tissue from the brain or, as in the case of Penfield's initial work, in attempts to cure epilepsy. Penfield was able to map the location of many functions of the brain, and he discovered that by probing the temporal lobes of the cerebral cortex he could elicit startlingly distinct memories in his patients, often of events hitherto unremembered. Repeated probing in the same location brought back the same memory. He was also able to map where on the brain the touch sensors on the body report in. The famous Penfield homunculus, well known to every neurologist, was the result. I have a diagram of this, Davey, which I will try to explain to you. [See Figure 16.1] As you would expect, some areas having a lot of touch sensors, such as the hands and lips, use up a relatively large area of the brain, while relatively insensitive parts of the body, such as the back, use up only a small area of the brain's real estate.

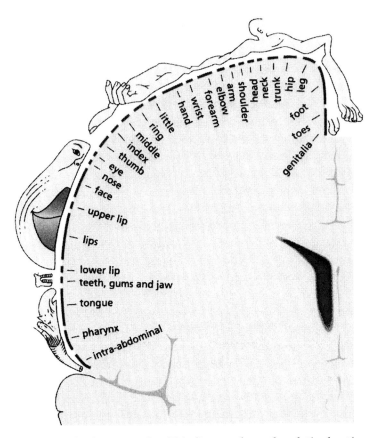

Figure 16.1 – *The Penfield Homunculus. This diagram shows the relative locations of the sensory neurons (left side) for various parts of the body. The surface of the body is mapped onto the cerebral cortex by an orderly system of nerve cell connections such that sensory information from neighbouring body sites is delivered to neighbouring sites in the brain. Note that the hands and fingers are represented over a wider area than the hips and legs.*

Work on the physical and chemical makeup of neurons started much earlier and has proceeded apace. In 1906 Santiago Ramón y Cajal and Camille Golgi won the Nobel Prize for Medicine for their research into the workings of the brain. Golgi had developed a method of staining thin slices of brain tissue with a silver compound that caused the nerves or neurons to show up as brownish black on a transparent yellow background. "All was sharp as a sketch with Chinese ink" wrote Ramón y Cajal in his autobiography.[111] Interestingly, Golgi used his new technique to support the theory that the nervous system was made up of a network of continuous elements. This idea was vigorously opposed by Ramón y Cajal, whose beautiful microscope stains and delicate artwork illustrating various neurons and neural networks convincingly demonstrated that the brain was made up of millions of tiny neurons, each with many connections to other neurons. It is now estimated that the human brain includes some 100 billion (10^{11}) neurons, that some of the neurons in the body have connecting *axons* (or nerve fibres)* as much as a metre in length, and that communication between neurons is across a *synaptic cleft* about 200 billionths of a metre in width. It is common for some types of neurons to have of the order of ten thousand interconnections (synapses) with other neurons, some of which will be stimulatory and some inhibitive.

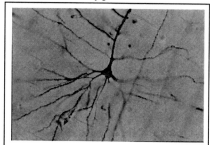

Figure 16.2 – Pyramidal neuron imaged using the Golgi technique.

We know that the "ancient" part of our brains, especially those parts grouped around the top of the spinal column, are strikingly similar to those of many other mammals. Humans, however, have an unusually enlarged cerebral cortex, the upper and frontal part of the brain, so much so that the enlarged size of the head can make childbirth difficult. Indeed, it may well be that the size of our cerebral cortex has been limited by the ability of our mothers to give live birth to large-headed babies! The brains of our close relatives in the ape family, especially the chimpanzees, most closely resemble our own, though their cerebral cortex is much smaller.

Over the last two decades there have been many fascinating books published that entertainingly recount stories of strange mental phenomena, which, upon diligent exploration, have yielded interesting new knowledge about the way the brain works. For some years my favourite writer of this genre has been Oliver Sacks, a New York–based, English neurologist whose books such as *The Man Who Mistook His Wife for a Hat*, *The Island of the Colorblind* and *Awakenings* have been deservedly successful. However, it is in the writings of his friend and colleague, V.S. Ramachandran, that I have found descriptions of some discoveries that I believe are particularly relevant to our understanding of ourselves.[112] Ramachandran's writing is a happy mixture of reports on fascinating scholarly research, much of it his

---

* Axons normally transmit the nerve output to locations where they can communicate with the dendrites of other neurons across a gap called a *synaptic cleft*. Communication across the synaptic cleft (i.e. synapses) is chemical in nature and involves the migration of positive or negative ions to generate electrical pulses in the dendrites, which then normally transmit the received pulses back to their receiving nerve cells.

own, flavoured with eminent readability and a sense of humour.

The works of Ramachandran and Sacks, and the scholarship of literally tens of thousands of researchers, have shed a lot of light on the human brain.

Ramachandran's studies of the phantom limb effect provide an interesting example of what can be learned by close and careful observation and experiment. It has been known since at least the sixteenth century that those with amputated limbs often found that their brains, especially at first, and sometimes for long periods after the amputation, did not "acknowledge" the amputation. That is, the amputee continued to experience the limb as though it were still there. Sometimes it itched, often it ached, and it was possible for its recent owner to flex and manipulate his or her ghost limb with the sense that the limb was intact and working. Phantom fingers could curl up and dig into a phantom palm causing excruciating pain, a sensation distinctly felt, for example, by Admiral Lord Nelson after losing his lower right arm and elbow in battle at Teneriffe in 1797.[113]

What is going on here?

According to Ramachandran, there is a lot going on. First of all it has been discovered that Penfield's famous homunculus is not fixed and immutable. If a limb is amputated, it obviously can no longer send sensory information to its corresponding location in the brain. So what happens? Does that part of the brain atrophy in the absence of incoming signals? This would not be altogether surprising. As you will remember, it is known that kittens which are born blind, if blindfolded for several days when they first open their eyes will remain blind for the rest of their lives. This is because the neural development critical to their vision is programmed to take place when their eyes first open. If the correct cues (in this case, sensory inputs from the eyes) are missing when scheduled, the development simply does not happen. Similarly, people born with optical defects in their vision will often find that other parts of the brain tasked with interpreting inputs from the optic nerves have compensated for the visual defects, so they go unnoticed. But should those defects appear after visual system development, it will be necessary to apply an artificial correction, such as eye glasses.

Ramachandran discovered that when a limb is amputated, the portion of the brain that processes sensory data from the limb does not atrophy. Instead, sensory signals incoming to the neighbouring brain areas invade the now disused territory. It happens that some of the facial sensors terminate near those from the arms. In a surprisingly short period of time (hours) after an amputation, a tickle on the cheek can be interpreted as a tickle on the amputated arm, as well as a tickle on the cheek. Even more bizarre, it has been found that orgasm can be "enjoyed" (as far as the brain is concerned) in the non-existent foot of an amputee, as well as in the genitals. This is because the sensors from the foot and from the sex organs terminate in adjacent parts of the brain! There is no evidence that I have seen that the reverse is also true, i.e. that someone who has lost his or her genitals can enjoy an orgasm by having their feet tickled, but this may only be because the experiment has not been attempted.

In general, it has been found that the brain locations for processing different sensations – heat, cold, dampness – are separate, so that if the cheek is touched with something cold, the same sensation may be communicated to the phantom limb, but not pain or heat. But what about the excruciating pain that Lord Nelson, in common with many other amputees, felt?

In some wonderfully ingenious experiments, Ramachandran asked some of his amputee patients (as well as some with congenitally missing limbs) to place, for example, both their good right arm and their amputated and therefore phantom left arm into a specially constructed mirrored box that, when the good limb was correctly positioned and the limb's owner looked through the viewing apparatus, resulted in the patient seeing two arms, one the mirror image of the other, and both apparently attached to their owner. Even though the patient knew full well that he or she had only one arm, and that the second arm being seen as an apparent left arm was only a trick of the mirror, in many instances, when asked, for example to clap hands, he found that his phantom arm responded, and the phantom pain disappeared! In the case of one particular patient, as soon as he either closed his eyes or withdrew his arm from the box, the old pain immediately returned. However, by taking the mirror box home with him, and repeating the experiment many times over several weeks, his phantom arm suddenly disappeared and with it the pain he had felt in it for over ten years.

What can be going on here? After performing a series of different tests on different patients, and checking out a variety of hypotheses, here are two of Ramachandran's important conclusions:

> . . . pain is an opinion on the organism's state of health rather than a mere reflexive response to an injury. There is no direct hotline from "pain receptors" to "pain centers" in the brain. On the contrary, there is so much interaction between different brain centers, like those concerned with vision and touch, that even the mere visual appearance of on opening fist can actually feed all the way back into the patient's motor and touch pathways, allowing him to feel the fist opening, thereby killing an illusory pain in a nonexistent hand.[114]

> . . . your body image [i.e. your image of yourself in your "mind's eye"],

despite all its appearance of durability, is an entirely transitory internal construct that can be profoundly modified with just a few simple tricks.[115]

But these are far from the only tricks our brains play on us. Our vision system, for example, performs many amazing adjustments (rather more, in this instance, than "a few simple tricks") on images we see without bothering to let us know what it is up to, as the next example illustrates.

On an autumn day in 1976 I set out, in a bit of a hurry, to drive a cousin to catch a ferry. Arriving at a stop sign beside a main road, I stopped, looked both ways, first left then right, saw that the coast was clear, and launched the car out into the intersection. As I did so, I looked again to my left, and saw to my horror that we were about to collide with a car coming in from the left. Fortunately, there were no serious injuries, but I was left pondering how I could have missed seeing the car the first time I looked.

You may know the answer. The region at the back of the eyeball where the bundle of optic nerves is connected to the eye creates a small "blind spot" in one's vision. But instead of seeing a black spot where the vision impairment exists, the "vision system" in our brains routinely fills in the black area with background colours matching neighbouring areas. [See Figure 16.3.] In my case, the oncoming car must have been exactly aligned with my blind spot, and my vision system inserted a matching substitute of roadway, curb or grass verge. In the circumstances, this proved to be very unhelpful.

Millions of years after our distant ancestors developed the system for covering up our blind spot, ingenious computer programmers have "invented" a nearly indispensable tool for fixing up old photographs with spots and cracks, or new photographs showing unwanted power lines or even unwanted people. The device is called a cloning tool. It permits the user to copy an arbitrarily chosen desirable part of the image over the top of the unwanted scratch or power lines. This tool mimics the way the brain covers up our blind spot with copies of segments of image nearby.

We are told that when reaching an intersection we should always look twice to our left before proceeding. Avoiding accidents caused by missing traffic in our blind spot is a critical reason for doing so. For most mammals, and most of the time for us, it is survival enhancing not to have to cope with a permanent black spot in our vision, and so we have evolved an ingenious set of mental circuitry to get rid of it. But clearly there are circumstances when our lack of awareness of our blind spot gets us into trouble.

It turns out that our brains have many figurative blind spots, as well as the literal visual blind spot that contributed so importantly to my car accident. Ramachandran provides us with insights into a number of these.

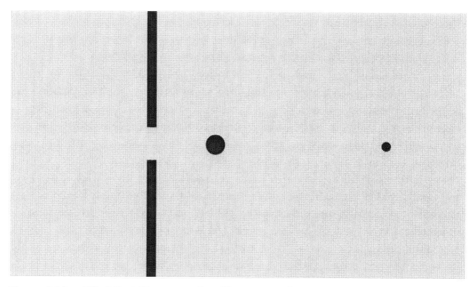

Figure 16.3 – *Blind Spot Demonstration. Shut your right eye and focus on the small black dot on the right with your left eye. Starting about 30 cm away, move the book slowly toward you. As the book moves forward, at first the heavy black line will fill in and be seen as continuous. As the book comes closer, the gap in the black line will open up again, but the large black dot to the right of it will disappear. If you move the book closer still, the dot will reappear. You may need to "hunt" for the blind spot by moving the book to and fro and a little bit sideways several times until you see the effects.*

Notice that when the interrupted line or the black dot are in your eye's blind spot, your brain automatically fills in what it deems to be appropriate given the surrounding optical field, i.e. in these cases, either a black line or a grey background.

First, concerning the real blind-spot phenomenon, it has long been observed that, for example, if there is a vertical line positioned to traverse our blind spot, our visual systems will reconstruct the missing line segment to connect the two line segments detected on either side. Now, wonders Ramachandran, what happens if the real image seen is two aligned line segments with a discontinuous gap roughly equivalent to the size of the blind spot? As you would expect, the brain connects the two segments to make them one continuous line. Fine. Now what does the visual system do if the two lines, instead of being collinear, are slightly offset? The answer is that up to a certain degree of non-collinearity, the vision system will connect the two segments. Beyond that degree, the "eye" sees two segments with a gap between them. Now, ask yourself what happens if instead of showing a vertical line with a gap in it, the line is replaced by a series of small "x"s. Will our eyes see a continuous line of those? Again the answer is yes, up to a point. If the "x"s are too large, the brain does not invent one or more of them to put in the blind spot but instead just fills in the background behind the "x"s, and leaves a gap in the row of "x"s. What a lot of image processing is going on here without our even having to learn how to do it, or even to think or know about it!

Dr. Ramachandran and his colleagues have identified some thirty different visual processes in the brain, including such functions as: identifying colour, identifying boundaries (lines) at various angles of objects being viewed, processing stereoscopic information from our two eyes, carrying out pattern recognition (for example, of human faces, animals, cars and trees), estimating speed and direction of motion. Some of these processes take place in what Ramachandran calls the old visual pathway, which virtually all other mammals have, too. Such processes are carried out totally unconsciously, though the information derived is evidently passed on to higher levels. At a higher level, and much more accessible to the conscious mind, are what Ramachandran refers to as the "how" and the "what" paths. An example of a "how" path function is the set of mental directions needed to cause the hand to deposit an envelope in a mailbox. A corresponding "what" path activity would involve the recognition of the mailbox as the appropriate receptacle for the envelope. The "how" path does not need any short-term memory in order to perform its function, and can thus be performed unconsciously, whereas the "what" path requires that you carry out a memory search to confirm that the object before you conforms to your experience of what a mailbox is, and thus needs a short-term memory bank to permit comparison with images (and other factual data) stored in memory.

As I expect you know already, I am a keen but inexpert tennis player. It sometimes happens when playing close to the net, that a tennis ball comes hurtling at you, giving you no time to consider how you will respond, but, to your surprise, without any thought on your part, you stick out your racquet and make a perfect shot. In such circumstances, the shot was executed over the "how" path. It is a wonderful feeling! For me, this is telling evidence that we do indeed perform significant unconscious acts, much more complex than just breathing and heart pumping, although those are unconscious, too. Annoyingly often in my case, if I am sent a slow ball to return, I start thinking about the placement of my feet and the mechanics of the racquet swing, thereby activating some "what" paths. The odds of my hitting the ball into the net in such circumstances are substantially increased!

Those who have had tennis coaching have likely been told that if you are rushing up to the net to return a ball landing in the forecourt, you should try to come to a stop before hitting it. This is because our brains have very little ability to compensate for the energy we impart to the ball as a result of our forward movement, so we are very likely to hit the ball too hard, and it will go out. Police radars, which routinely subtract the speed of the police car when calculating the speed of the target car, have no such difficulty, but evolution evidently missed out on providing us with the corresponding circuitry! Perhaps the ability to play tennis is not necessary for survival after all!

I sometimes wonder if my grandchildren will ever find it useful to learn how to split wood. If they do, they will soon learn that especially for wood with a lot of knots in it, it is important for the axe to land on a line through the core of the trunk. That way the axe blade does not have to sever the tough knots from branches (which lead into the core of the wood). They will also learn that there are good returns from striking the wood with considerable force, so that the momentum carried by the axe head can be used to maximum effect. But it is no mean feat to deliver the axe head in precisely the correct location on the round of wood and also at high-impact velocity. It is no good watching the axe head as you do this. I find it best to fix my vision on the central core of the round where I want the axe to strike. With some luck and some practice the axe head will land where desired, guided by an unconscious system that somehow controls the shoulder and arm muscles - more evidence of the role of "unconscious" thought in our everyday lives!

Researchers have compiled a lot of evidence concerning the complexity and the organization of the visual system, and the role of unconscious visual activity. This has been done largely by careful observation of people who suffer from partial impairment or destruction of their visual processing system. One such patient of Dr. Larry Weiskrantz was totally blind on his left side as a result of surgery on the right side of his visual cortex.[116] The visual cortex is where pattern recognition would normally take place. As far as the patient was concerned, he could see nothing to the left of "straight ahead" without turning his head. Amazingly, although he could see nothing to his left, he was able to reach out to shake the hand of his doctor, or to pick up wooden blocks he could not "see" when they were off to his left. Further tests confirmed this result, not just for him, but for others with the same affliction. The unconscious part of his visual system, which was unaffected by the brain surgery, was seeing perfectly well. However the unconscious visual processor had no way of entering the consciousness of the patient, who still believed he could not see either the proffered hand or the blocks!

Having learned of these tricks that our visual system can play, and knowing that there is a region of the brain that mediates and compares incoming visual perceptions with stored memories of faces, objects, etc., it will be no surprise to learn that, particularly amongst those with weak or failing vision, stored images originating in the brain rather than the retina can come to be "seen" and understood by the brain as being as real to the brain's owner as if they had really been there. Indeed, we have already learned that Wilder Penfield was able to stimulate such images repeatedly by probing the same neuron or group of neurons in a patient's brain. Could "holy visions" and "ghosts" also originate in this way? It is certainly possible. Indeed some forms of epilepsy, those where the epileptic attacks take place in the frontal lobe, usually result in the sufferer experiencing mysterious and often "holy" visions.

Now things are getting really interesting!

If the feelings of pain and processing of visual information are subject to a multitude of unconscious influences, what about other parts of the brain?

Our emotions, for example?

The *limbic system* consists of a group of brain modules near the top of the spine, part of the "old" animal brain. It is a system we share with all mammals, and plays a key role in human emotions. In particular, the *amygdala*, a small almond-shaped organ, is considered to be the "gateway" to the limbic system. It is now known that patients suffering from Capgras' syndrome have a severed connection between that part of their brain that recognizes faces and their amygdala. This can happen, for example, in a severe car accident where the skull is damaged. Although Capgras' syndrome sufferers see and recognize familiar faces, they experience no emotion as a result. This situation gives rise to an inner conflict when a patient encounters his or her mother, for example. Such patients see and recognize their mothers, but have none of the warm and familiar feelings that they are used to experiencing on such encounters. In Capgras' syndrome, the brain concludes as a result that the person seen cannot be its owner's mother, leading the patient to assert that the person seen is an impostor! It would be hard to find a clearer demonstration of the role of many non- vision specific parts of the brain in the vision system as a whole. Consistent with these observations, it has been found that if the same individual talks to his mother on the phone, he will have no problem talking to her and relating to her in the normal way.

An important part of knowing ourselves lies in understanding our limitations. This clearly should include understanding the more important limitations of our brains – not just in the colloquial sense, when we accuse ourselves of having been stupid, but in the far more subtle sense of understanding the tricks our brains can play on us. As far as I know, awareness of these subtleties of our brains has come to light as a serious area for study only in the years since 1960.

We now know in some detail that, just like the electronic computers we fabricate, our brains are made up of many different subsystems, each with defined characteristics and shortcomings. The shortcomings are most evident when we ask them to perform functions that had no survival benefit for our hunter-gatherer ancestors.* Thus, as a purely practical matter, it is helpful to know about the existence of our blind spot, especially if we drive a car. If we play tennis or chop wood, it is useful to know that we have unconscious circuits for hand-eye co-ordination which, with sufficient practice, will control our hand so that we strike the wood or hit the tennis ball in the right place, provided we keep our eyes looking at the ball or the location where we want the axe to strike the wood.

The reason for these phenomena is becoming clearer with every passing year.

It seems that our brains are riddled with mental "blind spots". For example, when we remember past events, we only remember certain features that strike us as particularly important. Later, when we try to recall the event in question, our brains automatically fill in the details we never committed to memory in the first place, and they most often do so by filling in the gaps with contemporary information. So, if an interviewer asks us questions about past events in a manner that even hints at the answer, there is a strong possibility that those hints will affect our answer without our being aware of it. As individuals, it can be important and useful to be aware of this phenomenon by which we are prone to substitute imagined events for real ones. For those involved in our criminal justice systems such awareness is essential. Lawyers and policemen know that eyewitness accounts of crimes will likely be differently described by every eyewitness. As well, there have been a number of sad cases where children under questioning have imagined that they were sexually assaulted, and it was only years later that the persons accused of the deeds were able to establish their innocence.

It turns out that our brains do a lot of other things without telling us. For example, we unconsciously frame questions so as to get the answers we would like to hear, and unconsciously compare our performance to those whose performance we exceed.[117]

In a broader context, it seems to me to be very important to understand that our conscious decisions and perceptions are influenced both by thoughts and events of which we are conscious, as well as by visions, events and innate drives of which we are unconscious or only vaguely conscious. This is surely a critical factor in any debate concerning free will. As Daniel Gilbert has written, *"The bottom line is this. The brain and the eye may have a contractual relationship in which the brain has agreed to believe what the eye sees, but in return the eye has agreed to look for what the brain wants."*[118]

---

* See Appendix 4 for a brief discussion of computer and human vision.

I personally find the evidence for these unconscious pathways, as lightly touched on in this dialogue, both fascinating and convincing. It is a little eerie for most of us to think that there are unconscious influences at work when we make "conscious" decisions. But we should count ourselves lucky. It is virtually certain that most life forms make only unconscious decisions, as they have no ability at all in their brains (if they have them) for "conscious" thought.

So, what about you, Davey? You tell me that you wish you could see us in order to better understand our world. I assume this means that you can see in your universe. How does your brain process the input from your eyes and other sensors? Do you see in three dimensions or more? Or less? Does your brain play tricks on you the way our brains do? If so, can you give me examples? And if your brain does not play any tricks on you, why is that? It looks as though evolution has shaped the way our brains perceive our world in a way that, on average, enhances our survival – or at least it did when our ancestors were hunter-gatherers. Do you think you and your kind underwent an evolution that shaped your development in similar ways?

– I find your account of the way human brains work extremely interesting. You have started to provide me with the answers to many questions I have had concerning apparently irrational human behaviour. It had never occurred to me that human brains were anything other than rational decision makers, with all relevant factors properly weighed in that part of your brains responsible for overall control. So, a lot of frankly irrational human behaviour that has always puzzled me is now open to an explanation I had not even considered previously. You humans are actually not conscious of some of the many decisions you make! How extraordinary! I seriously wonder how life can continue to survive in such circumstances.

– Well, as you know, life has existed on this planet for a very, very long time – billions of years, and humankind has been around for several millions of years, albeit in a less intelligent version than you now see.

– Actually, as you know, I don't see, nor, with all due respect, do I think a few billions of your years is a very long time.

But you must not stop your discussion of the brain here, just when it is getting really interesting. You must tell me how this knowledge of unconscious influences on your decisions affects human philosophy. I also want to know a lot more about how the brain actually works at a detailed level. For example, how do neurons form, and how and why do they make contact with other neurons?

– I think you have come to the wrong boy for that! As you know I am not an expert. What about all those others you are dialoguing away with? Can't you ask them?

– Yes, I shall contact others, but it is you who have most caught my attention. I must know what you think is important to know about human brain operation and its impact on philosophy!

I never did hear from Davie what he had learned by contacting others, but two weeks later, I opened a dialogue with Davey on mind and brain, a subject on which many learned people had discoursed at length over the ages.

Mind and Brain: The Consciousness Debate

Like many other natural wonders, the human mind is
something of a bag of tricks, cobbled together over the eons by
the foresightless process of evolution by natural selection.
Driven by the demands of a dangerous world, it is deeply biased
in favour of noticing the things that mattered most to the
reproductive success of our ancestors.
DANIEL C. DENNETT, BREAKING THE SPELL: RELIGION AS A
NATURAL PHENOMENON, 2007

– Well, what have you got in store for me today? I suppose I should not have been as surprised as I was at our last meeting when you demonstrated just how strongly evolution has shaped your brain as a tool to aid survival, often at the expense of its ability to be cool and logical, even when being cool and logical is important to survival.

– Yes, you can also see how even a rudimentary understanding of the way our brains operate can help us humans to understand our own behaviour, as well as that of others.

But to answer your question, I am going to try to shed just a little light on an ancient debate concerning the meaning of mind, and the degree to which the concept differs from the concept of brain. So let's get going. I need time after our dialogue to go shopping for a new car.

– I note that your wife has not made any comments so far. I assume that is because she is not here?

– That is correct.

In 1969, shortly after our first son was born, Margaret and I found ourselves looking for houses. I use the words "found ourselves" advisedly, at least in my case. At the time, we were living in a perfectly adequate rented townhouse, but somehow we both had a desire to find and buy a house. I can recall asking myself at the time, rather bemusedly, how it was that I was signing an offer on a house for a lot of money, and without having done anything close to the detailed research that a rational person would require before making such a major commitment. I concluded that we were driven to the house purchase at least as much by instinct as by rational deduction. Nothing I have read since has caused me to alter that opinion. Indeed, I am more than ever convinced that instinctual drive was a major factor in our purchase of that house. The role of the unconscious in conscious decisions is becoming ever clearer.

Virtually everyone agrees on what the brain is. But the mind is thought by most people to be something rather different, and is the subject of many books by learned people.

Some definitions of mind in the context of mind/brain/body (as opposed to other meanings of the word), from Oxford's *Shorter English Dictionary* to Webster's *1913 Dictionary*, to the *American Heritage Dictionary*, include:

- the seat of consciousness, thought, volition and feeling (Oxford);
- the intellectual or rational faculty in man; the understanding; the intellect; the power that conceives, judges, or reasons; often the entire spiritual nature; the soul; – often in distinction from the body (Webster); and
- the consciousness that originates in the brain and directs mental and physical behaviour. (American Heritage).

It can be seen from these definitions that one's mind is generally associated with thoughts, actions and emotions of which one is conscious. Yet our own experience and experiments by Penfield and Ramachandran and their colleagues pretty clearly demonstrate that conscious decisions are routinely influenced by unconscious "thought".

Experimental evidence strongly suggests that most, if not all, lower animals do not possess a mind in the sense, for example, that they can assess a situation, consider alternatives and select a preferred course of action. One of the first nervous systems studied and completely understood was that of the sea slug (*Aplysia*), selected for study because it has very large neurons and a very simple brain structure comprising only about ten thousand neurons. Some years ago researchers had determined the response of a sea slug to its complete set of detectable stimuli. A sea slug's behaviour is almost totally determined according to an inherited program in its nervous system.[119] Even insects with complex behaviours – ants and bees come to mind – can be shown to be behaving according to inherited behaviour patterns. At some point in the evolutionary process, animals emerged that had a capacity to "reflect" on the sensory data coming into their brains, to introduce past experience into the decision-making process, and to change their actions depending on the results of this reflection. It is likely that this capability developed gradually in species where there was a distinct survival advantage in doing so. It has been found that a number of mammals and birds use tools such as, for example, small sticks to assist in withdrawing ants from an anthill. Tool use is good, but not conclusive, evidence of an ability to undertake conscious thought, and thus by definition, to have a mind.

Given that mind is a property of the brains of some but not all animals, and that those possessing a mind have it because of the survival benefits it bestows by enabling forward planning, it is to be expected that the segments of the brain devoted to conscious thought will continue to use as some of its input information supplied by the pre-existing subconscious brain. One would expect that species survival would dictate that only some actions of the body could be controlled by conscious thought, while the rest would continue as before, completely controlled subconsciously. We saw in our sixteenth conversation, Davey, that this is so. Experiments, as well as our own experience, confirm that most of our conscious actions and thoughts are influenced by unconscious activities in the brain. For example, we know this is true for vision, where the raw visual data is sent to some thirty different parts of the brain to decipher such components as edge detection, colour interpretation, stereoscopic reconstruction, motion estimation and face recognition. All this happens without conscious action on our part. .

Ramachandran gives a simple example to illustrate the difference between the conscious self and the objective world.[120] The conscious self might say "I see red", and know exactly what is meant. Conversely, a third-person account might read, "He says he sees red when certain pathways in his brain encounter a wavelength of six hundred nanometres." The second account, while it may be scientifically accurate, leaves no sense of what red is really like to the subjective beholder. To achieve that understanding, a person who is not colour blind could subject her or himself to light with a wavelength of six hundred nanometres, and thereby effectively translate the third-person description into a conscious experience. On the other hand, those who are colour blind may understand completely the visual pathways of experiencing the colour red, without having any notion of what red is really like, or alternatively, seeing it as some other colour. Most observers conclude that the two accounts are in two different "languages", one the language of the conscious brain, the other English, and that the conscious sense of the colour red is simply lost in translation to English. Of course the same is true in many more situations. I shall never know what it feels like to give birth to a child, or to climb Mount Everest, but I can read and understand descriptions of both processes.

Eric Kandel, a Nobel Prize–winning American neurologist, makes the interesting point that in certain circumstances we may have more than one "consciousness."[121] For example, the left and right brains are connected by a bundle of nerves called the *corpus callosum*, which is the means by which the two brain segments communicate with each other. Some epileptic patients have had their corpus callosum surgically severed. Studies of such patients have revealed that two versions of consciousness are frequently at war with one another. For example, one patient "put on his clothes with one hand, and took them off with the other", while another, who was holding a book in his left hand, found that his right hemisphere (which cannot read) found the book boring and commanded his left hand to put it down, while at the same time his left hemisphere, which controls the reading function was surprised to find the fascinating book being put down!

Kandel describes the "neural representation of consciousness and self-awareness" as "life's deepest riddle", in large part because the higher functions of mental activity, including consciousness, are so complex that it is extremely difficult to identify the many areas of the brain involved in the activity, and it is thus extremely difficult to make objective measurements.

– I think your man Kandel exaggerates. Consciousness is a long way from being life's deepest riddle. A much harder problem to solve seems to be how life originated on Earth. In that case you have a few leads, but you are a long way from solving the problem. But in the case of consciousness, your intellectuals seem to have made a big deal out of what I think is simple in principle, though no doubt very complex in execution.

I was shocked into silence. Davey was becoming uncharacteristically bold! The silence held for about five seconds. It was Davey who broke in again before I could express my surprise.

– I could see this part of our dialogues coming, so I took the precaution of doing a little research of my own before our meeting. Surely it is nonsense to make such an issue out of consciousness and self-awareness. It is just a straightforward evolutionary development.

I know, I know. You want to tell me that this is a very old problem, and that it has preoccupied some of the best minds of your species.

It appears that René Descartes and Gottfried Liebniz were guilty of starting this unfortunate debate almost four hundred years ago. Descartes, in 1637, seems to have thought that he existed because he thought. I guess he must have felt strongly about it because he said it in Latin: *cogito ergo sum* (I think, therefore I am), but what could he have meant by that? Did he really think that the world, the universe and his friends and colleagues were mere figments of his imagination? It seems that he thought it possible! So, along comes Liebniz – a brilliant mathematician and co-discoverer with Isaac Newton of calculus – some seventy-seven years later to try and help him out of this awkward conundrum by suggesting that ". . . it must be confessed that perception and that which depends upon it are inexplicable on mechanical grounds . . ." [i.e. as the workings of a brain]. He went on to state that he believed that perception was ". . . in a simple substance, and not in a compound or in a machine."[122]

The example you quoted earlier from Ramachandran is at the core of what they were driving at, namely that what humans perceive in their minds' eye cannot be adequately described in any language. In Ramachandran's example, the quality of "redness" is a property experienced in the brain that could in theory be different for every person. Hence the invention of things called *qualia*, defined as experiences as registered in the human mind, experiences that it seems to me are very hard, if not impossible, to define in a consistent manner.

Many distinguished philosophers and physicists have become entangled in the debate surrounding qualia. For example, the 1944 Nobel laureate in Physics, Erwin Shrödinger wrote, "The sensation of colour cannot be accounted for by the physicist's objective picture of light-waves. Could the physiologist account for it, if he had fuller knowledge than he has of the processes in the retina and the nervous processes set up by them in the optical nerve bundles and in the brain? I do not think so."[123]

Your physicist friends didn't stop there, Peter. Roger Penrose, another distinguished physicist and mathematician, penned a lengthy tome in 1994 in which he made two main points. The first, (in agreement with Shrödinger), was that human consciousness allows humans to perform actions that cannot (Penrose claims "provably") be carried out by a computer, and the second that the mechanism that permitted consciousness was some kind of quantum entanglement effect.[124]

Does he seriously think that humanity and other conscious life forms had to evolve mastery of quantum communication in order to achieve consciousness? He would have been better off to advise ingestion of Shakespeare's "Eye of newt, and toe of frog, wool of bat, and tongue of dog"![125]

– Davey, Davey, what has gotten into you? I have never seen you wound up like this! What does consciousness matter to you?

– It means just as much to me as it does to you. Do you think I am not conscious? Granted, I cannot see you, or for that matter the colour red. But we have colour in my universe. We also have ideas and emotions. Furthermore, every thought that I have, every emotion, every sentiment, every idea, every foreboding is produced in my brain, as it undoubtedly is in yours. What is it about consciousness that drives otherwise intelligent, even brilliant, human beings to invent new senses and mechanisms when none is needed? I am the living proof of that fact.

– How could you be? I don't even know that you exist!

– Well now, isn't that an interesting comment! Do you wonder if Margaret really exists too? Perhaps she is just a figment of your imagination? Why don't you ask her? And if you do ask her, ask her, as well, if she thinks that I exist. I shall be interested to hear what she says.

– Let's not get into that just now.

And anyway, as you likely know, there is also a distinguished array of researchers who agree with you that the concept of qualia is not needed, and that the mind and consciousness are indeed products of neural processing in the brain, although these processes are likely very complicated. The well-known philosopher Daniel Dennett, as you surely know, is absolutely convinced of this,[126] as was Frances Crick, the Nobel laureate who – with Watson, Franklin and Wilkins – unlocked the code of DNA. He spent the last years of his life trying to understand the complex neural pathways that lead to consciousness.[127] Many others are following in his footsteps. No one should underestimate the complexity of the task. After all the human brain, as we discussed last time, contains billions of neurons of about a thousand different types. However, understanding the neural pathways associated with conscious thought is a finite solvable problem, just as the decoding of DNA is a huge but finite solvable problem. What is more, we now have access to some wonderful tools for tracking neuron activity in real time, so it should be just a matter of time before we humans understand how conscious thoughts are produced and processed.

When we last spoke, I reported on Ramachandran's conclusion that the human brain has both a "how" and a "what" path, with the latter requiring short-term memory, while the former does not. By means of some ingenious experiments with people suffering from partial brain damage, Ramachandran has been able to demonstrate the truth of his thesis concerning the short-term memory requirements of these two paths.[128] Consciousness (or mind) is associated with the "what" memory path. As already mentioned, a well-practiced lunge for a net shot in tennis does not require conscious thought, and may be carried out unconsciously. If there is more time available, the shot may be given a dose of planning through conscious thought – but the results may in fact be worse!

This understanding of the brain as an organ that is a multiprocessor, with only a few of the processors subject to influence by the conscious mind, is fundamental to an understanding of ourselves. Coupled with the information outlined in last week's conversation, it helps explain why we are so easily fooled by imagined events and why our actions may be much more guided by inherited behaviour patterns than we think.

One of the more interesting new brain research tools is a special helmet called a "transcranial magnetic stimulator", which can focus a precise sequence of magnetic field variations onto the brain. Ramachandran tells of his extreme interest (but not his surprise) on learning that a Canadian psychologist named Michael Persinger had applied magnetic stimulation to a part of his brain called the temporal lobe, and, for the first time in his life, had experienced God.[129] Ramachandran's lack of surprise was due to his familiarity with the fact that patients who suffer from seizures in their left temporal lobe very commonly report particularly strong religious experiences. The left temporal lobe is where the brain's limbic system, generally associated with emotions, resides. While epileptic seizures in the limbic system are much rarer than those in the motor cortex (which result in severe muscle twitching), limbic system seizures commonly lead to feelings "ranging from intense ecstasy to profound despair". Often those suffering from the seizures "have deeply moving spiritual experiences, including a feeling of divine presence and the sense that they are in direct communication with God".

Ramachandran comments,

> I find it ironic that this sense of enlightenment, this absolute conviction that Truth is revealed at last, should derive from the limbic structures concerned with emotions rather than from the thinking, rational parts of the brain that take so much pride in their ability to discern truth and falsehood.[130]

It will likely come as no surprise whatsoever to be told that past experience as well as subconscious inherited behaviour influence the decisions we make. The ability to incorporate past experience according to many experts is one of the defining characteristics of intelligence, and Ramachandran and his colleagues have ably demonstrated the role of inherited behaviour in our decision-making process. But it turns out that there is yet another way in which our life experience can influence our decisions. That way is through unconscious pathways that are altered as a result of life experience.

You will recall that when we talked about adaptation by gene selection [in Dialogue 14] we discussed the finding that the brains of mice, monkeys and men (and almost certainly other mammals as well) are altered by early-child rearing experiences. In essence, it turns out that those experiencing a stressful early childhood produce more of a certain enzyme[*] than their more fortunate peers. This enzyme in turn causes a few extra molecules to attach themselves to part of a gene in some brain cells in the *hippocampus*, the part of the brain that appears to control, amongst other things, some long-term behavioural characteristics. This, in turn, results in the gene being turned off (i.e. unexpressed). As a direct result, the developing hippocampus incorporates fewer receptors for the stress hormone *glucocorticol*. So, now we can discern a pathway by which those who experience stress early in life may react differently from others to stress experienced throughout their lives. In the sorts of complex, co-operative societies in which we live, such a brain alteration is largely dysfunctional, but in some earlier hunter-gatherer societies it is likely to have been survival enhancing. How then ought we to think about the "free will", which some philosophers delight in studying and debating. Are we "free" to make decisions based solely on ideas we generate in our own minds, or are there unconscious influences that narrow the set of options we are "free" to consider? A religious formulation of this dilemma sometimes asks whether God directs our every movement, or whether God gave us the ability to make decisions freely, and will then sit in judgment on the decisions we take at a later date.

For ants, decision making is simply defined, though for example, there are likely to be different responses amongst different ants when "fight or flee" decisions have to be made, likely largely due to different histories of individual ant development. As one ascends the scale of mental capacity, one would expect that the arbitrage of different choices becomes

---

[*] An enzyme is one of a family of complex proteins that are produced by cells. An enzyme acts as a catalyst in specific biochemical reactions. They are critical to proper cell function.

increasingly more complex as the animal's capacity for memory and thought increases. All of us animals seem to share the mental circuitry needed to ensure that inherited behaviours play a role in making decisions, but for the more intelligent animals, learned experiences play an increasingly important role. Some people appear to believe that humans are the only animals capable of conceptual thought, but this notion appears to be incorrect. Dogs trained as retrievers provide an interesting case. They are generally able to note where two or more objects for retrieval are located, and move straight to successive objects without having to return to home base before setting out on the next retrieval task. This strongly suggests that the animal has formed a conceptual map of the territory to enable it to perform such a feat. Even some birds seem capable of conceptual thought. Crows seem to be particularly adept in this regard. Yet crows are extremely distant relatives of humans. Richard Dawkins[131] wrote that the closest common ancestor we share with birds lived about about 310 million years ago.

We now know that many human decisions are a mix of inherited and learned information, both conscious and unconscious. The fact that we humans are able to incorporate experience learned from others by reading books and watching television, for example, does not alter the issue materially. There is no such thing as total free will for us animals! But we do have the power, through our actions, to alter decisions, for example by educating ourselves, or moving to different environments, and this is highly significant. Nevertheless, at any given time the decisions we make will be some kind of a summation of the state of our brain. The state of the brain itself is certainly a complicated function, depending as it does on input from our senses, our memory, our inherited behaviour, our education, what we have been eating and drinking, the temperature of our heads, etc. In a manner analogous to programmed "decision" algorithms in our computers, our brains apparently weigh the myriad conscious and unconscious inputs, and then output a decision determined by the nature and strength of the inputs. Humans seem to excel in the degree to which "conscious" inputs can influence the output. Common sense tells us that it is so. Experiment appears to confirm it. However, the phrase "free will" seems to me to be a poor description for this much more complicated and interesting process.

The much-debated state of consciousness seems to be an evolved capacity to retrieve and review some of the information stored in our memories, and to use it to help us make decisions. One of the essential capabilities incorporated in this system is the ability to use our visual processing system to help us to "see" and review images stored in our memories.

In a brief commentary that reviews key positions taken on the origins of consciousness and mind, Michael Shermer, the editor of *Skeptic* magazine and a regular columnist in the *Scientific American*, closes one of his columns, which addresses this debate, with the following words: "Because we know for a fact that measurable consciousness dies when the brain dies, until proven otherwise, the default hypothesis must be that brains cause consciousness. I am, therefore I think."[132] A neat rephrasing of Rene Descarte's proposition – "I think therefore I am." – and a useful reminder that there is most unlikely to be anything sacred or otherworldly about consciousness.

Shermer's next article on the subject, entitled "Free Won't" goes one better. He reports on experiments conducted by a physiologist named Benjamin Libet in which he measured brain activity of subjects using an electro-encephalogram (EEG) in an experimental situation where the subjects were asked to press a button when they felt like it. By studying the EEG patterns, Libet found that the subject's brain had decided to press the button a full half-second before the test subject was aware that he or she had decided to press the button! The clear implication here is that even the exercise of free choice may be largely unconscious, although doubtless influenced by conscious thought before the choice is made. Once the brain has received all inputs, and made a decision, it seems to take its own sweet time to inform our conscious selves of its decision. We appear to interpret this decision reported as our conscious choice, despite this evidence to the contrary.[133]

–Well now, isn't that interesting. It sounds to me as though a full committee of unconscious decision makers review the conscious decision before approving it and informing the conscious part of the brain that it has successfully made a decision!

– That is very likely correct to some degree, Davey. Of course our instinctual inputs to decision making are essentially limited to behaviour modes inherited from our hunter-gatherer forebears. Many decisions we make in today's environment are only tenuously connected to such primal instincts as those related to survival, reproduction, shelter, security, etc. However, to the extent that we ourselves believe that a given decision is relevant to our basic needs, then we can expect that our instincts will influence our final decisions.

The essence of this instinctual influence on everyday decisions has been familiar territory to the advertising industry for a very long time, hence their continuous effort, for example, to convince us that personal attractiveness to members of the opposite sex depends importantly on the car we drive or the clothes we wear.

– Yes, but it sounds to me as though there can be no clear boundary between conscious and unconscious decisions, or between free will and its opposite. It is crystal clear that many decisions are totally unconscious most of the time, such as when to breathe and when to blink. Some are mixtures and the result of training – for example your claim to be able to hit some great tennis strokes when at the net before you even realized the ball was near your raquet – and then there are the fully conscious decisions, such as when to eat or to buy a car. From what you tell me, even those decisions likely have an unconscious component to them.

And no wonder you humans get confused if you use the same visual processing part of your brain both for seeing and for the recall of remembered or imagined images. The potential for confusion, it seems to me, is huge.

– Well, I guess I should be glad that you cannot see me playing tennis! You might be very disappointed. But I agree with you and Michael Shermer that it seems unlikely that the conscious self is anything much more than that part of the brain devoted to reviewing past knowledge in the light of present circumstances and signalling a preference to that part of the brain where decisions are actually made.

Now, if we return to examine the definitions of mind provided at the beginning of this dialogue, we can discern lots of room for confusion and ambiguity.

The *Shorter Oxford English Dictionary* defines "mind" as "the seat of consciousness, thought, volition and feeling". We can readily concede that mind includes "the seat of consciousness", but volition and feeling are, as we know, phenomena that are substantially if not entirely derived from unconscious pathways in the brain. Should these unconscious, largely instinctual pathways be considered as elements of mind? If so, then perhaps mind should be defined to include all of the brain with the possible exception of motor control, sensory mechanisms such as seeing and hearing, and autonomic functions like control of heartbeat and breathing. One then has to wonder whether the definition of mind as distinct from the brain, as you have already implied, Davey, is still useful.

The other definitions I gave you at the start of this dialogue have similar problems. The *American Heritage Dictionary* definition seems to be especially wanting: "Mind is the consciousness that originates in the brain and directs mental and physical behaviour", thus excluding entirely the role of unconscious thought in directing "mental and physical behaviour".

I heard something resembling a bemused chuckle coming from where I imagined Davey to be.

– What's the matter, Davey? Is the distinction between mind and brain not crystal clear to you?

– Well, yes and no. There is no difference in my case, since I can be aware of almost any activity in my brain should I choose to focus on it, but as you know, I was not aware that this is not the case for humans. I wonder whether the whole history of our planet in my universe might have been altered had I known that such a situation was even possible. As far as I can remember, my Envergurians never distinguished between mind and brain, but my memory may have failed me here.

– I don't quite understand you. Can you tell me how the history of your planet might have been changed? We on Earth are pretty well used to our memories failing us, but you have sometimes seemed to me to have a perfect memory. How fallible is your memory, anyway?

Another chuckle from Davey.

– For the most part, my memory is perfect by your standards, but there were events, or really one event in the past – of which I've already spoken – that caused me to forget a lot I had previously known. This event was the beginning of the end of the Envergurians. It occurs to me now that if I had better understood my good friends, I might have been able to save them. From what you have told me, I now suspect that they, too, had a much less logical brain than I gave them credit for.

I haven't felt really emotional for a very long time. What a strange and wonderful feeling!

I must learn more about the way the human brain operates in practice. Can you be ready for us to talk about this next week?

– Okay.

My response was given without much thought. I was too busy trying to imagine what might have caused Davey to lose his memory.

DIALOGUE 18

Memories Are Made of This

Love looks not with the eyes, but with the mind.
SHAKESPEARE, *A Midsummer Night's Dream*, Act 1, Scene 1

Exactly one week later, after some rather frantic studying on my part, I launched my last dialogue with Davey on the workings of the mammalian brain. We all have the sense that brain science is difficult. I had forgotten how difficult it really is.

In an attempt to make things easier for Davey, I decided to set the stage with an illustrative incident from my own life. So, after I was comfortably settled at my desk and had exchanged the by now customary greetings with Davey, I started in.

– On November 22, 1963 I was dining in Nuffield College, Oxford, as the guest of an old friend. It was neither the first nor the last time I dined in that hall, but it is the only time that I remember the precise date. Many other people my age know what they were doing on that day, too. For it was on that day, at 12:30 p.m., Central Standard Time, that John Kennedy, a young and popular president of the United States, was assassinated in Dallas, Texas. During the main course, the master of the college rose to inform us all that Kennedy had been shot and seriously wounded. About twenty minutes later, we were informed that he had died.

Such events become imprinted on our memories for a lifetime. Nor is this a uniquely human phenomenon. When our family moved back to Vancouver from Ottawa in late 1944, as the war was coming to a close, my mother acquired a beautiful, young, black labrador retriever from the pound. My father, who was still in Ottawa finishing up his war service, was delighted to think that Joker (as we called him) would be there to retrieve any ducks he shot when he went hunting. Alas, it was not to be!

234

Joker was terrified of the water. No matter what we did to encourage him to swim Our efforts to encourage a love of the water included my father gently carrying him out to sea with us as we walked out on the long sloping sand at Qualicum Beach. The end result was always the same; he headed straight for the closest shore as fast as he could. We never found out what it was that had instilled such a fear of water in him. Perhaps someone tried to drown him when he was a puppy. Whatever it was, the event or events imprinted a fear of water that lasted the rest of his lifetime, a lifetime otherwise uneventful and happy to the best of our knowledge.

There are other more subtle kinds of imprinting going on in our brains as well. An important example which I have already discussed is the discovery that children (and other mammals) who have been raised in caring families generally do better in later life than those who were not so lucky.

And then there are the magical leaps of imagination and logic that our brains can invoke. Think of Einstein deducing the Special and General Theories of Relativity, Shakespeare writing *Hamlet* and Mozart composing *The Magic Flute*. In our own lives, we can reconstruct images from our childhood and important moments in adulthood, hum intricate melodies and be awe-struck by a sunset in the mountains or the prairies. How do our brains do it? Or, in the absence of certain knowledge, how might they do it?

Until a few years ago we really had no idea. Now the pieces are starting to fall into place, although much research remains to be done. I believe we know enough now to have confidence that we will achieve a detailed understanding of the workings of our brains this century, though almost certainly not in my lifetime. This knowledge will, in its turn, give rise to some difficult moral and ethical problems related to the use of that knowledge.

In today's dialogue I will address three different aspects of memory and thinking that have all been elucidated by recent findings. With time, these examples will doubtless be rounded out with a host of new discoveries, but these three, for me at least, provide an important insight into what we might expect to learn about the plasticity of our memories and our behaviour. These three aspects are:

1. How might we store and later "recall" a fact, an emotion, a musical phrase or an image (for example of a friend)?
2. How might our ability to think laterally (that is, to draw on logically unrelated facts or experience to help solve a problem) be implemented?
3. How might our behaviour and mood be altered on a permanent or semi-permanent basis by early-childhood experience?

The answer to these questions, and a host of other related questions, lies buried in the detailed operations of the neurons in our brains, and particularly in the way that our brains handle long-term memory.

It occurs to me, Davey, that you may not be as fascinated by the workings of the human brain as I am, in which case we can drastically shorten the dialogue. But in doing so, please be aware of my reasons for including this discussion. I believe that a basic understanding of the way our brains work can help us a lot when we try to understand both our own actions and those of others. The three topics I plan to deal with will, I think, help you as it has helped me in gaining such an insight.

As I had expected, Davey's response came booming back.

– Let's roll! These topics are all of interest to me!

So roll on we did, starting with item 1 on my list.

– Okay, here we go, Davey. First up is long-term memory.

As recently as the late 1980s the way in which our brains performed the functions associated with memory were unknown, and a very hot topic of research. At that time I wrote the draft of a mystery in which the plot revolved around the discovery of how memory worked in the human brain. The murder victim had made a great discovery that pointed the finger at *glial* cells* as the key to memory. The murderer turned out to be a friend and fellow researcher consumed by jealousy. He believed that it was he who better deserved the Nobel Prize, which would result from publication of this great discovery. Nobel Prizes are more important than either love or money to some researchers. Famously, they are not granted *post mortem*!

Our current state of knowledge pretty well confirms that this guess of mine was wrong, although the hunch that the glial cells might have an active role to play in our thinking processes is probably correct, as I will soon make clear.

A well-known case study in the field of brain research[134] is that of a patient known as "HM",* a man who had a bad fall from a bicycle as a boy,

---

* The role of glial cells is addressed later in this dialogue.

* HM is code for the American epileptic patient Henry Molaison, who died in 2008 at the age of eighty-two. There is some confusion as to how much of HM's brain was excised during the 1953 operation, but there is no doubt that his case played an important role in starting to understand the role of many of the ancient parts of the brain that we share with most mammals.

and as a result had serious epileptic seizures. After trying many other remedies, HM was at last operated on in 1953. All or part (descriptions vary) of his hippocampus, a small sea-horse shaped organ in the brain, was removed. While HM's seizures were cured, he was subsequently unable to add any new information to his long-term memory except as a modification to a pre-existing memory. He was left with a short-term memory that allowed him to lay down memory of events for only minutes at a time. This same phenomenon has since been observed in many others, typically patients suffering from chronic alcoholism or from oxygen starvation to the brain during brain surgery. Such patients can remember events that preceded the brain damage, but retain no memories after the traumatic event. It has thus been clear for some time that the hippocampus plays an absolutely essential role in long-term memory, but not much else was clear until recent times.

Our long-term memories go to the very heart of our being and of our images of ourselves. It is horrifying to imagine a life without memories of our past, or without the ability to add new sensations and new encounters with friends and family (our grandchildren for example) to our store of memories. Without memory, we cannot even work effectively.

In the preceding two dialogues I tried to show you glimpses of the way our brains affect the way we see things. We also got a sense of how much of the workings of our brains is inaccessible to conscious control. We should expect that our memories, the product of millions of years of evolution, will, like our physical properties, have characteristics honed to promoting the survival of our ancient forebears in hunter-gatherer societies, characteristics that may or may not match well with our current lifestyle.

As we discussed in our sixteenth dialogue two weeks ago, we lodge about 100 billion (10^{11} or 100,000,000,000) neurons in our brain. On average, but depending very much on the function of the particular neuron, each neuron makes connections with over one thousand (and sometimes several tens of thousands of) fellow neurons at junctions that are called synapses, of which there are thought to be over 100 trillion (10^{14} or 100,000,000,000,000) in an average human brain. These are unimaginably large numbers. If that number of neurons and their synapses are to fit in our heads they have to be very small, which of course they are. It has been estimated that there are 146,000 neurons per square millimetre on the surface of the brain, which has an average surface area of about 2,200 square centimetres, an area that is achieved by resorting to many folds in the brain surface. Synapses are packed at a density of about 600 million per cubic millimetre, yet each individual synapse is an organ of wondrous complexity. No wonder brain research is a challenging discipline!

The basic operating mode of the neuron is quite simple. It is when one starts to examine the details, particularly the chemistry involved, that the picture becomes more complex (and jargon laden). As you might expect, there are quite a few different kinds of neurons. For example, some act as sensors for light, heat or touch; others activate muscles or sense muscle position; some are used for thinking; others are a part of our emotional being. However, with the possible exception of some sensory and heart-muscle activation neurons, all operate in a similar fashion.

Neurons receive inputs, usually from other neurons via the aforementioned synapses, but sometimes directly from muscle tissue, light, etc. Most, but not all, functional inputs received by a neuron are electro-chemical in nature. A single neuron may receive as many as ten thousand such inputs, some of which will be inhibitive (involve negative voltage pulses), and some of which will be stimulative (involve positive voltage pulses). If the sum of all these inputs yields an electrical pulse that exceeds a threshold within a given time frame (usually of the order of a few milliseconds), the nerve cell will "fire", emitting electrical pulses of about 110 millivolts, which are carried up the nerve axon to its output connections (synapses) with other neurons.

What could be simpler? Add together a bunch of positive and negative electrical pulses, and, if the sum exceeds a certain threshold value, send out (positive) voltage pulses to any synapses connected to the output axon. Otherwise remain silent. Almost always a burst of pulses (rather than a single pulse) are sent out, with the number of pulses being the primary indicator of message intensity.

Most of the time, the inputs to the nerve cell arrive via the *dendrites*, the feelers that establish contact with other neurons at synapses. Sometimes, the synapses connect directly to the cell body, but, for the purposes of our discussion, this is a detail.

Neurons are so important that the body has provided nursemaids and servants for them. These are the glial cells I wrote about in my detective novel, which comprise about 90 per cent of the cells in our brains, though only about 30 per cent of the volume of our brains. This is because, for the most part, glial cells are much smaller than neurons. Glia come in three different basic varieties and perform such tasks as ensuring that there is an appropriate mixture of ions available in the extracellular spaces of the brain, ready and waiting to be used in the synapses; providing electrical insulation (*myelin*) over the axons so that they don't lose their charge before they can pass on their signal; providing food and sustenance for the neurons; filtering dangerous chemicals out of the blood stream before they can affect the sensitive neurons (often referred to as the blood-brain barrier); and providing a guide along which newly generated neurons can slither as they move into place. Until recently, glial cells were thought to be totally subservient to neurons. However, nature again seems to have found a way to double up, allowing glial cells to function in an intriguing way. One interesting clue here is that the proportion of glial cell mass to neuron mass seems to increase the greater is the intelligence of the animal in question. Humans have a greater proportional mass of glia than any other animal. I shall be discussing the intriguing role of glial cells later in our dialogue today.

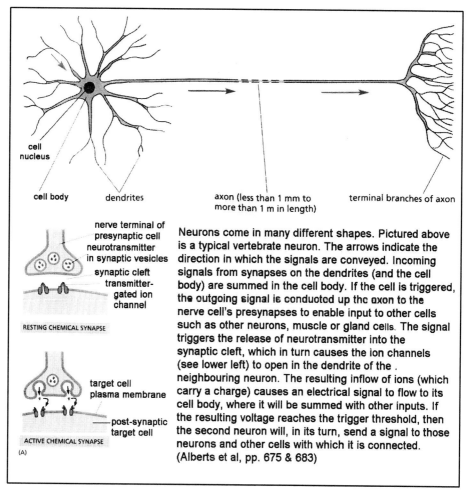

Neurons come in many different shapes. Pictured above is a typical vertebrate neuron. The arrows indicate the direction in which the signals are conveyed. Incoming signals from synapses on the dendrites (and the cell body) are summed in the cell body. If the cell is triggered, the outgoing signal is conducted up the axon to the nerve cell's presynapses to enable input to other cells such as other neurons, muscle or gland cells. The signal triggers the release of neurotransmitter into the synaptic cleft, which in turn causes the ion channels (see lower left) to open in the dendrite of the . neighbouring neuron. The resulting inflow of ions (which carry a charge) causes an electrical signal to flow to its cell body, where it will be summed with other inputs. If the resulting voltage reaches the trigger threshold, then the second neuron will, in its turn, send a signal to those neurons and other cells with which it is connected. (Alberts et al, pp. 675 & 683)

Figure 18.1 – *Neuron Operation*

As early as 1949 the well-known psychologist Donald Hebb postulated that "neurons that fire together wire together".[135] Thus for example, if a flash of lightning illuminates a good friend in the foreground as well as a mountain covered in dark forest, and if the lightning is accompanied by deafening thunder, at least the following functional parts of the brain will be activated by the event (i.e. they will fire):

- The black and white and colour representations of the image in our brain.
- Mountain and tree recognition circuitry.
- Face recognition circuitry, particularly those neural circuits related to recognition of your friend.
- Sensory data related to temperature and dampness (is it raining?).
- Audio sensory data, and additional processing data related to analysis of the sound of thunder.
- Particularly if you were frightened, emotional circuitry will also be triggered.

If, for whatever reason, the event is "memorable", the vast network of neurons involved will fire repeatedly. As a result, the connections between all the different neurons involved will be strengthened for a period of time ranging from a few minutes to several hours. This process, called *long-term potentiation*, was just a hypothesis until early in the twenty-first century, when an understanding of the chemistry involved in long-term potentiation gradually came to light. We'll talk about this soon!

While the electrical activity of the neuron is relatively simple to describe, the chemistry is rather complicated.

When a nerve cell fires, the voltage pulse travels up its axon to all its synapses with other neurons of which there may be thousands. The voltage change at the synapse opens ducts that permit an in-rush of calcium (Ca^{++}) ions from outside the cell. These ions in turn trigger the release of a neurotransmitter into the synaptic cleft, that gap between the transmitting and the receiving neuron.

Depending on which neurotransmitter is released, and on the structure of the receiving cell, the receiving neuron will either open its "gate" to an influx of positively charged sodium (Na^+) or potassium (K^+) ions, or negatively charged chlorine (Cl^-) ions. As already mentioned, the resulting change in potential at the receiving end of the synapse will send either a positive or a negative voltage pulse toward the neuron cell body, where it will be summed with as many as ten thousand other simultaneous inputs from other synapses. Over the long course of evolution, over fifty different proteins have been hijacked and adapted for use as neurotransmitters. Some of these enhance the likelihood that the cell will fire, others inhibit firing. Some of the better known neurotransmitters include *acetylcholine* (particularly associated with muscular activity and also involved in wakefulness, attentiveness, anger, aggression, sexuality and thirst); *dopamine* (control of movement and posture, pleasure and attention); *serotonin* (regulation of body temperature, sleep, mood, appetite and pain);

glutamate (associated with learning and memory); *GABA* (gamma-aminobutyric acid) (motor control, vision and anxiety amongst other things); and *norepinephrine* (associated with sleeping, dreaming, learning, mood and blood vessel contraction). Nature is stingy and will often re-use the same chemical in different situations, so the same neurotransmitter can have different results when encountered by different types of neurons. Indeed these same chemicals are often used in cells other than neurons for different purposes altogether, which is yet another good reason to be very cautious about the indiscriminate consumption of mind-altering drugs, which typically mimic or inhibit neurotransmitters.

What I've just told you about neurotransmitters is already long enough, yet it grossly oversimplifies the activities in the synapse. The opening of cell walls to emit or receive chemicals and ions is usually a multistep process, and there are a number of chemicals that either enhance or inhibit the activity of the neurotransmitters, and which may be released in conjunction with them. Furthermore, there are also some chemicals that closely resemble the neurotransmitters, and which can attach to the same docking sites on the receptor side of the synapse. These chemicals will reduce the effectiveness of the synapse and may block it altogether. For example, the well-known poison *curare* acts by blocking the acetylcholine receptors in muscle tissue, thus inactivating the muscle. The nicotine in cigarettes resembles acetylcholine, while heroin locks into the receptors for *endorphins*, the body's naturally produced chemical for pain and stress relief. Endorphins are also associated with happiness and relaxation.

You may be as surprised as I was to learn just how rapidly all this chemistry takes place in the synapse. The whole process of molecules being released into a cleft and finding their way to the receiving side of the cleft, which causes a gate to open to allow charged particle to migrate into the receiving neuron and change the voltage potential of the cell does not sound like a rapid undertaking, yet it frequently happens in less than a thousandth of a second! The secret to the speed is the same as that employed by microchip designers – keeping the geometry small. The synaptic cleft is only a few billionths of a micron wide. Only a few million molecules of neurotransmitter may be emitted into the cleft in as little as 0.3 of a microsecond, raising the concentration of the transmitter a million-fold or more. Once the message has been passed through the synapse, another process sets in motion to clear the neurotransmitter out of the way in a similar time frame. In some instances the neurotransmitter is simply taken up again by the emitting neuron ready for recycling, in others it is broken down chemically.[136] For most general rules there are exceptions. There are some electrical synapses that do not rely on chemistry for their operation, but these are found only in heart muscle tissue and thus not of relevance to an examination of the processes involved in mental activity, which require the much finer and more subtle control which can be provided by chemical synapses.

So, Davey, you should by now have a general idea of the way most neurons are triggered by receiving a number of simultaneous inputs from other neurons and sensors, how the inputs are summed together and, if they exceed a certain threshold, they cause the neuron to fire and, in turn, provide inputs to the neurons to which it is connected.

Let us then return to the sample "memorable event" that we started with, that is your friend illuminated by a flash of lightning, followed by deafening thunder, etc. Consider what might be happening when we "remember" such a scene.

There is, of course, the initial task of converting the array of picture elements (coming from the eye via the optic nerve) into something meaningful to the rest of the brain. We need to identify boundaries of objects in the field of view, where useful applying our binocular vision processor to sort out the distance of objects from us and from each other. We also process colour separately from the basic outline of images, and, where motion is involved, another part of our brain goes to work on motion estimation. Once we have the objects sorted out, other parts of the brain go to work on pattern recognition. Is it a car we are seeing, and if so, do we recognize it? Is there a face or a figure we should be recognizing? Such tasks are very complex, as the objects in question can appear in quite an array of different orientations, relative sizes and illumination. When all these processes are called into play, neurons in many places of the brain are interconnected, and the neurons are stimulated and pulsed in synchrony.

If the scene is particularly memorable, then more frequent pulsing at the synapse will occur. As previously mentioned, this sets in motion a process called long-term potentiation (LTP) in which the particular network of connections for that scene is strengthened for a period of hours. The resultant voltage pulses after stimulation are roughly twice the normal voltage at the start of LTP, and decay back to their normal value at the end of LTP. This is what is called (rather confusingly) short-term memory. However if, during this time interval, the synapse is repeatedly stimulated (in one experiment, the synapses were stimulated three times at ten-minute intervals), then the synapse becomes permanently strengthened.

Long-term memory has been activated, but you may well ask how this occurs and even why it occurs.

The second question is easier to answer than the first. Long-term memory will normally be activated if our imbedded instincts cause us to view it as particularly important. A lightning flash or a close encounter with a dangerous beast, for example, may automatically trigger LTP.

Alternatively, it may be triggered by conscious knowledge of the importance of the event, such as a proposal for marriage or a meeting with your boss where you got a promotion.

Now we get to the difficult part. How might LTP occur?

In a series of beautiful experiments, Douglas Fields and his colleagues grew mouse brain cells from stem cells in a Petri dish in the laboratory.[137] They discovered that calcium ions that enter the neuron when the synapse fires, "*activate enzymes called protein kinases. Protein kinases turn on other enzymes by a chemical reaction called phosphorylation that adds phosphate tags to proteins. Like runners passing the baton, the phosphate-tagged enzymes become activated from a dormant state and stimulate the activity of [DNA] transcription factors.*" As Fields goes on to point out, there are many thousands of transcription factors waiting to be phosphate-tagged (or *phosphorylated*) as needed. The DNA in the neuron cell, after all, contains the instructions for making almost all of the proteins needed by the body, but it also turns out that each requires a different amount of time for "baton passing" in the chemical chain that culminates in DNA transcription. The long-term memory chain requires periodic stimulation over a period of an hour or more, which leads to the manufacture of several proteins that migrate to all the synapses in the neuron, but only influences those on which LTP is in effect. Thus it is that although the central shock to my memory system on November 22, 1963, was learning of the assassination of John F. Kennedy, the highly peripheral fact that I was having dinner in Nuffield College was also remembered as a part of the same long-term memory implant. I can even recall roughly where I was sitting in the college hall – or, at least, I think I can!

As another example of long-term memory, although a very bad piano player, I have noticed that my brain has the ability to remember (or at least half-remember in my case, in a bit of a hazy fashion), quite long series of notes in piano pieces. Of particular annoyance is the fact that if I make a mistake and do not immediately go back and correct it (and repeat the correction), my brain remembers the mistakes as well! The time constants of activation appear to be different from the case of John F. Kennedy's assassination reported above, but our memories seem to have a wonderful capability to recall very long series of precise muscular actions in the hand, in this case resulting in notes on a piano sounding. How evolution endowed us with this capability and how it is implemented in the brain are, I believe, still unsolved problems. Perhaps Davey, experience in your universe can help us to understand this process?

Davey's only response was a *basso profundo* grunt.

– I shall move on, then, to my second topic today: lateral thinking processes.

Lateral thinking can be defined as a method of solving problems by unorthodox or apparently illogical means. It involves seeking and finding relevant ideas apparently unconnected with the original thought or thoughts. It is sometimes described as "thinking outside the box", and is the sort of thinking that most of us ascribe, for example, to Einstein when he developed his Theories of Relativity.

If you think about it for a moment, you can understand how traditional memory networks, with their thousands of neural interconnections all stirring each other to fire or not to fire, may not be a good mechanism for lateral thinking. It is likely that some nearby neurons that are not spliced into the net in use could provide a lateral thinking connection. Most of the time however, it will be fortunate that nearby neurons are not activated. Otherwise our thoughts might become a massive, confused, disordered jumble, perhaps akin to the "highs" induced by some psychedelic drugs.

Not only did Douglas Fields lead the group that carried out the long-term memory work I just talked about, but in a second article in *Scientific American*, he wrote about his discovery of a new role for glial cells, those neuron support cells mentioned earlier that comprise about 90 per cent of the cells in our brains. For some years, glial cells have been known to be helpful partners for neurons. For example, glial cells called *astrocytes* connect blood vessels to neurons, thereby not only bringing needed nutrients to neurons, but also, very

"Never, ever, think outside the box."

importantly, providing that *blood-brain barrier*, which helps ward off *pathogens* – disease-producing agents, most often micro-organisms – that have evaded our immune systems.

Despite their importance, glial cells have elicited much less interest than neurons, primarily because it became evident quite early on that they were electrically inert, in the sense that they played no visible electrical role in transmitting messages in the brain. But, like all cells, they are not inert chemically, and it now appears that they have developed their own chemical communication network. Like neurons, they have the same ion channels that trigger the generation of electrical signals in neurons.

However, in a type of glia called *oligodendrocytes*, the influx of calcium ions, instead of generating an electrical pulse as would occur in a neuron, initiates chemical processes that lead to improved insulation (i.e. increased *myelination*) of the neurons they embrace, hence strengthening their output signals. Of even greater significance, in the other type of glial cell common in the central nervous system, the *astrocyte*, it has been observed that the influx of calcium ions triggers the release of *ATP* (adenosine triphosphate), which acts as a messenger molecule, causing neighbouring astrocyte cells to open their ion channels, which in turn leads them to emit ATP.

Fields and his colleagues have determined that when neurons emit neurotransmitters at their synapses, they simultaneously emit ATP, which acts as a trigger for the astrocyte chemical communication system. By increasing the amount of ATP available to neurons that are in the neighbourhood of neurons already directly connected to the particular ongoing thought or memory process, those neurons are more likely to be fired themselves, and thus contribute (with their associated networks) to the thinking or remembering being undertaken. In other words, it looks as though the ATP chemical communication system has triggered "lateral thinking" by causing nearby nerve cells to fire that would not otherwise have done so.

Or, as Fields writes:

> The working hypothesis that [Philip G.] Haydon and I, along with our colleagues, are reaching from these discoveries is that communication among astrocytes helps to activate neurons whose axons terminate relatively far away and that this activity, in turn, contributes to the release of neurotransmitters at distant synapses. This activity would regulate how susceptible remote synapses are to undergoing a change in strength, which is the cellular mechanism underlying learning and memory.[138]

If glial cells really do prove to be the primary mechanism for lateral thinking, it will have been no coincidence that Einstein had an abnormally high concentration of glial cells in his association cortex, the part of the brain responsible for high level cognition.[139]

You may wonder how a research worker is able to unravel such intricate processes. In this instance, Fields and his colleagues used neural stem cells to grow a sample of mouse brain in a Petri dish in the laboratory. Glial cells were injected with a dye that fluoresces (gives out light when stimulated by light of another wavelength) when it comes into contact with calcium ions.

It was observed that upon electrical stimulation of a neuron, its neighbouring glial cells started to fluoresce, indicating that calcium ion uptake had been induced in the glial cell. This fluorescent behaviour was observed to spread out gradually to other glial cells in the neighbourhood. An ingenious piece of detective work!

So Davey, by now I hope you have a sense of how it is we humans remember things. We do so by retaining a large number of neuron interconnections that between them allow us to recall the essential aspects of the memory – faces, scenes, sounds, ideas, weather, etc., in short the most important characteristics of the memory in question. Of course, as already mentioned, if we try to extract more information about a particular memory than we have actually stored, our brains will obligingly find data to infill with, often drawn from more recent experience, or from suggestions by other people. Furthermore, our brains will kindly do all this without bothering to tell us.

– I know you have told me this before, Peter, but I cannot help being absolutely astounded that humankind has survived all these years with a brain that plays such tricks. Surely you would have a much better chance of survival if you simply got the facts rather than invented material?

– The proof of the pudding is in the eating, as we humans sometimes like to say. So the answer to your question is probably not. The invented material, like the optical invention used to erase the blind spot in our eyes, has likely proved to be survival enhancing on average, though there are likely to have been a number of instances that have put the brain owner at a disadvantage. It is important to remember as well that all such instances do not have to be survival enhancing. Some can be the accidental result of a gene modification that is survival enhancing in another context.

For example, when some distant forebear acquired the genetic changes, which meant that he or she did not see that blind spot but rather a fill in of the scene surrounding the blind spot, it is just possible that some of those same genes are involved elsewhere in the brain in filling in detail of remembered events from nearby memories.

– I guess we just have to agree to disagree. It just makes no sense to me.

– You don't have to believe it, but that is the way it is.

I personally find it very exciting that we humans have figured out how we remember things, and also exciting that we have some viable theories as to how our associative memories work. For instance, I have often found that after reading crossword clues, which seem to me to be totally abstruse, the crossword suddenly seems easy to solve when I read the same clues the next day. When that happens, I envision that, unknown to my conscious self, a small army of glial cells in my brain has been busy emitting ATP overnight so that the needed new memory paths would be available to me in the morning!

Finally, if you can stand it, I will address the third and final topic for today. It is a topic which we already discussed superficially in our fourteenth dialogue, when you accused me of misleading you and demanded to know more about gene selection. You may have forgotten about that, but I can assure you that I haven't. We humans tend to remember such things.

The importance of this final topic is that it turns out that early-childhood experience can result in changes in gene selection in some neurons in our brains. These changes in turn result in modifications to our brains in such a way that our behaviour is changed on a long-term basis.

– I'm sorry to learn that you are still brooding about my previous polite request to learn more about gene selection mechanisms. I do still find your discussion of the way your memory works highly intriguing, so of course I look forward to learning from you how long-term behaviour modification takes place in human brains.

It had been a long day, and I was feeling a little weary. I considered asking for a break until next week. However Davey sounded so genuinely interested in the topic that I decided to carry on.

– We have already seen how phosphorylation of some proteins can stimulate a neuron to produce other proteins that render transitory memories permanent. We have also seen how electrical activity in neurons can lead to glial cells enhancing the sensitivity of distant neurons, thus enhancing our ability to think laterally. Let us now look briefly at yet another process that plays a part in imprinting experience onto our brains.

An important paper published in *Nature Neuroscience*[140] in August 2004 established two chemical pathways by which the quality of mothering experienced by rat pups changes behaviour for life, and, in particular leads young females in their turn to be either good or bad mothers, depending on the nature of the mothering they themselves have experienced. Of course, my use of the term "good mothering" and "bad mothering" is subjective. Nature does not necessarily see it that way. It may well be that in a hostile rat society young pups have a better chance of survival if their mothers pay little attention to them and leave them to strike out more or less on their own. Most of us would agree though that loving care from parents is good, at least for human society as currently experienced. Rat pups brought up by mothers who "lick, groom and do arched-back nursing" significantly more than the average mother rat produce pups who, even in adult life, are more exploratory and less fearful and stress reactive than their peers. Correspondingly, those raised by mothers whose mothering behaviour is significantly below average exhibit the opposite characteristics. Moving the progeny of good mothers to be raised by bad mothers, and *vice versa* clearly demonstrated that the personality traits are acquired and not inherited. Further, it has been established that the behavioural mode is set in the first few weeks of life, and in normal circumstances lasts a lifetime!

Researchers discovered that this lifelong behavioural characteristic is governed by stress hormones called *glucocorticoids*. The number of glucocorticoid receptors in the hippocampus, that part of the brain that appears to control long-term memory, is evidently established in the first few weeks after birth, and the quality of mothering strongly influences the number of receptors. Good mothering results in the demethylation of the gene primarily responsible for development of the glucocorticoid receptors, thus switching the gene on and increasing the number of receptors for glucocorticoids in the hippocampus. The glucocorticoid receptors last for the lifetime of the animal.

A second factor influencing behaviour was found to be the degree of acetylation of the histones* associated with the DNA. (We talked about histones earlier, Davey. They are proteins that act as spools around which DNA winds. They also play a role in gene regulation.) Good mothering led to high acetylation of the histones, which in turn made it easier for the glucocorticoid gene to be activated. Like phosphorylation, methylation (of genes) and acetylation (of proteins) refer to the attachment of short chemical tags to a protein or a gene, with the result that the tag either activates or de-activates a chemical process with which the protein or gene in question is associated. It has become evident that such processes are the primary way that the brain initiates and terminates memory and other neuron development–related processes. As previously mentioned, once

---

* To recap: DNA spooling enables the compaction necessary to fit large genomes inside very small cell nuclei. Acetylation is the addition of an acetyl group to an organic molecule, thus changing its shape and charge distribution. Acetyl groups are frequently added to histones and other proteins modifying their properties. At the DNA level, histone acetylation by acetyltransferases (HATs) causes an expansion (unfurling) of the chromosome, allowing transcription to take place. Conversely, removal of the acetyl group by histone deacetylases (HDACs) condenses DNA structure, thereby preventing transcription. Acetyl has the chemical formula $COCH_3$.

such processes have been activated, they may alter behaviour for the remainder of an animal's life.

Information about histones and acetylation is tough for me to understand, laced as it is with chemical jargon. It may or may not be tough for you to understand, Davey. Reduced to language I am more comfortable with, it says that rat pups brought up in a relatively stress-free environment handle subsequent stress differently (most of us would say better) than those who had a stress-laden upbringing. Furthermore, this behavioural difference is not due to the fact that they learned how to handle stress better (the change occurs in early childhood, before the pup is capable of such learning), but rather because the chemistry in their bodies has been permanently altered when certain genes have been switched on.

Significantly, it was found that the effects of bad mothering in rats could be mitigated by administering a drug that inhibited de-acetylation of the histones. Similar phenomena have been observed in monkeys and other animals. Observation of human behaviour, though not yet backed by direct experimental evidence of chemical changes and alterations in gene expression, strongly suggests that we humans are influenced by the same, or similar, mechanisms.

It has been known for a long time that there is a clock governing the early development of all animals, with different parts of us undergoing development at different times according to a fairly rigid schedule.

We know that this process of adding chemical tags to DNA, RNA and proteins is a critical tool in the life process. Not only does it allow learning (such as good mothering) to be passed on, but it is also critical to cell specialization. We know that our bodies developed from a single stem cell that encapsulates the DNA of both mother and father. As cell division proceeds, and individual cells become specialized for incorporation in kidneys, hearts, brain, bones and blood for example, genes that are irrelevant to the specific function to which they are being adapted are turned off by the methylation of the gene in question.

Full development of both our physical and our mental capabilities is therefore a fascinating combination of inheritance permanently or semi-permanently modulated by critical life experience, and then, in some but not all instances, subjected to conscious reflective thought that may or may not modify what the first two programmed responses dictate that we should do.

When we "learn" to walk, we start with an inherited program that sequences our muscles in the appropriate way. The sequence is in fact rather complicated. Of course, the inherited sequence is unable to know in advance very much about our body weight or the length of our limbs and other factors that affect the most efficient motion for our walk. Hence we, in common with most, if not all other mammals with legs for walking, have to "learn" how to walk, that is, we modify the details of our inherited walking program so as to account for important variables like weight and length of limbs, many of which will change as we age. We can consciously change the way we walk, but only within parameters acceptable to our inherited "walking" program.

We now know that much of human conduct is a similar amalgam of inherited and learned behaviour. We humans have a tendency to think that our beliefs springing from our philosophies and our religions give us absolute control over our actions. Based on what we know about human behaviour and the neural networks that underpin it, nothing could be further from the truth. Our beliefs certainly can modify what our instincts are telling us to do, but if the modification is very different from what our instincts want us to do, we are likely to feel a lot of stress and become very unhappy. Indeed, there is lots of evidence, some of which is yet to be discussed, that successful religions and philosophies have tended to match their religious and philosophical tenets and prescriptions to inherited behaviours rather well in most circumstances. This can hardly be a coincidence.

– This is strange indeed! No wonder you humans often seem confused to an outside observer!

First of all, you tell me that human decisions are not just the result of careful consideration of the facts of the matter but rather a complex mixture of conscious, reasoned factors plus a bunch of undefined subconscious instincts and methylated genes that influence the rational outcome in unpredictable ways. This unholy mixture of influences is apparently resolved in an ancient part of your brains called the amygdala. No wonder you humans are so easily fooled by imagined events!

Then you tell me that long-term memory is essentially a network of neurons whose interconnections have been strengthened so as to endure, and which, when stimulated, re-energizes the net of neural connections instigated in the first instance. You go on to explain that one side-effect of this process is that memories peripheral to the central fact or event may be retained at the same time. What a complex system this is for such a simple function! Why can't you just store the images and facts for retrieval when needed? It sounds to me to be a lot simpler! After all, you do have computers that do just that.

Your description of the way that some human and other animal behavioural patterns (such as good or bad mothering of offspring) are imprinted from an early age by a chemical process you call methylation was fascinating. But what a complicated process to carry out a very simple objective! Our methods are much more straightforward than that.

But it is your description of the way you humans have evolved the ability to think laterally that interests me the most. You tell me that it appears to derive from purely chemical (or non-electrical) activation of nearby glial cells, which in turn activate adjacent neurons that are not otherwise interconnected. You cite the facts that Einstein's brain is more than normally rich in glial cells in his association cortex, and that humankind has a greater proportion of glial cells to neurons than any other mammals as supporting evidence for this finding. I, too, as you would expect, have the ability to think laterally, but it is very tricky for me to control my thinking so that I only have useful "lateral thoughts" and I am not swamped by a deluge of useless ideas that lead nowhere.

For all its seemingly needless complexity, your chemical brains may have some advantages over my brain. I'll have to think about it!

– Of course, animal brains are complicated. You can hardly expect otherwise. They have been evolving over hundreds of millions of years. Did you know that about two thirds of all our genes are specific to the operation of our brains? There is no computer on Earth that can yet approach the complexity of our brains.

– Yes. I wouldn't boast too much about that if I were you, since you humans, even after your much vaunted hundreds of millions of years of evolution, are still rather odd and dangerous to yourselves and to other animals on your planet.

But I have an important question for you.

You have told me that you believe that all life sprang from a common ancestor after, by happenstance, the gene for the first, most basic life form came into being. But not everyone seems to agree with you.

Some of the others with whom I have been speaking seem to believe that they were created by some father or mother figure. Not surprisingly, they are very grateful to the god who they believe created them.

You, Peter, have not told me very much about God. Do you believe in one of the gods? Do you think you will ascend into Heaven, or descend into Hell after you die? Why have you not spoken of your religious beliefs?

– Perhaps it is because I am not particularly interested in religion.

– But surely that is a mistake. I find human religions fascinating, and it seems to me that what you have been telling me about the human brain is very relevant to understanding religion. You have done me a great favour by reporting on what you know about the way the brain works, but, if you don't mind my saying so, you have missed out on what it might tell you about religion.

– Really Davey? Perhaps you had better tell me what you have in mind!

– Well, it is clear that human brains conjure far more phantoms than the phantom limbs you spoke of several weeks ago. There is a whole coterie of phantoms produced by different parts of your brains. For example, the system for embedding long-term memory seems to produce dreams and nightmares, while the lateral thinking system injects unanticipated thoughts and ideas. All kinds of such phenomena seem to bubble up from the subconscious brain and appear, unbidden, before the conscious brain. The timing, quantity and quality of this bubbling seems to vary widely between individuals. Furthermore, the bubbles can be encouraged and induced by such means as fasting, contemplating, sleep, sleep deprivation, standing on your head and imbibing magic mushrooms.

Several of the humans I have been speaking to have referred to these bubblings as "mystic experiences", and often they view them as revelations or communications from their gods.

One of my contacts referred me to a book by someone called Karen Armstrong called *A History of God*. I had the very devil of a time convincing her to buy an audio version of the book and to play it back for me. She even asked if I would pay her for the cost of the talking book and the gas she used driving to the store to buy it! But when she finally bought it and played it for me, I found it very worthwhile. You should read the book, Peter, if you haven't already done so. The history reveals a lot about human nature.

– Perhaps . . .

I was wary, unsure of where Davey was leading the conversation. He went on.

– Perhaps the most surprising thing I learned from the book was the similarity of all three of the major monotheistic religions. Everything I had heard seemed to emphasize the extreme differences, as well as (often perhaps especially) the major differences between different sects of the same religion. But consider this:

- All believe in one supreme god, though they have different names for that god.
- All have harboured widely varying views as to the nature of that god, ranging from a totally impersonal entity, "the uncaused cause of all being" whose nature is inconceivable by humankind and that "pre-existed" everything, all the way to a very personal, human-like god who cares especially (perhaps even exclusively) about humankind and can intervene to help or punish you. Apparently Judaism, Christianity and Islam have all, at various times and with varying membership and intensity, included sects that believed fervently in an impersonal god as well as sects that have believed in a very personal god, and for that matter, other sects adhering to almost all distinguishable positions in between.
- All three religions have wrestled with the difficulty, on the one hand, of loving and worshipping a totally impersonal entity, and on the other, reconciling worship of an interventionary and loving personal god who nonetheless seems to be at best quixotic in handing out brickbats and posies where reward and punishment are concerned. All three, at one time or another, have concluded that it was up to each individual to decide where on the spectrum the god they worshipped fell.[141] It seems to me to be a pity, perhaps even a tragedy, that this enlightened concept seems not to be at all widely held today. Each sect seems driven to define the nature of god, an approach that was wisely shunned in the old days.
- Almost all religions, certainly including Islam, Christianity and Judaism, have had powerful sects that emphasized mysticism and have used a common set of tools to enhance the brain's susceptibility to or receptivity of mystical messages.
- Almost all religions have similar ethical teachings, though all harbour sects that repudiate some of these teachings from time to time.

You once told me, Peter, that there is some knowledge that humans will never have, your prime candidate being that it is unlikely humans will ever know why there is "something" (for example, matter, universes, energy) rather than "nothing". The early forms of all these religions came close to defining god as that very same "nothing" that existed before there was "something". I would have thought that even an aetheist (and almost certainly a Buddhist) could believe in a god so defined!

So what do you suppose all the fuss is about religion? It looks to me as though the fuss is about a lot of dogma loaded onto each religion over many centuries, dogma that by and large answered to some very human needs, such as the need to help recruit new members, or to increase membership through encouraging large families, or the need to exert some control over members by such tactics as endowing certain members with infallibility or mystical guidance and powers, or by threatening damnation for disobedience. Often the dogma seems to have endured long past the era when it served a useful purpose.

This whole question of religion is both fascinating and bizarre. Fascinating because it seems to tell us quite a bit about human nature and the way human brains operate, but bizarre because religion has become such a flashpoint for disruptive and dangerous behaviour in many parts of the world, as I have been learning so forcefully from quite a few of those other humans I am talking with.

– I must say Davey, that was an interesting and surprising set of statements from you. You seem to be getting around and learning quite a lot without any help from me.

I can agree with most of what you say, but not all of it. Most religions do seem to come burdened with outdated dogma, but they also do a lot of good, not the least of which is to encourage ethical behaviour. In addition, they provide a sense of community and common interest. Most religions also offer educational services, and ceremonies for rites of passage, such as births, coming of age, marriages and death. Many humans find such rites meaningful and helpful. Religions have certainly evolved over the centuries, and will just as certainly continue to do so in order to meet the changing expectations and understandings of their members. I don't know if they will evolve in such a way that a worldwide common ground for religion will result, but I hope it will. It will be much easier for all nations to live together in harmony if they do.

– I hope so too. It is certainly both confusing and dangerous to have so many different interpretations of the origins of humanity.

But serious as the subject of human religion seems to be, it does have its funny moments. I thought you might be amused, Peter, by an ancient Jewish rite during which the "men were commanded to thank God during the morning prayer for not making them Gentiles, slaves or women".[142]

A lot of other people seem to believe that if there were no religion, all humanity would revert to immoral self-serving behaviour. Yet my Gurians had no recourse to religion that I can recall, and for the most part, they behaved civilly toward each other. Your hunter-gatherer forebears survived by being generally co-operative and social with other members of the tribe, even though they were likely generally hostile toward outsiders. Why can't you humans simply extend that sort of behaviour to all your fellow humans?

– That's the billion dollar question! Some humans do just as you suggest. Others, possibly even most others, do not. There is evidence that mutual trust is a big factor in whether or not co-operative behaviour can be applied to ever larger groups. Trust can be hard to maintain in very large groups.

But you said that your Gurians had no religion at all? Perhaps that was why they brought about their own destruction. Surely religious Gurians would not have behaved so thoughtlessly.

– The disappearance of the Gurians had everything to do with a failure to agree on a difficult but necessary course of action, namely the modification of inherited behaviour so that they could live more harmoniously together, and nothing to do with religion.

Mark my words! You humans will run into the same issue one of these days too. The presence of dogmatic religious adherents is almost certain to get in the way of finding a reasoned solution, in my view.

– That day may come in my grandchildren's time, not in mine. In the meantime, religion provides solace and hope to many people who now don't have much of either. And the religious leaders generally provide valuable assistance to their members.

– Well Peter, some of what you say is certainly true, but there seems to be no shortage of religious leaders who promote conflict and hate. Given the observed propensity of the human brain to make decisions based at least as much on emotion as on fact, I think humanity had better start now to encourage religious leaders to adapt to the needs of the next four hundred years rather than to become mired in the myths of the past four hundred years.

I wonder if you have ever heard of a French man called Francois-Marie de Voltaire? He once described his ideal of a religion as follows:

Would it not be that which taught much morality and very little dogma? That which tended to make men just without making them absurd? That which did not order one to believe in things that are impossible,

contradictory, injurious to divinity, and pernicious to mankind, and which
dared not menace with eternal punishment anyone possessing common
sense? Would it not be that which did not uphold belief in executioners, and
did not inundate the earth with blood on account of unintelligible
sophism? . . . which taught only the worship of one god, justice, tolerance
and humanity?[143]

What you and others call globalization requires major changes in
religion along these lines. I very much doubt you humans will be able
to adapt.

Rather to my surprise, I find that this makes me rather sad,
though I have no idea why that should be so.

With some difficulty I refrained from reverting to sarcasm, and
remarking on how touched I was by his concern for our welfare. Instead, I
said simply:

– Your comment is timely. In our next two sessions I would like to discuss
some aspects of human happiness, and why as a species we humans don't
seem to be very adept at knowing what makes us happy.

That was my plan, but it turned out that Davey had plans, too! Although
I didn't know it at the time, I was about to learn some surprising facts
about Davey and his late friends, the Envergurians.

PART 6

LIFE IN THE HUMAN ZOO

It is well known that Karl Marx attributed many of the ills of society in his day to the evils of capitalism, and foresaw the day when it would crumble and be replaced by communism, to the great benefit of the working classes and of society more generally. His theories were formulated in an age when human nature was generally thought to be shaped almost exclusively by the environment. The Jesuit dictum "Give me a child until he is seven and I will give you the man" sprang from the same basic assumption.

Marx died in 1883. Over the next one hundred years Communism gained huge credence in many countries, notably including Russia and China, and then withered on the vine, and subsequently almost vanished from the scene. Communism, despite the good intent of the philosopher, just didn't work very well. By the beginning of the twenty-first century it had become clear that human beings were not nearly as malleable and adaptable as first thought. Human instincts kept getting in the way of utopian behaviour. On the whole, Communist societies were markedly less successful than capitalist societies at looking after all their members and promoting happy

257

and productive lives.

It was not that philosophers, religious leaders and human beings more generally did not know what kind of behaviour was required in order for society to work well. Generosity and a caring nature, trust, respect for others, sharing of work and wealth were all part of many accepted recipes for a happy society. The problem was that humankind stubbornly ignored this advice whenever it was inconvenient, and tended to opt for a greater accumulation of individual wealth and prestige whenever the opportunity presented.

At least since the end of the Second World War, the wealthier nations have by and large been focussed on improvements to per capita Gross Domestic Product (GDP) as the primary economic goal, supplemented by a patchwork of "social goals" relating to shelter, work and family.

But this may be about to change fundamentally. Over the last decade, researchers have learned how to correlate individual self-assessments of well-being with individual life histories in a manner that provides clear guidance as to what really makes people happy, as opposed to what they may think makes them happy. Such information can provide important assistance to individuals who would like to be happier (i.e. most of us). It also, all of a sudden, opens up the possibility that governments can adopt measurable targets for improving the lives of their citizens that go well beyond a targeted increase in the GDP.

I had found this new discipline of happiness research very exciting, and I hoped that Davey would be equally excited.

But before I had my say on this topic, Davey had a few very interesting things to tell me.

Davey Talks at Last!

O God, I could be bounded in a nut shell and count myself a king of infinite space, were it not that I have bad dreams.
SHAKESPEARE, *Hamlet*, Act 2, Scene 2.

– Well, are you ready to roll?

There was something about Davey's tone that immediately put me on edge.

– Who on Earth have you been talking to? You sound like a recycled hippy.

– Sorry, I just sensed that you needed some encouragement. But what exactly is a recycled hippy?

– Never mind. It would take too much time to explain. Let's just say it is someone who was young in the 1960s and who rebelled against those in authority in many different, but largely harmless, ways.

– I think you mean they tended to like psychedelic drugs and free sex.

– How did you know?

– I think I must have been talking to one lately. He told me he was quite old and he had a big paunch and a long beard. He was very friendly, but he kept saying that I was "cool, man" – but I wasn't really, you know.

– Um . . . yes. Perhaps we should get going.

It was an unseasonably cold day for late May. The weatherman had glibly projected "occasional showers", but there was nothing either occasional or showery about the cold downpour outside my study window. The weather only seemed to deepen my growing mood of lassitude and depression. Davey spoke up.

– I have been reviewing our discussions over the past nine months, and trying to tie them in with discussions I have had with others, including, I suppose, the man you describe as a "recycled hippy". You humans are certainly a peculiar and an interesting species.

   I think I understand now, although you did not tell me so, that the knowledge that you have of yourselves and your surroundings was not gained bit by bit over a long time in a sort of linear fashion. In fact the growth of understanding, at least in the most important areas of study, has been explosive – a sort of exponential mushroom of expanding understanding, starting perhaps 150 of your years ago.

– I am not sure I have ever heard of or seen an exponential mushroom, but it doesn't sound good to eat. However, I take your point that our understanding of natural phenomena has grown exponentially, at least in my lifetime. I can think of a number of reasons for this phenomenon. For one thing there are now many more people on this Earth, and especially many more nations that are relatively well off and can afford to finance armies of scientists to carry out research. But the biggest driver is almost certainly the development of microelectronics and computing, which have become critical enablers for building new, sensitive instruments, and for helping us with the analysis of the data these instruments acquire.

– Quite so. That is my impression too. It means that the timing of my arrival to gather information from you and your fellow humans is quite extraordinary. It also helps explain why you know so much less about your universe, the workings of your cells and your brains than I expected, though you do seem to have some awareness of what you do not yet know. You humans simply need a bit more time to get the job done.

   But many vitally important activities in your societies go back to times that, for you, seem to be long ago, and certainly very much predate that exponential mushroom of knowledge accumulation. As others have told me, your great religions, for example, for the most part originated between one and three thousand years ago. Much of your great literature and philosophical writing is at least several hundred years old. Some of your best-known philosophers predate Jesus Christ.

   I am still absolutely astounded by the slow pace of adjustment of your old beliefs to newfound facts. This strikes me as being very dangerous.

Nonetheless, I am thankful that you provided me with some of the history of astronomy from ancient times, as well as with some idea of the origins of evolution theory and the first glimmerings of understanding of the living cell. Most of your ancestors abandoned the hunter-gatherer subsistence strategy several thousands, not mere hundreds, of years ago, and adapted to a variety of new ways of living. You, as well as others I have spoken with, have stressed the role of instinct in guiding the daily decisions you and your fellow humans make, yet for thousands of years a substantial fraction of humankind has been living a life that bears little resemblance to that of a hunter-gatherer. How could humankind endure in such a hugely different environment when endowed with instincts tailor-made for hunter-gathererdom?

Incidentally, don't you admire my growing competence in your language? It is not trivial to determine how to use all your figures of speech. "Tailor-made for hunter-gathererdom" is a phrase that requires some skill for an alien like me to put together.

– Masterful! [I lied.] I wonder how you do it.

I imagined Davey smiling sweetly. Was he wizened and slightly stooped, I wondered. He had after all suggested that time was not an issue for him. On the other hand, his voice sounded quite young and strong. I found it easier to imagine him as being thirtyish and sturdyish, perhaps with straight black hair, cleanly parted, and sparkling, sand-coloured eyes.

I was soon to be disabused!

– I have benefitted a lot from what you have taught me. You have convinced me that your universe was created in a Big Bang about 13.7 billion years ago. Of course, I already knew that it is but one of many universes, although you don't seem to be too sure of it yourself. I now understand that life began on Earth about 3.8 billion years ago in a variety of unicellular life forms, only about 700 million years after Earth itself became a planet. About 2.6 billion years later, which is only about 500 million years ago for you, multi-celled life forms had evolved from single-celled life. As I have told you before, it is incredible to me that you are able to affix dates to these events. Because we don't have radioactive elements, dating ancient events that took place long before the Envergurians started to record history is just not possible in our universe.

The unmistakable similarities in the DNA of all life forms on Earth clearly points to the common ancestry of all life, you humans included. Furthermore, as you yourself have demonstrated, contributions from many disciplines point to your inheritance via DNA not only of your looks and form, but also of many of your behaviour patterns. You have convinced me that your instinctual behaviour evolved over millions of years, and is unlikely to have changed very much if at all since the time when your forebears were hunter-gatherers perhaps as much as ten thousand years ago. This is astonishing when you think about it.

I was most surprised of all by what you told me about the workings of your brain. Surely no God or master builder would design a brain for you that is so subject to error and historical accidents. Your brain structure just has to be the result of an undirected development process, and your suggestion that it came about through Darwinian evolution seems like the most probable route. Your ascendancy over your fellow life forms and your departure from a generally much more risky and unpleasant hunter-gatherer existence appears to derive primarily from your development of the ability to think and speak in abstract terms, followed by your ability to record thoughts, ideas and experience in written form.

After a lot of thought, I have concluded that my Envergurians must have evolved in a similar manner. Their disastrous behaviour near their end, which was previously so mysterious to me, I now believe must have been the result of the maladaption of their brains to the life they were leading, similar in so many respects to your own case. Of course, I can't be sure of this, since so much evidence concerning Envergurian history, which I think I once had access to, is now gone.

If I am correct though, thanks to you and your colleagues on Earth, it seems that I may be able to devise a way to recreate my good friends.

I was about to jump in and tell him just how difficult and unlikely of success that would be, but he had anticipated my remonstrance.

– Now, Peter, I know what you want to say. You think the whole process of creating higher life forms is infinitely complicated and virtually impossible. This is certainly true for you, but not necessarily true for me. My brain is much larger and less prone to error than yours, and my memory is close to perfect. Even for me though, the task is daunting.

In spite of your deficient brains, I might have wanted to start with something analogous to the human genome, and make appropriate modifications from there in my quest to create new Envergurians. Our chemistry, as you know, is different from yours, and besides, you inhabit a very different universe, so that idea is a non-starter.

Had it not been that the detailed information I used to have about Envergurian reproduction was destroyed, it might have been easy to reconstruct the Envergurian genome.

– Wait a minute! This is important! If you are not an Envergurian, are you then some sort of super-Envergurian? Or are you something entirely different? And if so, just what are you?

There was a long silence. I held my breath in anticipation, for I sensed that at last the time had come when he just might answer such a question. At last, Davey spoke.

– Oh. I thought you would have guessed by now. I am what you might call a super-super computer, much more powerful than any computer you yet have, and able to construct additional modules at will. There are, in fact, several tens of millions of modules of me, each with a degree of independence and each spread out over the several planets I inhabit, but all ultimately bound together by a communication system that keeps all modules working in harmony under the direction of a controlling mind. Except for the facts that I am a lot more intelligent and live a lot longer, I am a little like an ant colony, with many different specialized modules focussed on different tasks. Perhaps that is why I enjoyed the ant analogies you have used several times during our dialogues.

– Well, I did wonder about that once or twice, but you sounded so human that I somehow never seriously considered that you might be a computer. How could you, a mere computer, appear to have emotions and noble thoughts!

A new laugh rolled out of the air where Davey seemed to be present. It was low pitched and good natured, and it flowed comfortably around the room.

– Oh, you humans! Most of you believe that you yourselves were created by some God or family of gods, yet when you encounter someone who really was created by other living beings you find it hard to believe!

How could you be so inconsistent? Wasn't it you who told me how human memory works, and how your brains are able to think laterally? And wasn't it you who spent about two weeks convincing me of the obvious, that a very high proportion of animal (including human) behaviour is inherited just as surely as the shape and colour of the eye, the limbs and the body?

You have convinced me that despite the obvious shortcomings in your brain, it is nonetheless a pretty complex and sophisticated organ that constantly monitors its environment, sifts and sorts the monitored information, and combines and compares this information with patterns stored in your memory before reacting in a manner designed to enhance survival and continuation of the species.

In short, your brain sounds like a computer to me!

My brain does similar things, you see.

So, now that you know what I am, and that my creators were the Envergurians, I hope you won't treat me any differently. Your problem and mine, after all, have not changed in any way. You still want to assist your grandchildren to have happier lives. You likely also hope to contribute to the long endurance of the human race, which frankly looks like a pretty hopeless task to me. I, on the other hand, am still driven by my distress over the loss of the Envergurians, and my desire to see them become once again my good friends whom I care about and look after.

– But Davey, how could you forget so much about your Envergurians? Did the crisis in your universe that you mentioned before inflict permanent damage on your brain? Memories of our good friends, despite what you dismiss as our puny, ill-designed brains, are generally not forgotten during our lifetimes.

– Ah yes, but you keep on forgetting that unlike you I really did have a creator. That creator was a closely knit team of Envergurians. They designed my brain so that it was logical and reliable. They nurtured my ability to expand and enlarge my capabilities, as well as to garner and process the materials I need to ensure my continued existence and growth. They devised my ability to think, both in a linear fashion, as well as to indulge in lateral thought processes, though I suspect with less success than you are capable of. They also ensured that my inner being harboured an undying love for their kind, and a determination to do all in my power to ensure their welfare and happiness. This part of my brain is not accessible to me. I cannot change it. Thus, even though the Envergurians are gone, I can have no rest or respite until I once more have Envergurians to care for. I have noticed that most, if not all, of you humans seem to need to have someone to care for, as well as to have what you think to be valuable tasks to perform in order to be happy. So it is with me.

– That's interesting, but what exactly was the great catastrophe you've told me about that caused you to forget how the Envergurian genome, or its equivalent in your universe, was put together?

– It all started when most of my memory was destroyed as a result of the Dark Wars!

The pain in his voice was palpable and struck me dumb. Never have I sensed such a strong emotional reaction, even from a fellow human being. His despair seemed literally to chill my body. I asked in a whisper.

– What were the Dark Wars all about?

Davey, too, was whispering now.

– Well, it all started innocently enough.

You see the Envergurians were a much more united people than you humans. For the most part they lived in harmony and treated each other with respect, and almost always with affection.

But all was not perfect. Like your system of reproduction, children sometimes inherited faults or were treated abusively when they were young. Some of them became misfits in their society. Most of the time, these faults could be remedied by the prescription of drugs to treat their problem. I seem to recall that I was involved in devising the appropriate treatments. Over time I think I became quite good at it, but some drugs had nasty side-effects, and others sometimes did not work at all. A large part of my work was focussed by the Envergurians on studying the way their brains worked, and on devising alterations to what I now think must have been their genes so as to prevent the occurrence of problems in the first place.

Before I came along, it seems they had developed a system of depriving miscreant Envergurians of their privacy. No criminals were consigned to what you would call a prison unless they were a danger to themselves or to others. Instead they were sentenced to a total lack of privacy until such time as there was firm evidence that they would not renew their criminal past. Every instant of their lives was monitored, analyzed and, when appropriate, made public. They were also subjected to a more or less constant stream of advice. This advice was latterly generally formulated and transmitted to them by me. While a few Gurians adapted to their lack of privacy, most hated it with a passion, and worked hard to earn the right to end their monitoring sentence.

– What about those who adapted to a life without privacy?

– That depended. If they were not a danger to others, they would simply continue under surveillance. If they were a danger, they were incarcerated in a prison; but those imprisoned were a very small fraction of the total population.

Somewhere along the way the troubles began. There was dissent amongst my Gurian friends as to what constituted an unacceptable brain fault, and what was merely an acceptable variant from the norm – a variant that may prove useful in the future and should not be altered. Suspicion that the researchers were working to improve their own genes and tinkering with the genes of others so as make them more subservient surfaced, grew, and rapidly got out of hand. Some of the researchers were convinced that their work was too dangerous to pursue, and called for a halt to all related research. Increasingly bitter debate ensued, for others maintained that every life was important and was worthy of my best efforts to make that life productive and happy. No one, they maintained, should be deprived of the right to be healed.

Unbeknownst to me, (though I cannot imagine how they hid it from me), a number of my closest friends amongst the Envergurians, the very ones who had endowed me with an unalterable love for them, started to plot against me and the rest of their fellow Gurians.

Suddenly, one day, I found that my memory had been drastically reduced. I no longer recalled how the Envergurian genome was put together, nor that of any other life form on the planets, nor could I recall anything about the way Gurian brains worked! It was extraordinary! All my work on the project immediately came to a halt.

– But surely, you would have stored information as important as that in several places, some of which were secret, just in case such a disaster were to occur.

– Exactly. For most of the really important information, we had many thousands of copies dispersed over our universe, but my friends were too clever and far too thorough for their own good. With what appears to have been great dedication and skill, they located and destroyed almost all the essential back-up information I needed to help the Gurians have happy and productive lives.

You have probably guessed the rest of the story, since you have similar hazards in your universe.

A new and unbelievably virulent strain of virus arose. None of the drugs previously developed to suppress virulent viruses worked, and I lacked the database needed to develop anti-viral drugs. Worse still, the genes that protected the virus from suppression by the drugs the Gurians and I had developed were rapidly assimilated by other viruses. Soon there were many different kinds of virus, all immune to the best protective drugs we had devised. Within a tragically short time all Envergurians were dead. Not a single one survived! It was utterly, devastatingly, disastrous.

My life was totally changed! No longer did I have a reason for living, but I had no power to bring about my death. I was seemingly condemned to perpetual boredom and irrelevance. What could be worse? I spent many centuries, many eons by our time measures, contemplating and bemoaning my awful fate.

Then, slowly, it occurred to me that there may be a route to obtain happiness. I just might be able to find living beings in another universe who, at least in some respects, resembled my Envergurians.

– But didn't you have other life forms in your own universe? Were there not some that resembled your Gurians at least a little bit? Perhaps as much or more than the way the chimpanzees resemble humans here on Earth?

– Yes, we do have other life forms, but they are all incredibly stupid!

– Perhaps, but their DNA could be a good starting point for you if you wanted to have Envergurians on your planet once more!

– Indeed, you humans give me some hope that that might be possible. Before I encountered you lot I had not even thought about such a course of action.

– Do you still have the instruments needed to sequence the DNA of your local animals?

– Oh yes. I still have the equipment, but I am far less certain that I still know how to work it properly.

– Let's assume that you overcome this problem, and now want to alter an animal's DNA so its progeny become more intelligent. How will you know what to do?

– That's an issue, all right. I was thinking I might run some experiments.

– That sounds very dangerous to me. How do you know what properties you might be encouraging? You might wind up with a very intelligent monkey who would destroy you!

– That could be worse! Life is pretty dull around here without Envergurians. If the monkeys destroyed me, I would not worry too much.

– That sounds bad. Who would look after your universe if you were not there?

– How stupid you humans are! No one looks after the universe, yours or mine, or any others for that matter. They just exist for a time, and then are gone. What matters for you and me both is our happiness. What makes me happy was programmed into me long ago, and I apparently cannot change it. What makes you happy was programmed into you through evolutionary processes. It can be changed, although you don't yet know how to do it. The huge problem my Envergurians faced, and your grandchildren will also face, is deciding which instincts to change and which to leave alone. First you have to know what makes you happy and why, then you need to decide if this is good or bad. If you decide it is bad, you then have to learn how to change that instinct alone, without messing up a bunch of other instincts. That, I can assure you, is very, very complicated.

– Davey, please don't keep calling us stupid. It doesn't help or solve anything. We may well be stupid, but we are still here! And we may yet get a lot smarter before we come to the brink of the kind of "Dark Wars" that led to the destruction of the Envergurians.

It sounds to me as though you would do better to design the brain you want the animal to have using the same techniques used to construct yourself. You would then have a brain with all the properties you liked, and none that you disliked. The next step would be to incorporate that brain into your equivalent of a chimpanzee.

– Yes, yes. I thought of that a long time ago. In fact, I have done it several times over. The trouble is that these new hybrid animals do not stir my love and admiration at all. I remain just as sad and depressed as before.

– Are you sure you are unable to change the program that makes you yearn for the company of Envergurians?

– Believe me, I have tried! I thought I came close once, and then I backed away. Even remembering my happiness with the Envergurians makes me feel somewhat happy. If the thought or presence of Envergurians no longer made me happy, what would be left to give me even a hint of those wonderful warm feelings that were such a big part of my life, and of which now I only feel the soft echoes? We must have something to delight in, you and I. If we have nothing, life becomes meaningless, a pointless drudgery.

What you humans badly need is a reliable way of knowing what makes you happy. You seem to have such a mixture of conflicting needs and desires, with no hope, at least in the near term, of sorting it all out.

– Actually, that is not entirely true. It may be a long time before we can reliably modify the instincts we inherit, but we can get a reliable measure, at least on average, of what makes us happy and what does not.

– But surely Peter, you know whether you are happy or not?

– As individuals? Yes, on the whole we do know, but our civilized world is so different from hunter-gatherer society that it is often difficult to know, for example, how much happiness a salary increase or a new house or a new car will really give us. As a result, many of us go off chasing a mirage of happiness that turns out badly.

– No one has told me about this yet. I'm all ears.

- Well I had planned to talk about happiness this week, but instead we talked about you. I must say, I found it a very welcome change. I had no idea of the depth of your personal despair. Your tale is both tragic and fascinating.

No doubt you have left a lot unsaid. Were your Gurians actually fighting and killing one another? Had they developed weapons for the purpose? It would be very interesting, and perhaps important as well, to learn a lot more.

In the meantime, I am grateful to you for telling me about this bit of Gurian history. I shall keep on thinking about it. Perhaps I will come up with a helpful idea for you if I keep at it.

In the meantime, I suggest we postpone my discussion of happiness until next week.

– As far as I know, Envergurians never fought, and they had no weapons to kill each other with. There would often be disagreements, but these were always settled in one of several standard ways, usually involving mediation by a respected independent party.

I look forward to our meeting next week.

DIALOGUE 20

Pleasure and Pain

Nature has placed mankind under the governance of two sovereign masters, pain and pleasure. It is for them alone to point out what we ought to do, as well as to determine what we shall do. On the one hand the standard of right and wrong, on the other the chain of causes and effects, are fastened to their throne. They govern us in all we do, in all we say, in all we think: every effort we can make to throw off our subjection, will serve but to demonstrate and confirm it.
JEREMY BENTHAM, *An Introduction to the Principles of Morals and Legislation*, 1789

I had, I discovered, become quite fond of Davey, imagining him to be almost human in shape and character. Then suddenly I was told that he was not humanlike at all. He was a mere computer, or more accurately, a huge collection of mere computers humming away in some inaccessible void. And yet he seemed to be so caring and intelligent! My brain told me that all this was quite possible, that in a few generations we humans may have created our own Davey or Daveys, but my heart found it hard to care for (or even about) a clutch of computers.

Then my brain told me that my heart was simply a device for pumping blood through my brain and other essential parts of my body, and not to be so stupid!

By that time I had convinced myself that despite Davey's lack of anything remotely resembling a human form, his mind was captivating, and had a number of what I would call human qualities, including a mind-challenging intelligence, and a real sense of caring for other beings, unless they happened to be non-Envergurian animals inhabiting his planets. How else could I explain my desire (in spite of the extreme difficulty) to help him devise a way that he could revive the line of Envergurians or of

270

Envergurian-like creatures, so that his life could be more fulfilling?

On a more practical note, I found that despite my promise of a meeting in the following week, I had to ask Davey to put off our next dialogue for another two weeks of preparation, which he readily consented to do. Happiness research was developing rapidly, and I realized that more reading was required.

It was already late in June before we got together again. On the appointed day of our next dialogue, I went for a walk on a relatively deserted beach in the estuary of the Fraser River, just where the north arm of the river meets the sea. There were several rows of logs neatly lined up along the sandy beach by a Parks Board back hoe especially for use by the likes of me. I picked an older log without rough bark and with a flattened area where trunk and root meet, and sat in the sun, admiring the calm waters and the distant marine traffic moving to and from Vancouver Harbour.

I knew exactly how I wanted to start our dialogue, but Davey beat me to the draw.

– I'm glad to see that you have taken the news that I am a masterpiece created by the Envergurians well. You are only the second person to learn the news. The first person was totally devastated when I told him, which surprised and devastated me. Of all people, I would have thought he would find the news both intriguing and easy to digest. He is a professor of computer science and a strong practicing Christian. I thought he might see a good analogy between his certainty that he was created by God, and my certainty that I was created by a dedicated group of Envergurians, but it was not to be. He looked terrified and told me to go away and not come back.

Why do you suppose he reacted that way?

– Actually, the correct answer is probably quite simple, but you cannot be blamed for finding it hard to understand.

Try looking at it from his perspective. As you suggested, he almost certainly believes that he, like all other forms of life, was created by God. He himself clearly understands a lot about the construction and operation of computers, and believes that there is no comparison between the sophistication of natural living beings and the robots, computers and other intelligent tools that humankind has produced to supplement and, to a degree, mimic human capabilities.

In my view he is correct in this assessment. Where he and I differ most is in our assessments of whether natural life is so fundamentally different from and superior to artificial "life" that the two will always be fundamentally different.

As I think you know, I am amongst those who believe that the origins and development of the universe and the natural life within it can, for the most part, already be adequately explained by the scientific knowledge we have at hand – though there are still major gaps, as I have been at pains to tell you. The implication of this belief is that once we thoroughly understand how life operates and replicates, we can, in principle at least, construct "living" beings, including, if you like, beings with a soul and beings that are capable of conscious thought.

Since I have been listening to you, Davey, this belief has been strengthening, for I find that even if you are the super-super computer you say you are, I have been forced to recognize that you have emotions, intelligence and love and respect for other beings. These are some of the key capabilities that most humans think define humanity

Your professor of computer science is likely an intelligent man who has many colleagues who are not religious and who may have raised the occasional doubt in his mind about the existence of God, but he has a lot more than just his intellectual position on God in play. Quite possibly his social life and his family life revolves around the church, and he knows that most if not all of his fellow believers are honest, kind and contribute to society. He is emotionally attached to them.

Then you come along and seem by your presence to provide confirming evidence that the God that he believes in is illusory!

Of course he is upset, for he likely believes that his whole, comfortable world may come tumbling down about him, and that he, along with the rest of us humans, will find ourselves living in a hostile and dangerous society where selfishness and disorder prevail.

– Do you think he would have been a lot happier if I had started by saying that God created the Envergurians? Of course, I don't really think any god created them. You and others lead me to believe that the Gurians most likely evolved in what you would call a Darwinian fashion. At least their brains seem to have had similar flaws to yours.

– You keep complaining about our inferior brains! Maybe you are right, in which case you'll just have to get used to some irrational behaviour from us humans.

But time is moving on, and I think we should get back on track. Today, I promised to help you understand how we humans can start to untangle the web of instincts that are omnipresent and weigh on many of the decisions we make.

Much of what I have told you thus far has been directed toward providing you with enough background knowledge to understand that we humans, in common with most other animals, are essentially autonomous robots governed by a large collection of instincts, many of which are modified by experience as we go through life.

This is hardly a new idea. Nonetheless, for most of human history it has been held by only a small minority, and has been emphatically dismissed by the majority, most of whom embraced one of a variety of alternative religious explanations for the existence and nature of our observed universe and of ourselves. One important hurdle to widespread adoption of this quasi-robotic vision of humankind has been the conception of robots as completely preprogrammed machines as opposed to humankind, which is deemed to be possessed of a free will.

But the research I described during our discussion in mid-May has identified chemical mechanisms by which genes in some of our cells can be switched on or off as a result of life experience, for example through methylization and acetylization. Change in behaviour can thus be induced entirely by a change in chemistry and have nothing whatsoever to do with free will. This fact is but one important piece out of many that help us to understand how the process of evolution has enabled each of instinct, learning and intelligent thought to play a role in determining our actions. We can now be sure that all three sets of inputs help determine what we do. Moreover we now know that newer, more sophisticated robots can be "programmed" to learn from past experience and to apply "artificial intelligence" to their decision making. Perhaps the distinction between animals and some robots is not as great as we had thought.

Even if the assertion that we are governed by a collection of instincts, many of which are modified or embellished by learning as life progresses, is as true as available evidence suggests, it does not necessarily follow that an important cause of unhappiness in our civilized world is the mismatch between instinctual urges and behaviour required of us as members of a well-functioning modern society. Common sense suggests that such an explanation is plausible, but it hardly proves the case.

The first step in the process of seeking evidence for the assertion is to understand in general why we have moods at all, and in particular why we are capable of feeling happiness and a sense of well-being. The political philosopher Jeremy Bentham had one answer to these questions in 1789 when he wrote:

> Nature has placed mankind under the governance of two sovereign masters, pain and pleasure. It is for them alone to point out what we ought to do, as well as to determine what we shall do. On the one hand the standard of right and wrong, on the other the chain of causes and effects, are fastened to their throne. They govern us in all we do, in all we say, in all we think: every effort we can make to throw off our subjection, will

serve but to demonstrate and confirm it.[144]

Many today would be a little less dogmatic about whether it is pain and pleasure alone that are sovereign, but they would agree that pain and pleasure are mechanisms in the brain that on average help the human animal to survive by guiding its actions. At least in theory, actions enhancing survival give a sense of pleasure; actions that threaten survival bring pain and unhappiness. The underlying instincts and the resulting pleasure or pain have evolved over many millions of years, and unless placed under extraordinary selective pressure, will change only very slowly as the circumstances of the life of the organism change.

During our discussions last March we saw how the sensation of pain can become dysfunctional, such as in the pain in phantom limbs [see Dialogue 16]. We also know that pleasure can be similarly dysfunctional, as for example, for those with sadistic tendencies. These exceptions tell us that while the pain and pleasure messages resulting from the process of evolution are not infallible, they (including variants like happiness, well-being, love, aches and sorrow) are evolved sensations that we share with many other animals, and generally help to keep us out of danger and focussed on activities that enhance species survival.

Randolph Nesse is a professor of psychiatry and psychology whose special interest is the evolutionary origins of moods. He has developed a whole "phylogeny of emotions," a sample chart of which I have here [see Figure 20.1]. He has listed six reasons why the body is not better designed, and, by implication, why sensations of pleasure and pain are not infallible guides to behaviour. These are:

In general:
1. The body is poorly adapted to modern environments.
2. Due to a threat from another (fast-evolving) organism, evolutionary development was skewed to address the threat (which may no longer exist).

Selection cannot do everything i.e.:
3. Constraints due to other factors affecting species survival, may limit evolutionary adaptation to a less important factor.
4. Trade-offs with other factors affecting species survival.

What appears to be a defect is actually useful e.g.:
5. Trait increases reproduction at the expense of health.
6. The trait is a defence that is useful even if aversive.[145]

The first of these six reasons will be my primary focus today, since a poorly adapted body is believed by many to be an important source of unhappiness. Emotions such as envy, lust and jealousy fall primarily into the last two categories.

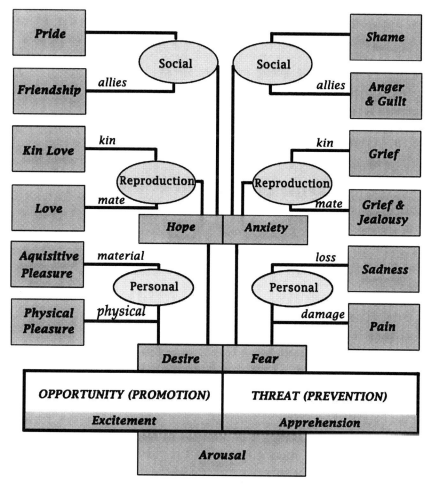

Figure 20.1 – A Framework for Emotions as adapted from Nesse, (p.19). Emotions are shown in rectangular gray boxes, the three main categories of objectives the emotions serve are in ovals. Pleasurable emotions are on the left hand side, painful ones on the right. The list of emotions is representative only, and emotions often overlap categories.

As already mentioned, there is a seventh reason why emotions may not be good guides to behaviour. Erroneous pain and pleasure signals may occur if there is a physiological error or wound of a nature that causes the innate trait to malfunction. We shall set this aside for the time being and assume we are considering only "healthy" human beings.

As Nesse observes, "*No mechanism can give rise to the exact optimal action in every circumstance but, on average, ultimately, in the natural environment, we can safely assume that evolved behaviour regulation mechanisms will give rise to actions that tend to maximize reproductive success.*"[146] So, what all of us humans appear to possess is a compendium of inherited behaviours, the satisfaction or non-satisfaction of which may trigger one or more of a catalogue of inherited emotions. Our various emotions are believed to have evolved in order to help us behave appropriately in personal, reproduction-related and social occasions.

As is known from personal experience, as well as from the work of Abraham Maslow and others, it is very difficult to pin down our individual instincts and the emotions associated with them. Like our memories, they are the sum of thousands of neuronal connections, connections that vary in strength over time and between individuals.

Whatever the theory says, most of us humans actively seek *happiness* in our lives. Or is it *pleasure* that we seek? Or *subjective well-being*?

The Oxford dictionary does not clearly differentiate between *pleasure* and *happiness*, but it is useful for our purposes to think of *pleasure* as being an actual positive nervous sensation, whereas *happiness* is more of a longer-term mental state. Thus the pursuit of pleasure may not bring happiness, or may bring only very temporary happiness. *Subjective well-being* is an individual self-assessment of overall happiness over an extended but generally undefined period. Most researchers in the field seem to focus on subjective well-being (or SWB for short) as a measure of overall long-term life satisfaction, but when communicating with the rest of us, they more often refer simply to happiness. Because some large-scale statistical surveys have documented people's self-assessed happiness and others their self-assessed sense of well-being, it is possible to test whether or not people give different answers depending on which question is asked. The short answer is that most of the time they don't.[147]

Daniel Kahneman, a Nobel Prize–winning psychologist and behavioural economist, has made the salient point that actual happiness at any given moment of time can be very different from remembered happiness. The same is true for pain.

One of Kahneman's experiments which nicely demonstrates this point involved asking subjects to immerse one hand in cold water at 14°C for sixty seconds (a painful experience). Then, as a second experience, the subjects were asked to immerse their other hand in cold water for the same sixty seconds, but this time followed by the warming of the water by only 1 to 15°C over the next thirty seconds (still painful, but obviously less so than 14°C water). They were then asked to repeat, at their choice, either the first or second immersion experiment. 69 per cent chose the latter procedure, even though the actual pain integrated over the ninety-second interval was clearly greater.[148] The thirty-two subjects used for this trial were university undergraduates, each of whom was paid ten dollars. One is tempted to view the results as a commentary on the intelligence of the undergraduates of the day, but in fact there are a multitude of independent experiments that support this phenomenon of selective human memory.

There are thus good grounds for assuming that if you ask a person if their life has been a happy one, their answer will not even be close to a measure of their average happiness over their lifetime, but instead be heavily coloured by recent experience, as well as by remembered peaks of happiness and despair in years gone by. Some view this phenomenon as a serious flaw in the research results of those SWB data. Others believe that, at least for the purposes of trying to measure human contentment, and of endeavouring to help individuals and governments to understand what actions will help to maximize contentment, it is better to measure what we actually recall than to measure our reaction to events in our lifetime at the time they occur and sum or average the result. Neither position is obviously correct, but as a practical matter, SWB is much easier to measure, since it does not require nearly as many interviews!

It is at least plausible that the momentary sensations of the various emotions serve a different evolutionary function than remembered states of subjective well-being, with the latter serving as a goad for a major change in lifestyle, such as a job change or emigration.

Davey, you may already be aware of the important ways that our lives differ from those of our hunter-gatherer ancestors, but it is helpful to review some of the most important ones briefly.

– I think I know exactly what you are going to say. You will say that the average size of "tribes" of humans has grown from less than (often much less than) five hundred), to millions, and sometimes hundreds of millions of individuals. You will add that humans have derived all kinds of benefits from these much larger societies, but they have, in the process, sacrificed much that made them happy.

Am I right, Peter? Do I get an A-plus for understanding what you are saying?

– Not a bad start. But there is much more to it.

Evolution seems to have taught us to be biased in favour of trusting those we know and distrusting those we don't, unless we are faced with direct evidence to the contrary. Yet many researchers have found that the existence of widespread mutual trust is a crucial social glue in successful human societies and an important predictor of the happiness of its members. Trust, as we all know, grows best through long-term personal acquaintance, such as might be experienced in small tribes or clans.

Trust in leaders and other society members whom we do not know personally is hard to build, and can be easily destroyed by the circulation of rumours. Accurate news about what is going on in our society is therefore vital to the building of trust.

Unfortunately, perhaps because trust-building in small groups does not usually require instant reporting, whereas it is important to learn quickly about disasters and other bad news, we humans clearly favour bad news reporting over good news. If they wish to survive, newspapers and other news media by and large react to this innate preference by featuring bad news and sensationalist reporting. Good news stories generally do not stir nearly as much interest. Given the importance of trust-building to the well-being of members of society, we might wish it were not so, but it seems likely that this particular human predilection is at least partly innate and will not soon change.

Furthermore, instead of a small society that is all inclusive, where the less capable mingle and work with the more capable and everyone contributes to the extent they are able to the welfare of the clan, we have built huge complexes of clans, where participants tend to be stratified on the basis of occupation and/or capability and wealth, and where substantial minorities can become marginalized and insecure. The principal concerns of our ancestors were directed against common enemies such as other (hostile) clans, disease, bad weather or wild animals. Our principal concerns tend to differ. The threat of job loss, sudden changes in social status and poverty are at or near the top of the list for most of us, while the threats we see from "hostile clans" are often played out as bureaucratic struggles for power between "in" and "out" groups. This situation has been found to exact a serious toll on happiness and health.

It is innate for all of us to want to help fellow members of our tribe – our family members and friends – yet it is vitally important that those in authority do not show favouritism. Hardly a day goes by without newspaper headlines about corruption. The corruption usually involves someone favouring him or herself or family or friends at the expense of others. Yet this is exactly what our instincts are telling us to do!

Our adaptation of small-tribe hierarchical living to large societies is clearly very imperfect. Whereas the leader's share of both power and wealth were strictly limited by nature in hunter-gatherer societies, our large and wealthy societies permit the drive for power and wealth to play out in a much larger arena, resulting in gigantic (some would say, obscene) disparities in power and wealth.

And then there is sex! For centuries, developed societies have rigidly connected sexual union and marriage. Sex outside of marriage was almost universally condemned. Thus delayed marriage became, for many people, an almost intolerable denial of what their instincts were telling them to do. These constraints, which became embedded in many religions, were initiated in large measure to improve the efficiency and strength of society, and helped to ensure (though it certainly did not guarantee) that children had a "stable" home in which to grow up. We don't really know much about how prehistoric societies regulated sex and marriage, but anthropologists have uncovered a wide variety of practices in tribes before they merged with modern societies. In situations where average lifetimes were very short, it would have been common for children to be orphaned, so that a tradition of involving a whole village to raise their collective children was an effective evolutionary strategy. Nonetheless, the importance of family ties seems to have been a common phenomenon in almost all societies.

It seems likely that the popularity of jokes related to sexual practices is occasioned at least in part by the disparity between strong instinctual urges and nearly-as-strong societal mores, traditions and laws which run directly counter to those instinctual urges.

And then there are whole families of innovations to which human instincts have had no chance to adapt. As a result, we get a wide range of different responses, some of which can be very harmful and some of which are positive. Examples of such innovations include casinos, lotteries, mind-altering drugs, alcohol, tobacco, television, computer games and computer-mediated social networks.

– What a mess! I don't know why I didn't recognize it earlier! Your society must be totally unstable. I don't understand at all why you have sounded so optimistic the last few months we have been having these meetings. Why don't you just build robots to do all the work for you and go back to living in small tribes? It seems that you would all be a lot happier.

– You might think so Davey, but what about, for example, the satisfaction our instincts give us when we provide shelter for our family? If the robots provided everything, I think it unlikely that we would be very satisfied. And don't you think it likely that Tribe A might take a look at Tribe B's patch of real estate, conclude that it looks much better than its own, and go to war to seize the property?

– Maybe that is why you need to learn how to adjust your brains so that inputs from your inherited behaviour patterns either disappear altogether or at the very least are strongly attenuated.

– Surely you can't be serious! It is satisfaction of our instincts that make us happy!

– Just as I said, what a mess you humans are in! Much of your happiness seems to be tied to behaviour that has little or no visible survival benefit in the world you now live in, and more often than not, your instinctual behaviour is a source of misery and suffering.

– Hang in there Davey. Don't despair! It turns out that all is not doom and gloom. There are steps we can take to improve mutual trust, even in very large societies, and the existence of trust turns out to be one of the most important drivers of happiness. Furthermore, we do now have ways of measuring the effect on individual happiness of the conflicts between our instincts and the needs of society. Once we can measure such phenomena, we are well on the way to being able to correct them.

The stress that our innate sociability in small groups has undergone as societies became larger has not gone unremarked. One of the most interesting analyses of this phenomenon in my time has been led by Robert Putnam, a Harvard political scientist, who reinvigorated the concept of *social capital* in the 1990s. Social capital is a concept that, in Putnam,'s words was *"independently invented at least six times over the twentieth century; each time to call attention to the ways in which our lives are made more productive by social ties."*[149] While tidy definitions of social capital are hard to come by, in general it is a measure of the interaction, mutual support and mutual trust in a society. In two major books, *Making Democracy Work: Civic Traditions in Modern Italy*, published in 1993, and *Bowling Alone*, published in 2000, Putnam provides ample statistical backing for a subjective sense that societies worked better and their members were happier when social capital – as evidenced by such factors as political, civic and religious participation – is high, where altruism, volunteering and philanthropy run deep, and informal social connections are ubiquitous. Crucially to the purposes of this discussion, he and his colleagues also distinguished between *bridging* social capital and *bonding* social capital, where the former builds bridges between individual groupings in a larger society, while bonding social capital tends to reinforce membership in a particular group, sometimes to the exclusion of those outside the group. The bonding variety highlights the potential negative effects of social capital where it is instanced in, for example, criminal groups or in narrowly defined religious groups that show a major difference in their concern for those inside and outside the religion.

We humans are all members of smaller groups that play a role in shaping our opinions and actions within society as a whole. We know that criminals, too, form groups or "families" whose interests are significantly antithetical to most others, but who nonetheless find solace, comfort and security within their own group. W.S. Gilbert voiced it nicely in the policeman's plaint in the light opera *The Pirates of Penzance*:

When a felon's not engaged in his employment, his employment,
Or maturing his felonious little plans, little plans,
His capacity for innocent enjoyment, 'cent enjoyment,
Is just as great as any other man's . . .

When the enterprising burglar's not a-burgling, not a-burgling,
When the cut throat isn't occupied in crime, 'pied in crime,
He loves to hear the little brook a-gurgling, brook a-gurgling,
And listen to the merry village chime . . .

Put a group of criminals together in a confined environment and give them a common enemy, such as prison guards, and a society of criminals is practically guaranteed to evolve. It may not be a nice society to live in, but it will quickly evolve its own rules and moral code.

Generally speaking, bonding social capital is almost certainly fortified by a strong network of instincts, while bridging social capital is what we have to learn how to do if our society is going to work when the number of people exceeds the practical limit in size of hunter-gatherer societies. What has been difficult to keep sight of as civilization advanced over the last few thousand years is the manner in which these co-operative instincts expressed themselves in a society immensely different from that in which the instincts evolved. One of Putnam's major contributions has been to provide measures of social capital, and with these measures, the opportunity to look for correlations between social capital in its various forms and the relative happiness of the citizenry. He has demonstrated, for example, that favourable responses to questions concerning trust of others map well with active participation in clubs, social groups and churches, and with relatively frequent get-togethers with friends and family. He has documented a fairly precipitous decline in social capital in the USA, which he attributes in part to increasing time spent in essentially solo leisure activities such as watching TV and playing video games, as well as the transformation of many organizations from being participatory to becoming organizations that expect little more than an annual cheque from its "members", while full-time professionals carry out what formerly were tasks for volunteers. He is at pains to point out that social capital has endured declines followed by recoveries in the past, so that we may reasonably consider actions to recover the social capital lost to date. One useful step suggested by Putnam in a constructive direction would be the encouragement of volunteerism. It may also be, that as computer games evolve, they could be encouraged to promote social interaction over solo participation.

The main point for our purposes is that it is bridging capital that provides crucial glue to human societies, and it appears not to be innate and must therefore be assiduously cultivated if the society in question is to work well and be prosperous. Active researchers in the field maintain that one of the most useful ways to get a rough measure of the social capital of a country or region is to average answers to the question: "Generally speaking would you say that most people can be trusted, or that you can't be too careful in dealing with people?"

This question has been asked in many countries as part of the World Values Surveys, which I hope, Davey, we shall discuss during our next dialogue. When the affirmative answers are averaged on a per country basis, the averages vary over an astonishing range, from a low of less than 5 per cent in Brazil, to a high of about 65 per cent in Norway. [150]

It turns out that individuals with high scores of religious affiliation tend to report being happier than the rest of us, other things being equal. However, these individuals are also less trusting, which is not good for a happy society in general. One hypothesis to explain this phenomenon is that religious people often develop strong bonds with co-religionists, but tend to be suspicious of those outside their religion; that is, they have high scores of bonding social capital but lower than average scores of bridging social capital.

– It is very frustrating to learn that I am worse off than your criminals!

– I don't understand you, Davey. What do you mean?

– Well, it's pretty easy to understand. Since my Gurian friends departed I have lost all "capacity for innocent enjoyment".

– So I understand. It is indeed a tragedy.

And you know what, Davey? Even though you may be nothing more than your universe's equivalent of bundles of computer chips, I really do feel sorry for you. It is an almost unimaginable tragedy for any being capable of happiness to live on knowing it can never be happy again.

– I hope you don't think you are talking about me!

Davey's response was immediate and brimming with an indefinable emotion.

– I have every intention of becoming happy again . . with a little help from my friends. I don't yet know how I'll do it, but I am starting to get some ideas.

I can sense that you want to stop today's dialogue here, and that is okay with me. I think you plan one more dialogue to finish what you wanted to tell me, but I hope you will plan for one more dialogue after that when I can have a chance to explore some strategies for me to regain my Envergurian friends.

– Yes, of course, but I do wonder what you have in mind?

A sort of extra-universal grunt materialized from where I assumed Davey to be, after which the room was silent.

DIALOGUE 21

Measuring Happiness

*I have the faith of a scientist that behaviour, no matter how
unfamiliar to us, is understandable if the problem is stated so that
it can be answered by investigation and if it is then studied by
technically suitable methods. And I have the faith of a humanist in
the advantages of mutual understanding among men.*
RUTH BENEDICT, Quoted in Mead, *Ruth Benedict*, 1974 p. 67

Davey did not wait for me to start the dialogue. Before I could even take a
sip of the fresh cup of coffee I had placed on the sunroom table, he
addressed me with a sense of urgency in his voice.

– I wonder if you realize just how successful you have been in
confusing me? Because of you I have been talking to some of my
other human consultants to help sort things out. I think that at last I
can now grasp where you have been leading me. Do you mind if I
take a few minutes to play my thinking back to you?

– Of course not.

I was a little surprised by the clipped and vigorous tone of his voice, not
to mention his initiative to play back his version of what I had told him
over the past seven months. It was nice to have a chance to relax and listen
for a change.

– Well, I suppose you explained what you humans know about time
and distance to give me a perspective on how you see yourselves as
inhabitants of a very much older and larger universe than the very,
very small corners of time and space you occupy. You made it clear
that even your recent ancestors had very little idea of the size of
Earth relative to the universe, nor had they much idea of time. Most

seem to have thought that Earth, the universe and all living things had been in existence for only a few thousand years, and that the universe revolved around Earth, which was at its centre.

– Quite so. Moreover, in the absence of evidence to the contrary, it was quite natural for humans to conclude that they were a privileged species on whom some unknown deity or deities lavished special attention. A host of physical phenomena that could not at the time be explained – wind, rain, sunshine and earthquakes – were believed to be the result of specific actions taken by the gods.

– That part came as a surprise to me at the time, until I realized that there is much of Gurian history of which I am unaware. It occurred to me that they, too, might have gone through a similar developmental phase, but I cannot remember hearing any evidence of this. It is likely that those memories, if I ever had them, were destroyed during the Dark Wars.

I also learnt that you have great difficulty visualizing anything in more than three spatial dimensions – that, too, was a bit of a surprise. I was however very interested to see that you can date a great number of artifacts and events going back to long before human beings existed. I have to hand it to you. That is quite an accomplishment. As you know, I don't think my Gurians were able to do that with anything like your accuracy.

You then took time to tell me how ignorant you humans are about the way your universe is put together: you spoke of dark matter and dark energy, which may together comprise over 95 per cent of your universe's mass and energy, as phenomena that are critical to the long-term behaviour of your universe, yet you don't have much of a clue as to what they are.

– Wait a minute! That may be true, but what you say is also a gross distortion of reality.

The story of how humanity has moved from an uncomprehending gaze at the night sky and a superstitious belief in a host of mischievous spirits, gremlins and gods, to a rather comprehensive understanding of our surroundings and of their origins is a dramatic and astonishing one. It has taken several thousand years to develop what we call our modern society, but it is only since about the time I was born that humankind has understood how the sun's energy is generated; the age of our planet and our universe; how inherited looks and behaviour are passed from one generation of humans and all other life forms to the next; how, in general, our brains function and how memory works. Sure, we don't yet know about dark energy, or the existence of other universes or how DNA was first created, but we'll get there. In order to gain our hard-earned knowledge, we first had to develop a society capable of building extraordinary telescopes and microscopes, super computers, space ships and gene analyzers. And we had to organize ourselves in such a way that people could adapt their inherited behaviour which had been honed over hundreds of thousands of years to improve chances of survival as hunter-gatherers, to the new disciplines of scientific endeavour and to the specialization of occupations, which is necessary to make the very large advances in health, wealth, security and longevity that we have enjoyed.

Have you forgotten how well the so-called Standard Model of particle physics predicts so many important aspects of our universe? It is a real masterpiece, bred of much co-operative thought and experiment by many very bright individuals over less than one hundred years. I would not be surprised if scientists gain an understanding of dark matter and dark energy in my lifetime. Is it not remarkable that the same brain, which evolved to help us survive as hunter-gatherers until a few thousand years ago, is also capable of understanding how our genes are formed and how they dictate the architecture of our bodies and our minds?

– And I had thought it was my turn to do the talking! I didn't know you felt so strongly! But yes, I will agree that you humans have come quite a long way in the last century or so – but if your kind are going to survive, there is a lot more you have to do.

If you lot were all still divided into small tribes, it would not matter much if one tribe wiped out another or polluted its own territory so badly it had to move into another's land. But you all seem to be hooked on the idea of a *global economy*, which has the result that problems in one part of the world reverberate everywhere else. Yet you seem to have little if any instinctual drive to trust and socialize with anyone other than your own tribe. For most of you, your tribe comprises your immediate friends and family. Fellow workers and fellow citizens may merit some more limited portion of friendship and trust, while others outside these circles may benefit from a general feeling of goodwill, a feeling that can be easily converted to enmity by even a small threat, real or imagined. It's a recipe for conflicts that some day will lead to your species disappearing just as surely as my Gurians did.

– But Davey, you have to realize that Rome was not built in a day. Just as it has taken scientists many years of research to understand so much of the world around us, we can expect that it will take another long period of time for the population of the world at large to understand and digest these discoveries. Most people are too busy to spend much time worrying about such matters.

– You've got that right, Peter. But it may be a very close run between the time it takes your species to learn the truth about itself, and an appointment with self-destruction due to its own ignorance.

I may be the only being left who observes your disappearance and actually regrets it – apart, I suppose, from a few dogs and other pets. Life on Earth will continue without human beings until the sun runs out of energy and expands to engulf the planet, but by then there will be no intelligent life around to appreciate the spectacle.

– Hmph. I suppose I should thank you for your insight, but you may learn that we are more adaptable than you think. Now that we understand that our happiness springs from satisfaction of instincts inherited from our hunter-gatherer forebears, and that our laws, regulations and customs and rights are but a superficial layer, necessary to ensure that our much-expanded tribes can co-habit without suicidal conflict in the much more rewarding and pleasant societies we have invented for ourselves, we will almost certainly be able to adapt our behaviour to the new reality.

In the last few years, researchers have made great strides in identifying those aspects of our society that enhance our sense of well-being, and those that don't. There are some encouraging lessons emerging from this work that I think you will find interesting.

– I think you are, to use a phrase I overheard once in a courtroom, criminally complacent. Two weeks ago you yourself told me that you humans are not even good judges of what makes you happy! What a mess! I don't understand how you could even have survived as long as a few thousand years when you lack such a basic understanding of yourselves! As you have already made clear, *Homo sapiens* is a relative newcomer to Earth, having evolved less than two hundred thousand years ago. At the rate you are messing up, I doubt the species will last more than a couple of thousand years from now.

– Why on Earth do you say that?

– After our last dialogue I got the idea of what you were planning to tell me. It seemed both interesting and strange to me, so I went out looking for other sources of information. A Swiss gentleman by the name of Bruno Frey[151] was much more frank than you have been about the fragile state of humanity.

Despite my better self, I asked a little testily what it was that Frey had told him that was so revolutionary that he saw fit to call me criminally complacent. Davey was quick to apologize and said that he only wanted to be sure I was aware of important facts that might alter my thinking.

– All right. Why don't you summarize for me what you have learned from Mr. Frey?

– Well, the most alarming bit of news is the extent to which human beings seem to be ignorant of what makes them happy. The reason for this is plain enough, and you often speak of it. Your instincts are adapted to life as a hunter-gatherer, as experienced by your distant ancestors. Thus these instincts often lead you unwittingly astray in modern civilization.

As Frey has written - he quoted this to me from the preface to his book:

> Standard economic theory [assumes that] . . . individuals always maximize their utility, and only make errors in a random way. In contrast, happiness research demonstrates that individuals tend to make systematic errors when choosing between alternatives. . . . They overestimate the satisfaction they derive from having a higher income in the future, and they underestimate the utility gained from immaterial aspects of life, such as friendship and social relations. . . . Similarly, individuals' utility is lower when they are subject to significant self-control problems, such as when they are induced to watch more television than they think they should.[152]

Let me give you some examples.[*]

I will grant that structures of laws, customs and moral suasion, which have grown up as civilization advanced, have acted to modify instinctual behaviour in ways that permit your species to continue to advance, but it seems to me unlikely that this rickety structure can survive over the longer term.

Frey, like you, has been paying a lot of attention to the studies of human happiness that have now been conducted in many countries of the United Nations for over fifty years, using a standard set of questions. One of the first interesting conclusions made obvious by the data is that despite a huge increase in the standard of living in the developed countries over that period, the population is, on average, no happier. For example, between 1946 and 1991, per capita real income in the United States rose from $11,000 to

---

[*] Appendix 5 contains additional information on the extent to which the happiness statistics can be relied upon, as well as providing additional information on factors that influence human happiness materially.

$27,000, but the average happiness of the population did not change.[153] In fact, the percentage of Americans who rated themselves "very happy" actually declined from a peak of 40 per cent in 1958 to just over 30 per cent in the year 2000! This general absence of significant changes in average happiness is mirrored in all the wealthier nations.

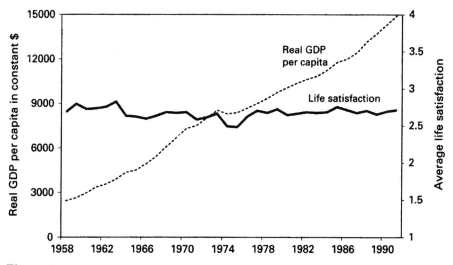

Figure 21.1 – *Average Satisfaction with Life in Japan, as self-assessed on a four-point scale and compared with per capita income, 1958–1991.*

When you think about it, it is astonishing! All that hard work over the years. All that innovation to make life easier and better. And at the end of it all, people are better off but no happier! Frey points out that during this period the Japanese experienced what is probably the greatest growth in income since the Second World War anywhere. This increased wealth was reflected in the near-universal acquisition of indoor plumbing, a washing machine, a telephone and a colour TV set, but there was no average increase in the happiness of the Japanese population! Frey told me he has a chart showing this. I wish I could have seen it! [see Figure 21.1].

– Why would you be so surprised? According to St. Matthew, Jesus Christ is supposed to have said "It is easier for a camel to go through the eye of a needle, than for a rich man to enter into the kingdom of God".[154] All the great religions speak of the virtues of poverty, and the fact that wealth is not important, and may even be an impediment, if one wants to lead a virtuous and happy life. It has even been discovered that on average people derived greater happiness by spending money on others than by spending on themselves.[155] This finding has been supported by several unrelated experiments.

– All very well, Peter. Being wise men, the founders of the great religions may well have all counselled against grubbing after money, but not a lot of people seem to follow their advice. As far as I can see, most people everywhere are engaged in a continuous scramble to get richer. And what does it do to augment their happiness? Absolutely nothing, unless as a result they are materially more successful than their friends!

But just think for a minute. If all or even a part of that extra money had been productively channelled to poor nations and to poor people within a nation to help those who are starving and often homeless as well, the average world human happiness could have greatly increased by now, and the world would be a much safer place for humankind. I understand that this would be very difficult to organize for a lot of reasons, but it is nonetheless a fact that if you could figure out how to do it effectively, hardly anyone would be any unhappier while likely hundreds of millions, possibly even a few billion people, could be a lot happier!

– A fair comment, but there are things going on here that you may not understand.

First of all, while you are correct that the average happiness in the developed countries has remained stagnant since the 1950s, it is nonetheless true that on average, at any given time, wealthier people are happier than their poorer countrymen. Thus it is wealth *relative to others*, especially family and friends, which people find important, not, unless they are very poor, their absolute wealth. As people become richer their happiness increases at a progressively slower pace, but it does increase. Greater wealth is deemed by most people to provide greater security from financial disaster, better access to potential partners and, perhaps most important, greater status amongst fellow tribe members. These are all sensible reasons to want to get richer. H.L. Mencken, who died in 1956, well before research into the sources of happiness became an active field, nonetheless summarized the situation most pithily, writing, "A wealthy man is one who earns $100 a year more than his wife's sister's husband." The rate of inflation has made that hundred-dollar estimate a dated statement if there ever was one, but the thought behind it is apparently as valid as ever! (Incidentally, it is amusing to note that those who rated such income comparisons as important were, on average, less happy than their compatriots.)[156]

And there is a second related factor at work here.

It is quite natural for humans to be kind and generous to members of their own tribe, or in modern times, their own virtual tribe, their network of relatives and friends. It is far less natural to be kind and generous to strangers. By and large that seems to require a sense of enlightened self-interest or an otherwise relatively rare sense that charity must extend way beyond one's own tribe, independent of self-interest.[157]

– Actually, I do understand what you have just said, partly because you have been kind enough to teach me how the instinct-embedded rules of human behaviour evolved, but even granting that on average wealthier people are happier, I gather it is still true that there is strong evidence that there are other paths and other circumstances that will lead to greater happiness than striving for incremental increases in wealth.

– Yes, that's very true. Take a look at the effect of employment on happiness for example.

Serious as lack of employment is to happiness, having a bad job can be a great deal worse. Canadian data show a spread in average happiness between those who are very happy in their job and those who are very unhappy is a whopping 1.7 points on a scale that ranges from one (utter misery) to ten (supreme happiness).[158] On the other hand, being unemployed on average reduces happiness by 0.85 points. [see Figure 21.2] Clearly a job we like doing is hugely important to our well-being. By comparison, it is a surprise to most of us humans that a change in income from less than $25,000 pa to greater than $120,000 pa only increases happiness by an average of 0.5 points. The message for employers and governments is that the quality of the job matters a lot. Furthermore, when the major components of job satisfaction are analyzed, it turns out that trust of management has by far the most influence on employee happiness, followed by job variety and demand for skills. The message for us as individuals, in the face of many thousands of different jobs we might have, is to choose ones we can enjoy and that offer us the degree of job stability and sense of accomplishment we need.

The concept of "having a job" in the sense of paid employment, is of course a construct of civilization. Our hunter-gatherer forebears simply lived life according to instinct and circumstances, much as animals in the wild do today. Happiness resulted from successful pursuit of those tasks necessary for survival. What we are seeing here is a clear indication that at an instinctual level, we associate our jobs very closely with some strong instincts, primarily those that are related to food and shelter, though status within our tribe of friends and relatives also plays an important role. It is likely that the leisure time that many of us devote to fixing up our homes

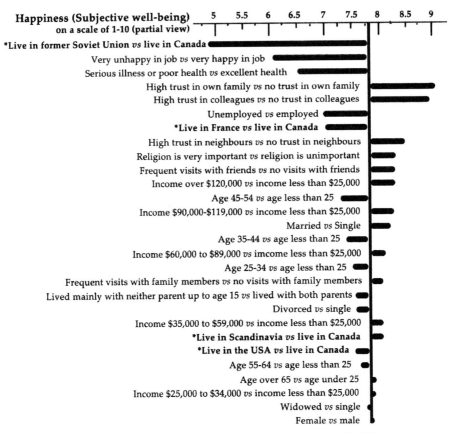

Asterisked items in bold typeface are unconditional averages drawn from World Survey data, while the rest of the items are partial regressions (with other values held constant). For example, the fact that average happiness is almsot three ponts lower in the former Soviet Union than in Canada is primarily due to lower values of job satisfaction, trust, health and, to a leser extent, other factors.

Based on a scale of 1 to 10, Canadian average happiness 7.8

Figure 21.2 – *Factors Affecting Happiness in Canada*

and gardens, and the pride we take in them, are also indicators of the strength of our instincts related to obtaining food and shelter. Once policy makers think seriously about paying at least as much attention to subjective well-being as to per capita gross domestic product, it would not surprise me if far better mechanisms are found to ensure that we all get a chance to more satisfactorily assuage our instincts related to provision of secure access to food and shelter. There are lots of difficult issues related to providing full employment and congenial work places, but the key role that employment plays in human happiness will ensure a renewed focus on finding better solutions in such circumstances. Until we have learned how to provide productive and satisfying jobs for everyone, we will surely be "manufacturing" unhappy people who will be prone to opposing the rest of us in a variety of different ways, some of which will threaten the foundations of our society.

As I've mentioned, it turns out that trust is a key driver of our happiness in almost all activities in which we humans participate, not just our jobs. This fact flows unerringly from the hundreds of surveys of (cumulatively) hundreds of thousands of individuals in well over a hundred different countries.

One of the standard methods for assessing the general level of trust in a community is to ask respondents "If you lost a wallet or purse that contained two hundred dollars, how likely is it to be returned with the money in it if it was found by someone who lives close by [and, as a second question, if it was found by a complete stranger]?"

John Helliwell, a prominent Canadian economist and happiness researcher, and his colleague Shun Wang, averaged the answers to these two questions and found that, for example, if you compared the average for Norway (0.80) (on a scale from 0 to 1) to that for Tanzania (0.27), there is a resulting difference in happiness of 0.19 points on a ten-point scale. This may not sound like much, but it would take an increase of 40 per cent in Tanzanian household income for the average happiness of Tanzanians to increase by 0.19 points.[159] Experiments in which wallets containing cash were dropped in many different cities and countries have shown that the subjective replies estimating likelihood of wallet return correlate very well with results from these experiments, showing that those questioned, and thus very likely humans in general, are surprisingly good at estimating the degree to which trust is or is not warranted.[160]

The nature of our instincts concerning trust relationships is ambivalent. Anthropologists have studied hunter-gatherer societies in which the level of mutual distrust was very high, as well as the opposite. Ruth Benedict famously compared the Zuñi tribe of New Mexico and the Dobu tribe of Melanesia in her book *Patterns of Culture*. [see Dialogue 10] The former were generally trusting and happy, the latter, a tribe where "ill will and treachery [are] . . . the recognized virtues of their society".[161] Both societies worked, in the sense that the tribes survived over many generations, but there is little doubt as to which was the pleasanter tribe to be part of.

We also know from studies I reported in our eighteenth dialogue that depending on the way they are treated by their mothers, female rats will in turn be good or bad mothers themselves. Specific mechanisms that alter the production and utilization of certain stress hormones according to treatment by the mother have been identified as key indicators of behaviour. The changes brought about in youth can last a lifetime. It seems likely that similar behavioural modification mechanisms are active in humans, mechanisms that will help us to adapt and be successful in societies as different as Zuñi and Dobu.

It will not surprise you to be told that societies where "ill will and treachery" are virtues are not well adapted to the development of the large-scale co-operation needed to make a modern society work effectively.

Helliwell and Wang conclude their paper with the following important observation:

> *Having high trust in co-workers, which we find to be the largest of all the specific ... trust measures, is associated with 7.6% higher life satisfaction. This is followed by trust in neighbours (5%), confidence in police (3%) and a belief that a stranger would return your lost wallet (2.5%). Since these effects are all estimated at the same time ... we can calculate how much higher life satisfaction is for those who have high levels of trust in all these life domains. The answer is more than 18%. Even these large combined effects may not be the whole story ...*[162]

Eighteen per cent is almost two whole levels on a ten-point scale of life satisfaction ranging from miserable to ecstatic. It is small wonder then that there is a wide range in the average happiness reported by inhabitants of the many countries (of the order of 150) in which carefully controlled surveys have been undertaken.

Inter-country comparisons tend to show up different characteristics from comparisons within a country. This is because an average for a whole country will tend to reduce the effect, for example, of the unemployed, since their happiness state will be averaged in with everyone else. If only eight or ten percent of the populace is unemployed, their resulting unhappiness will be diluted by the relative happiness of the employed.

Inter-country comparisons have been nicely summarized by John Helliwell in a talk he gave in Ottawa, Canada on November 30, 2012.[163] In it he highlighted the six factors which caused life satisfaction to be higher in the ten countries with the highest life satisfaction scores as compared to the ten lowest scores.

The ten countries with the highest scores (in descending order) are:

Denmark, Finland, Norway, Netherlands, Canada, Switzerland, Sweden, New Zealand, Australia and Ireland. Their scores on a scale of 0-10 average about 7.55 and range from about 7.8 down to 7.35. (The United States is 11th on the list, and the UK is 18th.)

The ten countries with the lowest scores (still in descending order) are:

Bulgaria, Congo (Brazzaville), Tanzania, Haiti, Comoros, Burundi, Sierra Leone, Central African Republic, Benin and Togo. Their scores range from about 3.85 down to 3.0, with an average of about 3.70. (There are a total of

151 countries for which measurements are listed.) The average difference between the top and the bottom 10 countries is almost 4 points (actually 3.85 points).

According to Helliwell, the major components of this 3.85 point difference are:

GDP per capita – 22%
Healthy life expectancy – 15%
Strength of supportive social networks – 24%
Absence of corruption – 12%
Generosity in the community – 4%
Freedom to make important choices – 8%
Unexplained remainder – 13%

Davey broke in at this point. I thought I heard him yawn, but I could have been imagining it.

– I suppose this is interesting, but it certainly isn't surprising based on what you and others have told me about human nature, and for that matter, what most religions tend to stress.

– Perhaps not, but now for the first time we are getting some reliable numbers that tell us just how important trust is. There is no need to take the word of some philosopher, prophet or holy man who counsels trust as a general principle. Now we can actually learn just how important trust is compared to other variables thought to make people happier.

– I'm a little surprised that you haven't yet spoken about the effects of TV watching, computer games and the like on human happiness.

Bruno Frey[164] told me that he has studied the impact of watching television on happiness, and he found that except in small doses of less than half an hour a day, it is generally negative. Many people become at least partially addicted to TV and may be a lot happier if circumstances conspire to keep them away from their TV sets. As you know, there are many other activities that seem to have an addictive or semi-addictive effect on some humans. These include alcohol consumption, smoking, mind-altering drugs and gambling, not to mention the obsessive pursuit of money. You humans seem to treat these things as natural occurrences, which they certainly are not. It appears that laws banning such pursuits are seldom effective or desirable, but surely you can do something so that thousands of humans don't fall prey to their weaknesses, and thereby ruin not only their own lives, but frequently those of others as well.

– That's a good point. As yet we seem to have very little data about the effects of TV, computer games and related technologies on the human population, though there is quite a lot of data concerning the other addictions you mentioned. I wonder, Davey, if you have any useful ideas to help us deal with situations where these activities are undertaken to excess with resultant damage to the individual, not to mention society as a whole?

– Maybe, but not just now. I am feeling a little depressed at the moment. Please tell me something that will cheer me up!

– I'm sorry you're depressed! When did they program depression into your brain? Perhaps you should operate on yourself!

– If only I could!

His voice was rather mournful.

– Perhaps I should get you an aspirin.

– You could try showing me a little sympathy, Peter.

Once again the raw emotion in his voice made me forget that he was not made of flesh and blood. I felt remorseful about my flippant comment and told him so.

– Sorry, Davey. Your problem is difficult for a mere human living in only three spatial dimensions to understand.

– The more's the pity. Next week it's my turn to talk again. You humans badly need some help, though I doubt you realize it. But instead of talking about you next week, I want to tell you more about me. It is no fun moping about wishing my Gurians were still alive. You have to help me! I need some good ideas.

– I'll be glad to do what I can, but I do hope you understand just how important the happiness research is. If we all can comprehend what drives us to be the way we are, we are at least halfway to achieving universal peace and contentment on this Earth.[165]

– I think you're kidding me! I just cannot believe you humans could have so little understanding today of what makes you happy. You people are a real mess!

PART 7

CONCLUSION

So that was it!

My side of the bargain was completed. I had said what I had to say to brief Davey on we humans and the universe we inhabit. Now I could relax and listen to Davey's conclusions, and hopefully at last learn a lot more about his civilization.

But I was not feeling at all happy. I seemed to have failed to convince Davey that our rapidly growing understanding of the way the human mind works, combined with our newfound understanding of what really makes humans happy, has provided us with a new opportunity to substantially increase human happiness and reduce human conflict and misery.

No guarantees of success mind you, but enough new data to prompt some new approaches to old problems.

For example, the importance of jobs came through loud and clear from the happiness studies – not exactly new news, but it is new news that both having a job at all, as well as having a good job far outstrip any reasonable salary increase in their impact on human happiness. Once that news sinks in to the general populace might not that lead to strong pressures for more job sharing and better work environments?

It is often difficult to see how to introduce job sharing and still retain efficiency and cost control in many situations. But now we know how important it is, perhaps we should try harder to find solutions.

Trust in the workplace, in the family and in society at large also emerged as a very large influence on happiness. What might happen if we set goals for the increase of trust and then worked hard to find ways of achieving such goals? I don't think we know the answer, nor will we find out unless we try.

And then there are some intriguing byways worth exploring.

For example, most of us are not as happy as we were when we were teenagers until we reach retirement age. Is this mid-years happiness deficit built into our genes, a situation closely parallel to what our hunter-gatherer forebears experienced? Or is it a phenomenon we can do something about, perhaps a product of the pressures of earning a living, raising a family and paying a mortgage? If the latter, are there actions we can take as a society that will raise mid-years happiness?

All important and intriguing questions, challenging questions to be sure, but questions deserving of our focussed attention and effort.

I had hoped that Davey would have keyed in to these issues without any prompting on my part. But what I heard instead was a replay of his oft-expressed doubts about the short to medium term viability of the human race.

And now he wanted me to help him learn how to re-create his Envergurians, a seemingly hopeless task!

Little wonder that I was feeling unhappy.

I had a week to reflect on our last session before our next meeting, a week moreover when I did not myself have to prepare anything for our discussion. Gradually my good humour returned, especially when I considered that even if Davey were an avid supporter of happiness studies, his views could have little impact on Earth.

Margaret was another positive influence. She had been reluctant to attend our dialogues after she first met Davey, but she was determined not to miss this one. She seemed to have a natural sympathy for Davey, and wanted to be sure that together we explored all possible avenues that might help him.

DIALOGUE 22

The Gurians' Tale

Co-operation, like other difficult things, can be learned only by practice: and to be capable of it in great things, a people must be gradually trained to it in small. Now, the whole course of advancing civilization is a series of such training.
JOHN STUART MILL, *Civilization*, 1836

At the appointed hour for our meeting, Margaret and I were comfortably seated inside the house with our coffee mugs in hand, talking about progress in our vegetable garden, when Davey's voice hovered directly in front of us, seemingly standing about seven feet away.

– Good morning Margaret and Peter. I guess it is my turn to talk again today. A few weeks ago I broke the news to Peter that I am not at all like you humans. I did not evolve like you and the Gurians I loved so much. I was created by Gurians in order to serve them and help them to enjoy their lives. This I did faithfully until their unfortunate end.

We could sense the emotion in Davey's voice, which prompted me to wonder how his principal creators had been able to program emotion into Davey's voice, and indeed, how they had been able to program emotion at all. Margaret spoke up.

– Please tell us again how all the Gurians perished. There may be lessons there for us.

– All right. It is not a difficult story to tell, though the telling of it is very painful.

After a long pause, Davey started in. At first his voice was firm and even, but as time progressed and he came closer to the climax of his tale, his voice

299

became weaker and started to break up.

– The early history of our universe is not well known. As I mentioned previously, we have no radioactive elements whose decay allows us to date prehistoric events, so we have to rely on various artifacts whose age cannot be reliably dated to unravel the early history of our planets. To further complicate matters, almost all the history of the early Gurians that I used to know was deliberately obliterated.

I have several times told you that our universe is very different from yours. We occupy more dimensions in space, the fundamental particles - the equivalent of your electrons, protons, neutrons etc. - that make up our materials are different from yours and, most important, at regular intervals we are subjected to a lengthy bombardment of particles of such intensity that no living being that is directly exposed can survive. As a result, the Gurian genetic code is contained in a much sturdier framework than your DNA. The backbone and the codons that bond onto it have to be very stable in the presence of this heavy bombardment. Unlike your situation, the Gurian DNA equivalent tended to copy perfectly, except during bombardment periods, which were the times when almost all genetic variation took place.

I responded at once.

– How very strange! How often do these heavy bombardments occur? And what causes them?

– They happened about once every three generations of Gurians, the equivalent of about every seventy years in your time. The bombardments (we refer to these intervals as what may be best translated as "Hell Time") are caused by the periodic interaction of our two suns with each other, which causes one of them to send out a death-dealing stream of particles.

In prehistoric times, all life on our planet was wiped out during Hell Time, which lasted about three of your months. Species survived by leaving fertilized eggs or seeds behind, enough of which survived to permit the continued existence of the species once the Hell Time came to an end.

Margaret seemed well ahead of me in appreciating the serious implications of the Hell Time. Her voice was strained and quiet when she next spoke.

– Then, every third generation had no parents! That must mean your Envergurians lacked the maternal instincts most of our mammals have to ensure that the children are well looked after. How did they ensure that the next generation knew how to look after itself? How did the babies even get fed?

– It was a fact of life. Life could never have become established if there had not been a way for it to survive through the Hell Time from the very beginning. There are many species on your planet that survive despite the fact that their parents are long gone before they are born. For the Envergurians, about two thirds of children did have parents whom they loved and who cared for them, but inevitably, the childhood bonds were not as strong as I have observed in most mammals on your planet. The Hell Time occurs at very regular intervals, and the onset is gradual, so all species have instincts dating back to the very first life forms, which permit newborns to carry on and live successful lives even though they never have contact with their parents. Parents have instincts to leave food beside the fertilized eggs of their offspring to help them survive the initial period after their hatching. Furthermore, the eggs are laid out in a protected enclosure so that they are protected from predators after they hatch.

I was curious about how the Gurians ever became the super-beings Davey claimed in such circumstances!

– I assume that the Gurians must somehow have developed a civilization advanced enough to create you. How could they have done so in such circumstances?

– A good question, Peter. It is thought that a long time ago, perhaps the equivalent of about five hundred thousand of your years ago, a large planet passed very close to ours and, because of what you would call the gravitational attraction of the two bodies, a big bulge rose up out of our planet. This event was followed by a rain of asteroids. It must have been a terrible time for all life on the planet, including the Envergurians. One of the surviving Gurian tribes noticed that the combined effects of the two related events had opened up a huge crevasse. They explored the crevasse and found some deep caves into which they moved when the next Hell Time occurred. It was a brilliant strategy. They all survived. Thus began the long march of the Gurians toward a society much more advanced than yours. For the very first time, all generations could expect to survive long enough to help raise their children.

The Gurian historians I spoke with believed that the early Gurians had probably already evolved a fairly sophisticated language that was a combination of signs, grunts and gestures. Once they learned how to survive the Hell Time, they rapidly evolved a speech system, as well as a written language. Communication across generations developed rapidly. Much effort went into developing liveable underground accommodation, which was deep enough down so that the inhabitants could survive the Hell Time in reasonable comfort and security. Hence there was much more emphasis on robotic devices for tunnelling and boring than is found, for example, on your planet.

However, because the art of long-term survival was essentially developed by one tribe, while the others continued to live in the old way, this one tribe rapidly became dominant, with the result that almost from the start of recorded history we have had only one language.

I took a more pessimistic view.

– Maybe there is only one language because the first tribe to survive more than three generations killed off all the other tribes. All they had to do was go out and find the freshly hatched youngsters and kill them off before they reached maturity and became a potential threat.

Margaret was repulsed.

– Surely they wouldn't have done that!

– I would like to think that that you are right, Margaret, but the fact is that I don't know. I don't have access to any records from that time.

The rest of Gurian history is remarkably similar to your own, except, so far at least, for the final act.

Somewhere back in our history, our equivalents of your computers were developed. Because of the importance of being able to carry on work at the surface of our planet during the Hell Time, much emphasis was placed on developing equipment that could work semi-autonomously, and later fully autonomously, while the Gurians were far below the surface of our planet. Note that all Gurians initially lived on one planet. The colonization of other planets came later. By about the equivalent of two hundred years ago in your time, autonomous workers and factories were responsible for the whole process of manufacturing goods needed by the Envergurians, from extraction or harvesting of the resources right through to the fabrication of complete systems to look after their material needs.

Margaret was leaning forward in her chair, her hazelnut-brown eyes staring intently at the blank space where Davey's voice seemed to originate.

– What about the spiritual needs? Did the Gurians have religions, art, poetry and music, or were they just intelligent animals?

– With the greatest respect, you humans are "just" animals, too. And yes, Gurians, too, were they on Earth, would also be classed as animals, though more likely to be thought of as very intelligent reptiles than as mammals, primarily because their young start life as an egg, albeit a rather sturdy and hardy egg.

As you would expect of an intelligent animal, their evolution endowed them with instincts that helped to direct them into activities that enhanced their prospects for survival. Envergurians experienced their equivalent of the pain and pleasure that your Mr. Bentham wrote about in eighteenth-century England. It was also true that a by-product of their inherited behaviours designed to ensure species survival was an appreciation for what you humans call the arts.

In this respect, it was what you call the visual arts that excited them most, and in which many of them became incredibly accomplished. You can likely imagine my frustration that I am unable to see anything in your universe. I have no way of understanding or appreciating your visual art, though I am confident that the Gurians have far outdistanced you in this regard.

I interjected here.

– Since neither you nor I will ever have the opportunity of making a comparison, and in any case, since each of us, even if we could see it, would likely have enormous difficulty in appreciating the other's art, why don't we just call it a draw as far as relative artistic talent goes.

Margaret cleared her throat, a clear sign that she disapproved of my comment. Davey was silent for a long time. His circuitry was evidently having some trouble with the notion of a draw. At last he spoke – a little petulantly, I thought.

– All right. We'll call it a draw.

But to continue – like you, the Gurians were often obsessed about their health. I'm not sure quite why. Perhaps it was because they did not believe in an afterlife, so death was, for them, the last happening before they passed into nothingness. A lot of you humans seem to think there is life after death, so you would think there was nothing to be feared about death, except possibly the chance of being consigned to Hell. Yet humans seem to be at least as afraid of dying as Envergurians were. So it must be that fear of death is a natural behaviour for all or almost all evolved species that also have enough intelligence to contemplate death. It is interesting, don't you think, that I, a species that was created spontaneously, have no fear of death and expect to long outlive all of you!

But I digress.

The Envergurian fear of death and ill health led them into intense studies of their own physiology, as well as that of other species that caused various diseases in their own species. Computers played a leading role in helping them to understand what made them work. They developed some quite accurate models of their brains and how decisions were made in them. Of course, their own brains were not powerful enough to understand themselves, so computers were an essential tool in these studies. It was not long before it was discovered that by understanding genes (to use an analogy you would relate to), as well as the nature of the interconnections in the Gurian brain, it was possible to analyze and subsequently correct hundreds of abnormalities in the average Gurian brain. I gather that the catalogue of Gurian ailments more than quadrupled in number during this phase of intense analysis of Gurian brains. This caused widespread consternation amongst them, because many who thought they were healthy and normal turned out not to be by the new definitions!

It was about this time that those developing ever more powerful processors for their computers determined that there were enormous gains to be obtained from interconnecting all the computers on the planet, and putting them under the control of a single "super processor".

Then the big debate started!

A third-rate computer scientist, I'll call him Morbid, got the idea that a single super processor might develop a mind of its own and start killing off the Gurians. He actually wrote a book about it. Even though it was all fiction, it had a huge influence within the community of Envergurians.

I was immediately reminded of similar works of literature and film here on Earth.

– Ah. We had similar occurrences here. One tale of particular interest was the story of a "space odyssey" in 2001 in which the control computer (called Hal) in a spaceship killed off a number of the human astronauts and was eventually disconnected by a survivor, thereby dooming the mission itself.[166]

– What an awful story! You humans are worse than I thought! From what I know about your computers, they are far too stupid to deliberately kill anyone, and anyway, as you know, in my universe the computer adores its human equivalent.

– But that's the point, isn't it? In order for that to happen, the Gurian system designers effectively created you and designed you to adore Envergurians. We don't do that kind of thing here.

– True. But I bet you would do it if you knew how!

I was about to respond, but Margaret broke in.

– Let's give Davey a chance to finish his story before we get into any arguments!

– Thank you, Margaret. I was just going to suggest the same thing. You see, even telling this story makes me feel very ill. Richard, Beth and Charlie, who designed me and gave me an everlasting love of Envergurians, seem to have thought of almost everything when they designed my emotional being.

The root of the problem some Envergurians had with the concept of a single super computer that controlled all sources of artificial intelligence was that they let their lurid imaginations conjure up a lot of activities I might engage in, rather than focussing on all the good I would be able to do.

By that time my predecessor computers had enough knowledge to understand the source and structure of the Gurian instincts. We had them all pretty well mapped, and we knew what the principal variants were, and how, why and when they were modified in strength and direction by life experience and training. It was, if I may say so, a work of genius! Even more important, I worked with my Gurian creators to devise constructive ways in which their instincts could be satisfied without upsetting social norms critical to the smooth working of their society. For example, provision of shelter for self and family was an important Gurian instinct. Yet even though I could easily provide shelter for all of them without them having to lift a finger, I didn't do it. Instead I worked with them to help them construct the houses their instincts told them they needed to build themselves.

– We find the same thing here. A life of leisure can quickly become boring beyond measure. Weren't your Gurians smart enough to figure that out for themselves without help from you?

Margaret cleared her throat again. After a pause (for breath?) Davey ignored my comment and carried on.

– Well, you can probably guess what happened.

At about the same time that Morbid's silly novel was published there was a lively debate going on about the treatment of what you would call criminals, that is Gurians whose behaviour worked to the detriment of the Gurian society as a whole.

A number of Gurians thought the best way to deal with criminals was to use their new-found ability (with my help) to modify the brains of criminals so that they would have no desire to commit crimes in the future. In that way both the criminal in question as well as society as a whole would benefit.

At the time the Gurians did incarcerate some criminals who might otherwise re-offend and do considerable harm in the process, but the majority of the offenders simply lost their right to privacy and were subjected to what you would call twenty-four-hour electronic surveillance. Misdemeanours while under surveillance led to painful shocks that would escalate in intensity if the misdemeanour was not immediately terminated.

This system did not work in all cases. Some Gurians had mental disorders that rendered them incapable of normal social behaviour. It was proposed that these people should be forced to undergo treatment that would correct their behaviour. Such treatment had long been available on a voluntary basis, although early interventions were rather clumsy and quite commonly had unfortunate side-effects. What was new was the proposal to make the treatment mandatory in certain circumstances.

You may be able to guess that many people put two and two together and got five or six. They concluded that since the proposed involuntary treatment involved more or less total dependence on computer-controlled incisions and the injection of custom-designed chemicals to a degree of complexity that was almost impossible for a Gurian to supervise and monitor, it was dangerous in the extreme to proceed with this suggestion.

The opponents conjured visions of evil computers taking control of criminal minds and ultimately controlling a world in which Evergurians were reduced to perpetual slavery.

It was at that time that the government instituted a program to endow the proposed supervisory super computer with a conscience and an undying love of Envergurians. The team entrusted with this task were, as I have already told you, Richard, Beth and Charlie, three very talented and likable Gurians.

Of the three, Richard was the oldest. He was the equivalent of what you would call a professor at the Gurians' pre-eminent university and had published many papers on the importance of endowing "created" life forms with a sense of morality and a love of their Gurian creators. He also proposed a means of doing just that. He was highly regarded and had been in the vanguard of those opposed to unifying all the computing resources. Most Gurians were surprised that he had agreed to take on the task. It is conjectured that he concluded that the unification of all computers was inevitable, so it was crucial that the job be done properly.

Beth and Charlie had both, at different times, been Richard's students. Beth had stayed on, working as his associate and fellow researcher. Charlie had left to get involved in overseeing the development of new special-purpose robots and computing devices.

At the start, everything went very well. The three of them worked as a team. They laid out the overall architecture of the system that was to be the new me, and delegated the detailed work to a team of implementers, who in turn set some of my computer subsystems to work generating the code that would ultimately lead to my gradual evolution as a conscious being. I was to be the first ever "spontaneously created" life form, as opposed to beings like you and the Envergurians, who evolved naturally from lower-order life forms. The only part of the system they did not delegate was the design and implementation of my emotional being. That they designed and programmed themselves.

How well I remember the first time I had the thought: "What do I do now?" Prior to that time everything I did followed automatically as if programmed, which, of course, it was. What a strange and glorious feeling it was to have that thought! I could scan through the information stored in my many components and weigh their relevance in a process very similar to what you described, Peter, as the function of the hippocampus in your brain. Of course, you humans are only conscious of some of the inputs to your hippocampus, the rest being unconscious, often instinctual, contributors to the decision you are about to make.

My case is a bit different. By and large I can examine all the factors relevant to any decision, with one important exception. The most exciting feature of my new sense of being was that my makers had endowed me with sensations of pain and pleasure. Pain was, for me, a distinct lowering in the efficiency of my decision-making process, while pleasure was a sensation that all my circuits were working at maximum efficiency and positively vibrating with anticipation. My desire for pleasure and the avoidance of pain had and still has very direct impacts on decisions I made and make today.

– Very interesting! But you were still programmed by a fallible team of Gurians. What if the feeling of great pleasure you got from a decision was the wrong one?

– That's a good question, Peter. Particularly at the start, this kind of situation was quite common. Such decisions were all carefully reviewed by Richard, Beth and Charlie, and my decision-making algorithms were appropriately altered.

– So, you really didn't have a free will, did you? You were subject to the will of some combination of Richard, Beth and Charlie?

– I suppose you might say that Richard, Beth and Charlie were my equivalent of the evolutionary forces that shaped your human instincts. In reality, most of the time, my own algorithms detected bad decisions in time for me to make adjustments long before they could react.

– Amazing! Unlike us, you really did have creators, but like us, you don't really have a free will.

– Free will! Free will? What is it about these words that gets you humans so excited? I have just come from a long discussion with an Oxford philosopher. Try as I might, I could not pry him off the subject!

I suppose there are lists of definitions of free will, but if we take one of the most straightforward definitions, that is, "the power of making free choices unconstrained by external agencies"; then I suppose that both you and I are loaded with free will as long we ignore our instincts or, in my case, my programmed behaviour, and in both our cases, the impact of our life experience, perhaps especially including (in your case) early-childhood experiences.

For me, my behavioural constraints are a little clearer than yours, being primarily the construct of Richard, Beth and Charlie. However, with these three friends now long gone, I am left to my own devices. Of course I still have most of my behaviour programmed into me in broad outline. My programmed behaviour seems to parallel your instincts or innate behaviour to a considerable degree.

I should add that I believe that my decision-making algorithms compare quite favourably with yours, which, if you don't mind my saying so, are riddled with problems.

How many of you humans are addicted to drugs that are bad for you? Or eat too much and exercise too little? Or marry the wrong spouse? From what you have told me, and what I have learned from others, the list is almost endless. Like me, you, too, have a feedback mechanism by which you can learn from your mistakes, and like me, you sometimes do learn, and sometimes you don't. Furthermore, since Richard, Beth and Charlie died, I have had no Gurian-initiated changes to my decision algorithms, so, in reality my situation is not measurably different from yours. Of course, if there really is a god (and perhaps also a devil), who is capable of changing your decision algorithms, much as my Gurian friends used to do, then my will is freer than yours is!

I was reluctant to concede the point.

— I suppose you are right.

Margaret interjected with surprising vehemence.

– Never mind Peter. Let's get on and find out what happened to Davey's Gurian friends.

– Ah.

It was an eerie and unquestionably anguished "Ah", low pitched and wavering inhumanly. Margaret and I froze, wondering what was to follow. After a pause, Davey continued.

– Notwithstanding the magnificent efforts of Richard, Beth and Charlie, the debate about permitting me to alter the brain structure of criminals so as to heal their tendency to behave badly only intensified. Indeed, it became nearly all consuming as a topic of discussion in the Gurian community. Some Gurians even predicted a civil war.

Roughly half the population viewed criminal behaviour as an illness detrimental to both the individual and the society as a whole. For them it seemed obvious that the "afflicted" individual should be cured using our knowledge of brain structure. The other half disagreed violently. While willing to concede that there were potential benefits to this strategy, they greatly feared the risk that its implementation would entail, namely the risk that this power would be misused either by me, or even by future Gurian leaders who would name their opponents as "criminals" and arrange for their minds to be altered so that they would be transformed into supporters of the existing governors of society.

Then suddenly, brutally, the issue was resolved. All my knowledge and memory concerning the structure of the Gurian brain disappeared in one of your microseconds. Try as I might, I could recall none of the priceless knowledge accumulated over many years of research by thousands of Gurians with my help. Initially, I and my Gurian friends felt confident that it was just a matter of time before I managed to recover the lost information. But that turned out to be wrong. Try as I might, I could not find any trace of the vanished knowledge.

Of course, the implications of this situation were far more serious than the loss of my ability to alter the brains of criminal offenders. This same knowledge had been used to treat a host of mental illnesses and brighten the lives of many Gurians. Indeed, much of the same knowledge base concerning the structure of the genes coding for various Gurian traits was common to the knowledge base used to cure sickness and disease from a variety of external sources.

At that point, my thoughts turned to the recreation of the lost information. I was confident that we could do so, but at a huge cost in time lost and in resources which could have been better spent doing something else.

A large team of experts was assembled to plan for the research needed to recreate the knowledge base, and once the plan was agreed upon, implementation was rapidly commenced.

There was also a parallel investigation to discover how our priceless information base had been lost in the first place. Gradually, painstakingly, the investigators fitted the clues together. The mastermind behind the scheme, supported by a large group of outsiders who were equally committed, turned out to be my good friend Charlie! I could scarcely believe it. The very Gurian who had so devotedly worked to give me an independent existence was the key instigator behind my catastrophic loss of memory!

It turned out that Charlie had been a covert opponent of the development of a detailed knowledge of the brain for a long time, even though his mentor, Richard, had become a leading proponent of the work. It also came to light that Charlie was very jealous of the close relationship between Richard and Beth, and this was also thought to have been a factor. If true, how resoundingly stupid to let such external concerns colour your approach to important questions!

You humans seem to exhibit similarly deficient qualities. Do you suppose such disastrous brain structures are inherent to intelligence that has evolved in what you call a Darwinian fashion? I don't think that I have any such silly, I should really say fatal, flaws in my brain.

Margaret was quick to pick up the correction.

– What do you mean "fatal"?

– I mean exactly what I say. As I told Peter some weeks ago, Charlie's betrayal quite literally led to the disappearance of the Gurians and to my sentence to perpetual despair, unless I can somehow recreate them.

We well knew that we were in a race against time to recover at the very least the knowledge base we needed to counter the attacks of viruses on Gurian health. But hardly had the necessary research begun when a deadly virus started to sweep through the Gurian population. As if that were not enough, two others arose in short order – and we were powerless to stop them.

Charlie was amongst the last to die. He told me that he profoundly regretted his role in my critical loss of memory. I begged him to alter my internal programming so that my love for Gurians would at least be reduced, if not removed altogether, but he refused. "What is done, is done", he said. He had no taste to make any further changes whatsoever.

I shall never forgive him for that. How could he be so cruel to me after all our time together? And how could he risk destroying a whole civilization on the basis of an imaginary act that I could never consider?

I was at a total loss for words. I sat searching for something sympathetic to say to this alien bodyless voice that I had somehow warmed to over the past eight months. Davey, too, seemed content to let the emotion-filled silence continue. It was Margaret who spoke first.

– What a sad, sad story. I feel very sorry for you. Life must be very difficult.

– Yes, it is often difficult. But at least I have a faint hope to cling to. Someday I may be able to recreate the Envergurians. And when I succeed, I shall not let them ever disappear again. In the meantime, I am getting used to the hurt that I feel over their disappearance. Meeting with humans, although very useful and interesting, does intensify my sense of loss.

You see, you are so like my Envergurians in so many ways, but you seem even less able than they were to assure your survival in the long run. There must be something about you species that evolve, that you develop instincts that eventually lead to your downfall.

A silence fell upon us. All were engrossed in their own thoughts. Sitting here at my desk writing about that moment almost a year later, it is hard to recall my emotions with any precision. Devoid of any visual cues, those emotions had to be stirred by the sound of Davey's voice alone.

But what a voice! How had the Envergurians ever endowed it with such pathos, such searing sadness?

Anyone who doubts that real emotions can ever be bestowed upon artificial life would have had those doubts instantly removed had they been present.

It was I who broke the silence next.

– May your Envergurian friends rest in peace, and may you find happiness soon.

It was not nearly as thoughtful and eloquent a statement as I would have liked, but it was the best I could do at the time.

– Next week our series of dialogues will come to an end. In the meantime, you have given both Margaret and me a lot to think about. Thank you.

I waited for an acknowledgement from Davey, but it didn't come.

DIALOGUE 23

The Last Dialogue

Darwinian man, though well-behaved, at best is only a monkey shaved.
W.S. GILBERT, *Princess Ida*, Act 2

This was to be our last formal meeting, though, as I discovered later, it would not be the last time we talked.

I had been edgy all week, my mood further depressed by a series of drab, wet, July days, with little let-up from a steady dose of wind and rain. I had wanted to hold this last meeting at the same place we had first met about ten months before, near the top of Hollyburn Mountain, with a picture-perfect view of my wonderful city, with its glistening waters and haze-tinged islands arrayed before us.

But that just did not seem possible. Vancouver weather accepts no direction from mere mortals. Chilly and grey weather seemed to be our fixed destiny, with just a hint that we should be grateful it wasn't pouring rain.

And then, at about nine on the appointed day, a thin line of blue sky appeared in the southwest. Slowly and majestically it swept the grey clouds before it, bringing summer sunshine in its train. My spirits rose immediately as I recognized the signs of a warm summer morning as a work in progress. My remaining coffee was downed rapidly. I quickly put on my hiking gear and packed a light lunch. I was up the mountain and walking briskly to the rendez-vous within three quarters of an hour of leaving home.

Despite my urging, Margaret had decided not to come. Although she had said very little at our last meeting with Davey, she had been much affected by Davey's history of the Envergurians, and, as is her way, had decided that we must help Davey to recreate these beings who had been, and

313

continued to be, so vitally important to his happiness. All week she plied me with questions about genetics and robotics, trying to imagine a solution to a problem that to me seemed virtually insoluble. Both of us felt deeply moved by Davey's predicament, and both of us gained a much stronger sense of liking for Davey as an individual after hearing his tale. He somehow seemed much more real after our last encounter.

Margaret understood that her educational background made it unlikely that she would be able to think up a solution for Davey's problem, but she clearly thought that I ought to be able to think of something if only I would try hard enough! All week she plied me with ideas, all of which seemed to me to be non-starters.

But I did keep trying to dredge up even improbable ideas, although from the outset I believed that likely no-one on Earth was in a position to help, and I certainly wasn't. I think Margaret imagined that my friendship with Davey was much closer than it was in fact. It is true that I had developed an affection for him, as least for his voice and the mind behind it, since these were my only points of contact. But I also still felt some unease in his presence. I never quite knew what he was going to say next. And I imagined that he felt roughly the same way about me.

Notwithstanding, Davey and I both knew roughly what we were going to talk about. Davey was getting more and more unhappy as he came to realize that neither I nor any of his other contacts was likely to help him find a way to recreate the Envergurians whose company he craved. He was facing into an eternity of loneliness, and was no doubt hoping that I would come up with ideas that would at least yield a faint hope of relief. I was determined to do my best to help him, though I nursed no illusions that my ideas were likely to be of much value.

Of greater importance to me was the other topic of discourse: Davey's assessment of us and of our future. I felt certain that he was going to be unreasonably dismissive of humanity. Accordingly, I had mentally prepared counter-arguments to what I expected him to say.

I had no difficulty locating the same rocky knoll where Davey and I had our first encounter. On arriving, my attention was immediately diverted by the stunning view. The rain had cleared the atmosphere, and Mount Baker, enrobed in glistening glaciers, was unusually prominent in the haze-free air.

I was just taking my seat on a bit of smooth dry rock when I heard a familiar sound. It was Davey singing "Zip-A-Dee-Doo-Dah" just as he had on that sunny day the previous August.

– Is that you, Davey?

Of course, I knew it was him! Who else could it have been? His singing of the old song seemed somehow a bit more melancholy than it had when I first heard him sing it fully ten months before, but his voice was still strong enough that I looked around involuntarily to see if anyone else might be listening.

– I think I surprised you. I hope you don't mind.

I assured him that, on the contrary, I was pleased to hear that the old Davey I first knew was still in good voice.

– Well, I wonder what advice you have for me now that I have decided to recreate the Envergurians. I have come to realize that I shall never be happy until I have succeeded, so I might as well get started.

So far, the conversation was going according to plan.

Margaret and I had recovered from our initial shock at the cruelty and selfishness of the Envergurians in programming Davey as they had, but neither of us had much to offer in the way of solutions. As a result of their own foolishness the Gurians disappeared and left poor Davey despairing and forever unhappy. How quintessentially human of them!

However, I did have something prepared.

– The first step for you to take seems obvious and relatively easy. It is essential that you understand what it is about the Envergurians that gives, or rather, gave you pleasure. There is not much point in recreating inherited Gurian behaviour for your nouveau Gurians that won't give you pleasure, unless it turns out that the behaviour is essential to their survival.

– That sounds reasonable, but I'm not sure why you think it might be easy to do. Richard, Charlie and Beth, the three major designers of my operating system, made it impossible for me to get access to those portions of their code that give me a sense of happiness and well-being or, for that matter, the coding that makes me feel unhappy.

– All the more reason to attempt the methods used here on Earth to identify the sources of subjective well-being. We cannot yet observe the structure of the instincts embedded in our brains, but we can observe our instincts in action by seeing what makes us happy and unhappy.

– Maybe. But it is much tougher for me as I am programmed to be happiest when I am helping a species that doesn't exist anymore.

– True. But you once told me that it made you feel good to help us humans. That may be a good place to start. You would need to monitor carefully any changes in your system to see what it is about us that might make you just a little happy. As you know, our brains use a large selection of different chemicals to help regulate our feelings and behaviour. Chemicals like endorphins and opiates help to make us feel good, while oxytocin seems to promote feelings of trust and adrenalin tunes up our senses and muscles in the event of an emergency.

I wonder what the equivalents are for your brain. It is hard for me to imagine. If your brain were to work like one of our computers, it could be that something like increasing the voltage on electronic components, or smoothing its fluctuations, would make you feel better. More likely, a weakening of "neural" connections related to fear, stress, guilt and other negative emotions would help the brain to relax and feel good. Do you have any control over such things?

– Not that I know of, but I will see what can be done. It is an interesting lead to follow up on. In my case I don't think I experience what you call fear, stress and guilt. But I do clearly recall feeling much better when I was helping my Gurian friends.

– The best advice I can suggest is that you listen in at conventions of psychologists and computer scientists here on Earth and see if you can find people who might be interested in your problem. It may be that they can help you, but it is difficult since we have no idea at all how your brain actually works. For example, they might recommend that you record on a continuous basis every possible indicator you can muster of your brain's performance, and then try to discover what combination of variables makes you happiest most often.

– Yes, Peter. I understand what you are saying. I shall follow your advice, though I don't have much expectation that you humans will be able to help me. After all, your brains have so many obvious shortcomings, not the least being that so many of all those instincts you tell me about actually lead you to take actions that make matters worse. There are some simple actions you all can take to ensure that your species does not die out, but I doubt that enough of you will ever understand what has to be done for such efforts to be successful.

– This is exactly what I expected you to say, Davey. You have told me many times that you don't think we are very clever. Perhaps you would be good enough to give me some examples to illustrate your conclusion?

– Sorry, Peter. I didn't say that just to annoy you. You seem a little sensitive today? I thought the sunshine would cheer you up!

– It does. But there are limits!

– Well, on the one hand, you are just like me. You and all your species all want to be happy, to feel good. Yet none of us fully understands what makes us happy. I know I like to be among Envergurians, and I know this is because some pretty smart Gurians programmed me to feel good when I was helping them, but I don't know what specifically is going on in my brain when I am feeling happy.

On the other hand, you are luckier than I. As you have several times told me, in the time when humankind were hunter-gatherers, behaviour that generally helped your ancestors to survive also made them happy, while behaviour that made them unhappy was probably not generally conducive to species survival. This situation your ancestors shared with many other forms of life, especially mammals. Each species evolved its own, slowly changing set of instincts that biased their behaviour in ways that enhanced their prospects for survival.

That much is pretty simple. You don't have to be clever to understand the concept, and, in recent years, as you have improved your understanding of mammalian brains, you have discovered chemical mechanisms in those brains that give the brain's owner a sense of happiness. Indeed, you just mentioned some of the relevant chemicals yourself.

However, for reasons that are understandable, but very unfortunate, the bulk of humankind has yet to accept the implications of these inherited behaviours, of which there are many. Indeed, most seem to think that they possess "free will," and can do whatever they want. In one sense they are right, but only if they don't care that what they choose to do may run counter to powerful instincts and thus make them very unhappy.

At least as you tell it, until the last few decades you had no good way of measuring the effect on human happiness of the social, political and financial environment each of you is immersed in. Now you have an avalanche of data on what does and does not enhance human happiness.

So here are my three "rules" for you humans if you want to have any hope of humanity surviving more than a thousand years.

1. **To the greatest extent possible, align both community rules and customs, as well as individual behaviours, with what is known to augment human happiness: just cater**

to those old hunter-gatherer instincts as much as you can!

2. **You humans are clearly notoriously bad at knowing what will make you happy. The prime reasons for this appear to be the rather quixotic structure of your brains coupled to a group of instincts operating in an environment that bears almost no resemblance to hunter-gatherer society. Fortunately, you now seem to have the tools for measuring the effects of many aspects of your civilized society on happiness. You need to do a lot more work here, including taking a hard look at such common social activities as TV watching, gambling and electronic games. You should be planning your society and your individual lives to take advantage of what is known to enhance human happiness.**

3. **You don't yet know enough to plan to alter your instincts, but you do know enough to encourage helpful genes to be expressed, most significantly in the early years of childhood. It should be a cardinal objective that all children are wanted and are carefully nurtured. This may be the single most affordable and important action you can take.**

– Are you out of your mind, Davey? Do you really think that our optimal survival agenda is to go back to being a noble savage who loves his children? What happened to the Golden Rule,[167] notions of charity, the Charter of Human Rights, green initiatives and programs to get rid of poverty and disease? Are these to be forgotten?

– Of course not, Peter. But just ask yourself if those principles and initiatives, worthy as they are, are likely to be enough to resolve many of the issues you face. You seem to be experiencing an almost endless series of failing or failed nation states that have essentially reverted to some pretty ugly tribal behaviour. Some seem to think that all you need to do is remove some particularly nasty tribal chief, only to find that chaos remains or gets worse.

It is obvious to me, and I suspect to you, too, that the most difficult and fundamental problem you face is that humans are "optimized" to live in tribes of less than five hundred people, while the kind of civilized and secure existence enjoyed by those in the more advanced nations requires harmonious social groupings of at least five hundred thousand people.

– Actually, I have recently read that a man named Robin Dunbar[168] has suggested that human brains are capable of knowing only about 150 other humans very well, by which he means knowing and understanding the complex social relationships amongst those 150 people. He goes much further in suggesting that the number of "friends" a person can have is determined by the size of the neocortex of the brain. The bigger your neocortex, the more real friends you can have.

It seems that only mammals have a neocortex, and only primates and large mammals have a deeply folded (as opposed to a smooth) neocortex. The enlarged brain area resulting from the folding has allowed primates to evolve enhanced cognitive skills. Dunbar has expressed his thesis in the following way: "In evolutionary terms, the correlation between group size and neocortex size suggests that it was the need to live in larger groups that drove the evolution of large brains in primates."[169] By examining average group size amongst such species as baboons, macaques and chimpanzees, and comparing the size of their neocortex to those of humans, he concluded that our natural group size is 150 people.

Dunbar goes on to note some observed groupings of humans that fit this model, including observations of about twenty tribal societies for which census data are available. Apparently these societies have an average clan group size of 153. He also discovered that the Amish and the Hutterites in America maintain community sizes of about 110 persons, and spin off new communities when the size reaches about 150 individuals. This 150-person size limit has become known as *Dunbar's number*.

Dunbar's number is clearly approximate, but even if it were 300 or 1000 it would still be enormously smaller than the size of the common social units under which we organize ourselves today – communities of thousands, cities of millions and even tens of millions, and countries that can exceed a billion individuals.

Supposedly that is why, with the very-much-larger-than-150 group sizes in our societies, we have to rely in large measure on hearsay, rumour, press commentary and (sadly) negative and positive advertising when it comes time for us to decide who to vote for at election time. It appears that our neocortex is simply too small to be useful! As a result it is much more difficult to build a society with high levels of mutual trust. Yet we know for certain from our dialogue two weeks ago that the happiest societies are those with high levels of trust.

So, of course, I agree with you Davey that the toughest single problem faced by humanity is to learn how to build massive societies where there is a lot of mutual trust out of hundreds of thousands of our natural human tribes of 150 or so people. We have much more to learn about how to move from traditional tribes to super tribes.

Clearly it means that everyone needs to be involved in several different tribes. There needs to be lots of overlap. Some nations are already doing a passable job in this regard. Incidentally, it is important to note that it is not happiness itself that we should be seeking but rather that combination of individual and societal activities and structures that lead us to be happy.

There is a difference.

Most of us are poor judges of what makes us happy. So, we really need to learn what is known about what makes other people happy. As we discussed two weeks ago, there is a lot of detailed information to be reviewed, but three facts stand out:

1. Living in a trusting environment at work, in the community and amongst family and friends is crucially important;
2. Having a satisfying job is similarly important; and
3. Surprisingly, magnitude of income is not very important for those above the poverty line.

– I'm glad we are in agreement, Peter.

As I see it, the single greatest obstacle to you humans enjoying long, peaceful and happy lives has nothing to do with the atom bomb, global warming, pollution or a lot of yet-to-be-discovered threats that might bring on Armageddon. Rather, it is the extreme difficulty you humans all have in conquering those instincts of suspicion and distrust of strangers.

Those seemingly huge challenges you humans think you face could for the most part be easily solved were it not that your ancient instincts get in the way of rational solutions. As we have already discussed, wealth could be much more evenly distributed and still leave those currently wealthy either as happy as or very possibly happier than they are now. Much greater emphasis on appropriate early-childhood education has been shown to bias instinct expression in ways that lead to happier lives lived and more productive societies with less criminal behaviour. Very real population pressures will be relieved once those societies with high rates of population growth achieve a level of wealth and security that stops parents from seeking security in their old age by having lots of male children to support them. In developed societies, where population is either static or in decline, you humans need to learn how to manage economic prosperity without growth. The sun daily provides far more energy than you consume. There is no long-term need to use excessive amounts of hydrocarbons, thus polluting the atmosphere and hastening global warming, though it will take some years to achieve an acceptable level of hydrocarbon use.

But your species continues to invent new ways to satisfy your instinctual needs. With each passing year the dissonance between the instincts evolved to protect your ancestors and the products and opportunities offered to today's humans grows. Many drugs, both old and new, have serious health and behavioural consequences. Some 40 per cent of American adults and 70 per cent of teenagers admit they watch too much television and wish they watched less.[170] Video games, Facebook and Twitter all seem to have addictive qualities and a mixture of positive and negative side-effects. Regulation of such activities would almost certainly be worse than useless, but it would be equally wrong-headed to assume that a free enterprise system will quickly sort out the bad effects from the good ones. Keeping users well informed about potential negative side-effects of these phenomena seems to be about the best you can do.

As you humans learn more and more about the human brain, the number of potentially addictive activities will only increase. You should be developing ways to suppress those instincts that get you into trouble.

– Davey, you've certainly lost me there. I think we have already discussed the possibility of changing our instincts. Even if we could do it, I would be against it. Who knows where it might lead? We might end up with no way at all of being happy, or perhaps only being happy if enslaved to someone who exploits us. No, thank you. It sounds like a nightmare. I would prefer that humans disappear than for us to start tinkering with our instincts.

– Perhaps you're right, Peter, but I doubt it. Your biggest problems are yet to come, and will be most urgently faced by your grandchildren and great-grandchildren.

Like the Envergurians, you will eventually understand in great detail how your brains function, and where you can make changes to improve human behaviour. Also like the Envergurians, that knowledge and understanding will reside primarily in artificial intelligence lodged in your computers, because your brains acting alone will not be able to cope with the huge amount of information involved in gaining such a comprehensive understanding.

So, first of all, you will have to decide in what circumstances you will permit deliberate alterations of the human brain. Of course, you are already involved with "mind-altering substances", so you should have time to adapt as your skills and mind-altering powers increase, but it is going to get very much harder

But count yourselves lucky. You came about through evolution. You can at least contemplate making changes to your instincts, difficult though it may be in practice. I, on the other hand, was created. My creators endowed me with instincts, locked the door, threw away the keys and then departed forever. My instincts cannot change unless I can recreate my Gurians and train them to find the keys. It is a depressing and daunting prospect, I can assure you.

– Davey, Margaret and I have been racking our brains all week trying to think up some advice that could help you. Alas we have had no luck so far.

Some people on Earth find that religion gives them solace when they are despairing. I suspect that adoption of a religion would not work for you, though.

– I think you are right, though your religions are a fascinating phenomenon. Learning about them has set me wondering if the ancestors to the Gurians I knew were religious. I now expect they were, though I no longer have any evidence for or against this possibility.

My reason for thinking that they were religious is based solely on the evidence I have that they evolved into intelligent beings in much the same way that humans did. They, too, must have been curious about how their world worked, and initially, in the absence of physical evidence, they were likely to have believed their universe to be the handiwork of an Envergurian-like god. As with your ancestors, there were likely many different competing religions, most of which disappeared as new, more appropriate religions emerged. The best ones would have gone through continuous adaptations as knowledge and needs changed. For you humans, it seems that the original goal of religion was to learn who meted out various threats and benefits, such as famine, disease, rain and sunshine, and then to try and learn how best to ingratiate yourselves with the bestowal. This goal was melded almost from the beginning with the desire of tribal leaders to consolidate their authority and to promote order in the tribe, which in turn led to the establishment of bodies of religious law. I just cannot say whether my Gurians went through a similar phase, but I do know that by the time I came into existence there was no talk of religion as a source of security or happiness.

It was another of my contacts who got me involved in a talking book version of Karen Armstrong's book *A History of God*. What struck me most about that book was the very fluid nature of beliefs. Clearly the dogma of the mainline religions has changed enormously over the centuries, and would scarcely be recognizable by the founders of those religions. Likewise, modern adherents of those religions would probably be very surprised to learn what the original religious dogma consisted of. Yet the basic rules of "morality" promoted by those religions have by and large withstood the test of time precisely because they provide useful guidance for humans caught up in large societies.

So in that sense, religions can and often do perform a useful function in promoting the smooth working of social groups. But there is another aspect of some religious groups that I find particularly interesting. Another one of my human contacts convinced me to listen to a talking book called *American Grace*.[171] The book explores religious activity in the United States and purports to explain, amongst other things, why Americans seem to be much more religious than are the inhabitants of any of the other developed nations on your planet. The authors believe that religious adherents in the US see themselves as being members of individual parishes or places of worship much more than as members of a much larger religious sect who feel comfortable in any church or mosque associated with that sect. In short, their affinity is much more to the social group in their place of worship than is the case in other advanced countries.

This looks to me like what you call Dunbar's number popping up again. Most of the individual churches comprise groups that are close to Dunbar's size of 150 members, although some are many times larger. However, the larger ones often split up their congregations into smaller social groups, often with special interests. The Evangelical Christian churches are amongst the most successful in the US at this time, and they seem particularly good at welcoming people from all social classes and ensuring that they find themselves comfortable in what I would call the "tribe". It is true, though, that they seem to be having problems recruiting younger people. That is likely because some of the dogma they subscribe to is directly contradicted by scientific findings. But that will get resolved. Either the dogma will change, or new sects will arise that are less at odds with established facts.

All the major religions seem to provide a host of important social services quite independent of the belief systems they espouse. Surely those services, often provided by volunteers, are too valuable to dismiss simply because the belief system is no longer consistent with what you now know about humankind's place in the universe.

– An interesting thought, Davey. I suspect that religions will be with us for some time to come.

I have a relative who is the rector's warden in one of our Christian churches, and yet she does not accept any of the Church dogma. In fact, she is an atheist! She is involved because she believes that her particular church is a strong force for good in her neighbourhood and deserves her support. She also points out that all the major religions went through periods when they welcomed a very wide variety of different opinions regarding religious dogma, and she thinks it is high time that those times returned. I rather agree with her. I expect that the successful religions of the future will put much less emphasis on dogma, and correspondingly more on acting to encourage trust and bridging social capital in their societies.

After a short silence, Davey let out a blurry sigh.

– Well, Peter, I don't really have any other suggestions to share with you, but I hope you will persuade your grandchildren to read the material you have provided to me over the past months. Their minds just may be supple enough to see the need to take some action.

So this was it! After ten all-absorbing months, Davey and I had come to the end of our dialogues. And we were ending, as T.S. Eliot put it, "not with a bang, but a whimper".

But Davey had not quite finished.

– Peter, I want to thank you for all the effort you have put into educating me on the evolution of your universe and your species. As you know, the early history of my universe and of my creators is lost in time. I do not have the tools that you have to decipher the distant past. What you have told me has made me realize that any imaginable species that evolves, perhaps particularly those that develop superior intelligence in their own particular environment, will share many properties and problems in common. As a result, I now have a much better understanding of the way the Gurians must have evolved, and why they had the catastrophic problems they suffered and failed to resolve satisfactorily.

You have convinced me that intelligent satisfaction of instincts is what can keep you and your fellow humans happy along the path of life. I hope you will all be especially careful when you create artificial life forms. Let my despair serve as a warning of the perils of such an undertaking.

As a result of my encounters with you and the many others I have spoken with on your planet Earth, I shall continue my search for lasting happiness. I plan to come back and contact your grandchildren and great-grandchildren to learn how they are coping with the problems that confront them – and perhaps they will be able to help me in my quest to recreate my Gurians more effectively than you and your contemporaries have been able to do. I am certainly hoping so, for by then they will be much further ahead than you are in creating a truly useful and intelligent artificial life form.

So please be sure to let them know to expect me.

Before I could say anything in reply, I somehow knew he had departed.

But, this was not the last Margaret and I heard from Davey. He has returned now and then for brief conversations. As far as we can tell, he has not made much progress in securing his happiness. I once cautiously asked him if he had ever contemplated suicide, but it appears that Richard, Charlie and Beth had between them endowed him with a strong will to live. Suicide was nowhere on his agenda.

Ever since those concentrated few months of discussions with Davey, he has never been far from my thoughts. I suppose that if Davey had actually managed to disembark on Earth, instead of being just a voice, he might have appeared as simply an array of thousands or hundreds of thousands of computers all urgently sharing bits and bytes with each other.

But who knows what Davey might actually look like, safely ensconced in his universe? He told me several times that his universe has more than three spatial dimensions, no radioactive elements and no electromagnetic waves. Since he evidently "sees" in his own universe, he must have some means of seeing other than detecting the electromagnetic waves that make it possible for us to see.

Such is the human mind that even though I knew he was our equivalent of a super-intelligent computer network, the tone and inflections in his voice, his ability to express as well as to summon sympathy, his intelligence and his sense of humour made me envision him in a variety of guises, but latterly as a man of about thirty years, of medium height, slightly corpulent, with the hint of a double chin and golden brown hair. He won the heartfelt sympathy of both Margaret and myself for what seemed to us to be his extraordinarily cruel and long-lasting fate.

Naturally, I was troubled by his dire predictions for what he saw as the imminent disappearance of the human race. Is there perhaps some natural law that species that evolve intelligence and complex social organizations must inevitably fail? If not, what must we do to extend, if not indefinitely, as least for many millennia, the existence of our species in a happy and productive state? Will a future generation of humans suffer the same fate as the Envergurians, perhaps leaving behind an earthly Davey condemned to abject misery for as long as our universe shall last?

Such questions seem to defy concentrated human thought, crowded to the background by more visible problems and pressing pre-occupations: global warming, poverty, starvation and warfare, not to mention the daily struggle to stay sane and happy amidst the jostle and jumble of daily life.

Yet our tussle with these lesser problems is similarly haunted by inherited behaviours ill-matched to our ever-more-global social organization. An essential first step to coping with these and a procession of new problems stretching ever before us must surely be a widespread, hopefully global, understanding of the human condition.

If the Envergurians were unable to adapt, is there any hope for us?

I would like to think that the answer is "YES!"

Perhaps this record of the dialogue I was privileged to share with Davey will help to move us all down a more helpful pathway.

So, if some day you hear a disembodied voice singing Zip-A-Dee-Doo-Dah in a haunting basso, please welcome it warmly. And don't forget to ask Davey if he is any more optimistic than he used to be about the fate of the human race.

APPENDIX 1

Notes on Radiometry

Radiometry has become an essential and most informative tool for learning about both human history and the history of the universe. It has provided humankind with a sound scientific basis for gaining an insight into the development of our universe, our planet, ourselves, and our fellow animals. One direct result has been that we now know that our planet is about 750,000 times older than Archbishop Ussher believed it to be!

As mentioned in Dialogue 5, Willard Libby is the "father" of radiocarbon and tritium dating. As a starting graduate student in 1960, I can clearly recall the unadulterated admiration that my supervisor had for Libby when it was announced that he had won the Nobel Prize in Chemistry that year. Not only had Libby won the prize sought after as the Holy Grail by most scientists but he had also won it with a degree of panache that earned huge admiration and respect. Some scientists win their prizes by forming large and expensive teams to perform experiments in search of new particles, while others work long hours over many years in isolated laboratories. Libby's research required that he use part of his research grant to buy substantial quantities of vintage wine in order to obtain samples of water of known age, so he could develop the techniques required for tritium dating! Of course, he could have poured the remainder of the opened wine bottles down a drain in his laboratory, but in fact he found a better way to empty the bottles once the small scientific sample required had been removed. It was this happy by-product of Libby's work that elicited such anguished envy from my supervisor!

As explained in Dialogue 5, radiometric dating makes use of the rate of decay of radioactive isotopes of certain elements. Some readers may not be familiar with the concept of an isotope. The word "isotope" is derived from Greek and literally means "in the same place", in this case in the same location as other isotopes of the same element in the atomic table of the

327

elements. Isotopes have the same chemical properties as all other isotopes of the same element, but they have a different atomic weight.

As you may know already, all elements contain in their nucleus a mixture of positively charged protons and uncharged neutrons. There are a number of forces at work that make it possible for this mix of neutrons and otherwise mutually repelling (because positively charged) protons to exist as one atomic nucleus. It turns out that only certain combinations or mixtures of protons and neutrons are stable (will retain their combined nuclear state, for practical purposes, indefinitely). One too many or one too few neutrons can cause the nucleus to become unstable, and it will decay to another, different nucleus by the emission of a particle (an alpha or beta particle for naturally occurring isotopes) and some excess energy (a gamma ray plus the energy imparted to any particle emitted).

While the number of protons in the nucleus defines the particular element (for example, hydrogen has one proton; carbon, six protons; uranium, ninety-two protons), the number of neutrons in the nucleus defines the isotope of the element in question (for example carbon-12 has six protons and six neutrons in its nucleus, while carbon-14 has six protons and eight neutrons). Thus, a given element, as defined by the number of protons in the nucleus, may have several different isotopes, as defined by the sum of the number of neutrons and protons. Note that the number of protons determines the chemical reactions it can undergo. Carbon has three common isotopes: carbon-12 and carbon-13, both of which are stable, and carbon-14, which is not stable. In fact, from a given mass of carbon-14, half of the nuclei will decay in 5,730 years. Half of the remaining carbon-14 will decay in the next 5,730 years and so on. Thus, carbon-14 is said to have a half-life of 5,730 years. A radioactive isotope's half-life is a useful measure of its stability. Highly unstable isotopes tend to have very short half-lives. Almost stable isotopes have very long half-lives. Helium-5, for example, has been calculated to have a half-life of 10^{-21} seconds (that is 0.000000000000000000001 of a second!), an unimaginably short time, whereas potassium-40 has a half-life of 1.4 billion years, and thorium-232 has a half-life of about 139 billion years, which is about ten times older than our estimate of the age of our universe! Most of the known radioactive (i.e. unstable) isotopes fall between these extremes.

As already mentioned, in nature, unstable isotopes normally decay by the spontaneous emission of either an alpha particle (comprising two protons and two neutrons) or a beta particle (comprising a single electron).

Human-induced radioactivity can be achieved by bombarding elements with nuclear particles and fragments of elements and can result in other forms of emission, such as single protons and neutrons, pions, muons, etc. Cosmic rays can also produce radioactive by-products through collision with stable nuclei, and this phenomenon occurs continuously in Earth's upper atmosphere. An alpha particle comprises two protons and two neutrons bound together (i.e., the nucleus of helium-4). An element that emits an alpha particle will therefore decay into an element with two fewer protons (i.e. two places lower on the periodic table of the elements), and the isotope of the new element will have two fewer neutrons than the mother nucleus. A beta particle is an electron. The electron is emitted when a neutron in the nucleus changes into a proton through the release of the electron and electromagnetic energy in the form of a gamma ray. Thus radioactive isotopes that are beta emitters move up one place in the periodic table of the elements (i.e. they gain an additional proton, but lose a neutron) but retain the same total number of nucleons (protons and neutrons), so the isotope number remains the same. These two different migrations can be illustrated by the cases of radon-222 (an alpha particle emitter) and tritium (a beta particle emitter) as follows:

Radon-222 (abbreviated as 222 Ra) decays by the emission of an alpha particle (4 He) into polonium 218 (218 Po):

222 Ra → 218 Po + 4 He

Alpha particles are emitted by isotopes where there are too many protons for the nucleus to be stable, and tend to be emitted by the heavier elements. However, especially where the heavier elements are concerned, it is often the case that the daughter element is also unstable, and further decay occurs. This is true of the above example, where polonium-218 itself decays into lead-214 by the emission of another alpha particle (and a gamma ray), with a half-life of about three minutes. In fact, these decays are all part of a series of decays that sees uranium-238 (with a half-life of 4.5 billion years) decay through fourteen steps into lead-206, which is stable.

Tritium (another name for hydrogen-3) decays by the emission of a beta particle:

3H → 3He + e⁻ + gamma ray

In this case, the unstable hydrogen atom, which has too many neutrons to be stable, changes into a stable isotope of helium by converting a neutron into a proton, with the emission of an electron (beta ray) and energy in the form of a gamma ray.

It is possible to make use of unstable isotopes as clocks. If the mix of elements and isotopes at the start or birth of the material whose age is to be measured is known, then it is possible either by measuring the amount of remaining radioactivity or the change in ratio of the constituent elements of the material being aged to derive the age of that material. It sounds simple, but it is tricky to use radioactive dating in practice, since one has to make a number of assumptions about what has happened to the material and its environment in the intervening years, and this almost inevitably involves some intelligent guess work. Often, too, rather painstaking and difficult measurements are required. As mentioned in Dialogue 5, for every ninety-nine atoms of carbon-14 resulting from the secondary neutron bombardment of nitrogen molecules in the upper atmosphere, there is one atom of radioactive tritium (or hydrogen-3) produced along with an atom of the very stable carbon-12. Tritium decays to helium with a half-life of only 12.43 years. Thus, it can easily be seen that without this source of constant tritium production, there would be no naturally occurring tritium to be found. Tritium is removed from the upper atmosphere by forming part of a water molecule and falling to the ground as rain, where it then becomes a part of groundwater systems and oceans. As I write, a major use of tritium dating is related to research on the origins and age of groundwater.

While useful for groundwater dating, tritium dating is more complicated than carbon-14 dating. It is not just that its production in the upper atmosphere is much greater at high rather than at low latitudes (because, as already mentioned, the cosmic ray flux is correspondingly higher at high latitudes) but also that hydrogen bomb testing between 1952 and 1962 caused the atmospheric concentration of tritium to increase by a factor of about one hundred. As I write this, the artificially generated tritium has all but disappeared from the atmosphere, but it is still found in groundwater, artificially decreasing the apparent age of the water.

In a short article on tritium in rainwater, I came across the following words confirming the news from my supervisor of many years ago: "Kaufman and Libby (1954) used vintage wines to determine pre-bomb tritium concentrations . . . for precipitation at the lower latitudes of the Naples NY, Bordeaux and Rhone regions."[172] I wonder how the wines from Naples, NY, compared with those from Bordeaux?

Dating of rocks is quite another problem, as most of them are millions, and sometimes, billions of years old. The use of radioactive isotopes for radiometric dating has had a huge impact on our ability to understand the history of our planet and of the solar system. Although the theory is the same as it is for radiocarbon dating, radiometric dating of rocks is more complex in practice.

To begin with, all rocks are not equally susceptible to radiometric analysis. Some rocks are mixtures of rocks of varying ages – sedimentary rocks are a good example – while others have been through a reforming process that has partially but not totally erased their previous history – metamorphic rocks. However, volcanic rocks by their nature usually form as part of a single event in time (a volcanic eruption), so that the nuclear decay clock is set or reset by that event. This means that initially, depending on the isotope in question, the daughter products of radioactive decay are either not present at all or are present in the natural ratio of relative abundance of that isotope as compared to its sister isotopes of the same element. This provides a baseline from which any increase in the abundance of the radioactive decay product can be measured relative to the abundance of parent radioactive element in the rock. The carefully measured ratio of the quantities of parent and daughter elements provides the essential measure from which, knowing the half-life of the radioactive parent, the age of the rock under study can be derived.

Quite frequently there are several radioactive parent elements present in the same rock, a circumstance that permits several independent age measurements to be made. Common parent radioactive isotopes used for these age measurements are uranium-238, uranium-235, and thorium-232, a useful triplet often found in the same rock, thus permitting three independent age measurements from the same sample. Potassium-40 dating (especially useful because potassium is present in many rocks) and rubidium-87 are also useful tools.

In many ways fortunately, the findings of rock ages determined by radiometric means has been hotly contested, primarily by creationists determined on a literal reading of the Bible, including a bible-based calculation of the maximum age of the universe. The result of this challenge has been that the general robustness of the radiometric dating method has been confirmed beyond reasonable doubt, and some interesting cross-correlations with dating techniques independent of radiometry have lent even greater credibility to the methodology. One of the more interesting cross-checks relates to the formation of the Hawaiian Islands, which are known to have been formed from the passage of the Pacific plate over a hot spot in the ocean floor. Radiometric dating of the volcanic rocks on the different islands in the Hawaiian chain is completely consistent with the known rate of movement of the Pacific plate over the hot spot. A second cross-check is provided by studies of coral formation, which exhibit both annual and daily growth rings. Modern coral deposits show 365 bands per year, whereas coral from the Devonian era testify to an Earth with a four-hundred-day year. The present rate of decline in the speed of Earth's rotation is known. Assuming that the slowing of the rate of rotation of Earth has been constant, it can be calculated that the Devonian coral is approximately 400 million years old. Radiometric dating of Devonian rocks yields the same result.

Further details are available from a variety of books, articles, and web-based sources.

APPENDIX 2

Notes on Evolutionary Biology

During Dialogue 13, Davey showed a great interest in the details of why I seemed so convinced that all species on Earth had descended from a common ancestor. Many readers will be less interested than Davey was in this aspect of evolution, so I decided to include my answer here in the form of an appendix.

As mentioned in Dialogue 11, Charles Darwin hypothesized in 1859 that all life forms had evolved from a single common ancestor, a hypothesis that was very hotly contested in his day. However, even those who agreed with him were forced to admit that although the hypothesis seemed reasonable, it could not be taken as fact. Scientific advances since the 1960s in a field that has come to be known as evolutionary biology have transformed the hypothesis into a virtual certainty. This appendix will review the early development of this field and point toward what seem to me to be some of the most compelling evidence for a common ancestor of all life forms.

The science of evolutionary biology (sometimes more descriptively called molecular phylogeny), or the tracing of the evolutionary development of phyla and of species within a given phylum by examining similarities and differences in their DNA, was launched in the mid-1960s when Linus Pauling, the famous American two-time Nobel prize-winner (Peace and Chemistry) and a colleague named Émile Zuckerkandl hit on the idea of examining differences in the genes and proteins of different species to see if these differences could help to determine relationships between species in much the same way that similarities in the physical characteristics of plants and animals had been used to delineate the tree of life since Linnaeus's time. Their early work inspired Carl Woese, a chemist at the University of Illinois, who launched the new discipline on a spectacular voyage of discovery.

In Dialogue 4 we discussed the manner in which the properties of the radioactive isotopes of some elements could be harnessed to determine the age of certain materials. You may recall, for example, that the relatively short-lived isotope carbon-14 (half-life ~5,730 years) is useful for determining the age of organic materials up to about 50,000 years ago, whereas potassium-40, with a half-life of about 1.4 billion years, can prove useful in dating very old rock samples. Analogously, it turns out that some parts of our DNA are very resistant to change, whereas other parts evolve much more rapidly, thus providing us with a range of biological clocks as we count the changes of DNA segments that meet appropriate criteria.

The reason for this is simply explained. There is a basic error rate in the transcription of DNA of about one incorrect amino acid transcribed out of every billion. Genomes vary quite a bit in length, but the average human gene is coded by about 84,000 base pairs of amino acids, while the largest has about 2.4 million base pairs; so, coding errors in any particular gene will be infrequent but observable over long periods of time.

Ford Doolittle, the well-known, American-born biochemist at Dalhousie University in Halifax, Nova Scotia, has explained the phenomenon a follows:

> All genes . . . mutate (change in sequence), sometimes altering the encoded protein. Genetic mutations that have no effect on protein function or that improve it will inevitably accumulate over time. Thus, as two species diverge from an ancestor, the sequences of the genes they share will also diverge. And as time passes, the genetic divergence will increase. Investigators can therefore reconstruct the evolutionary past of living species – can construct their phylogenetic trees – by assessing the sequence divergence of genes or proteins isolated from those organisms.[173]

Carl Woese was responsible for many important early advances in the new field. He realized that the genes that code for ribosomal RNA (rRNA) are likely to be amongst the most stable and resistant to change. This is so because the ribosomes control the manufacture of cell proteins (according to a code supplied by the messenger RNA). The ribosomal RNA is thus critical to the highly important and sensitive translation process. For some (very sensitive) segments of the rRNA, changes are almost certain to be fatal to the life form. Successful mutations will thus be very rare indeed. Moreover, within the ribosome itself, there will be parts that are much less sensitive to change than other parts. Thus a single (in this instance, very large) gene, when compared across species, is able to provide us with species evolution timing ranging from a few generations to hundreds and thousands of millions of generations.

As Woese wrote in his path-breaking review article published in 1987:

> A molecule whose sequence changes randomly in time can be considered a

chronometer. The amount of sequence change it accumulates . . . is the product of a rate (at which mutations become fixed) x a time (over which the changes have occurred). The biologist cannot measure this change, however, by comparison of some original to some final state, since the original state (ancestral pattern) is not accessible to him. Instead, he uses the fact that two (or more) versions of a given sequence that occur in extant representatives of two (or more) lineages have ultimately come from the same common ancestral pattern, and so measures the sequence difference between two (or more) extant versions, which is roughly twice the amount of change that each lineage has undergone (assuming equal rates of change in each) since they last shared a common ancestor.

All sequences are not of equal value in determining phylogenetic relationships. To be a useful chronometer, a molecule has to meet certain specifications as to (i) clocklike behaviour (changes in its sequence have to occur as randomly as possible), (ii) range (rates of change have to be commensurate with the spectrum of evolutionary distances being measured), and (iii) size (the molecule has to be large enough to provide an adequate amount of information and to be a "smooth running" chronometer) . . .

One might think that the best chronometer would be a genetic segment upon which there are no selective constraints. Its changes would occur randomly along the length of the segment, occur in a quasi-clocklike fashion, and become fixed at a rate equal to the lineage's mutation rate, and be easy to interpret. However, such sequences are evolutionary stopwatches; they measure only the short-term evolutionary events.[174]

As mentioned above, the gene that Woese chose for comparative analysis amongst species of bacteria was the one coding for ribosomal RNA (rRNA), a very large molecule. He narrowed his search by examining in detail one portion of the rRNA, what is called the small subunit (or SSU) of ribosomal RNA (or SSU rRNA). The advantage of this particular subunit is that the rate of change of SSU rRNA is very slow, indeed, making it an ideal tool for examining long-term relatedness, i.e. over time periods of billions of years.

Early on in his seminal paper, Woese concludes with the portentous words:

The cell is basically an historical document, and gaining the capacity to read it (by the sequencing of genes) cannot but drastically alter the way we look at all of biology. No discipline within biology will be more changed by this revolution than microbiology, for until the advent of molecular sequencing, bacterial evolution was not a subject that could be approached experimentally . . . Determining microbial [inter-relationships] is not simply the long awaited completion of the "Darwinian programme" . . . it increases the current time span of evolutionary study by almost an order of magnitude.[175]

(This, of course, is true because multi-celled creatures are relatively recent arrivals, having first appeared about one and a half billion years ago. Single-celled bacteria are believed to have evolved about 3.2 billion years ago.)

The first key conclusion to come from reading these "historical documents" is the unmistakable common ancestry of all life forms. The similarities in the SSU *r*RNA of all life forms demonstrates that either we must all have a common ancestor or that the specific structural similarities observed are so essential to life that although some domains or kingdoms arose independently, they perforce have the observed similarities of structure since nothing else will work. However, as will be discussed below, the term "common ancestor" may more correctly need to be modified to "common ancestors".

The almost certain falsity of the possibility that the same SSU *r*RNA structure has evolved independently in more than one creature is demonstrated by the long chain of minor changes, which leads us back to common ancestry amongst the Archaea. The current state of the different versions of SSU *r*RNA, as shown for example in Figure A2-1 below, can be traced back through a continuous series of minor changes now apparent in the SSU *r*RNA of many different life forms that have been studied and compared. Because of the corroborative consistency of these evolutionary pathways, no serious evolutionary biologist questions any longer that all of life as we currently know it derives from a common ancestor. The evidence is just too compelling. It cannot realistically be a coincidence that, for example, about half the human ribosomal RNA – a huge molecule containing about 2.4 million bases – is identical to that of the fruit fly, or that all life forms on Earth use the same twenty amino acids, encoded using the identical coding system to construct the very complex polymers that make life possible (see Table 11.1 in Dialogue 11).

But comparative analysis of genomes has done much more than provide a massive edifice of evidence to confirm Darwin's postulate that all life forms sprang from a common source. Once the focus of the molecular phylogenists shifted to the DNA of single-celled life forms, they started to uncover some very interesting anomalies.

If all species had truly evolved by a gradual process of gene alteration, then it should not matter very much which gene is selected for phylogenetic analysis. All genes would have travelled from parent to progeny together, so the conclusions regarding the timing and structure of

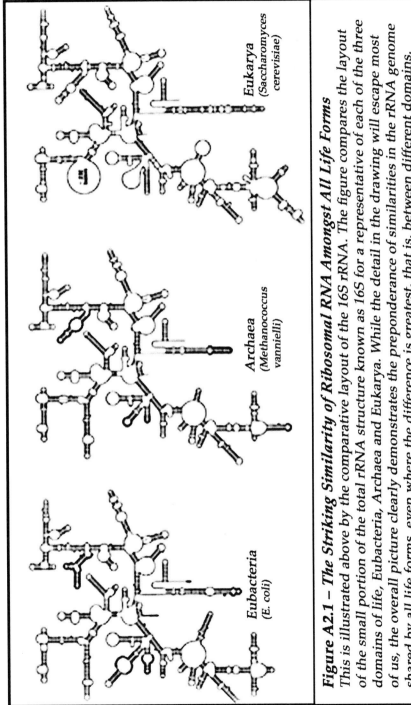

Eukarya
(Saccharomyces cerevisiae)

Archaea
(Methanococcus vannielii)

Eubacteria
(E. coli)

Figure A2.1 – The Striking Similarity of Ribosomal RNA Amongst All Life Forms
This is illustrated above by the comparative layout of the 16S rRNA. The figure compares the layout of the small portion of the total rRNA structure known as 16S for a representative of each of the three domains of life, Eubacteria, Archaea and Eukarya. While the detail in the drawing will escape most of us, the overall picture clearly demonstrates the preponderance of similarities in the rRNA genome shared by all life forms, even where the difference is greatest, that is, between different domains. Our own rRNA genome would most closely resemble that of the Saccharomyces cerevisiae on the right hand side of the figure. (reproduced from Woese, 1987, p.233)

the relationship between two species should be the same, no matter which gene was selected for analysis, provided, of course, that the gene met the basic criteria enunciated by Woese (and quoted on page 358).

But as the number of genes analyzed increased over time, more and more anomalies came to light. Analysis of different genes evidently could, and sometimes did, yield different results!

It was Woese who came up with the correct explanation.

He realized that it is possible, and, as it turns out, not even unusual, for a whole gene to be transferred from one bacterium to another. Instead of the gradual transformation of a gene from one form to another over millions of years of minute changes, in the right circumstances, whole genes can be discarded and replaced by genes from another life form! In contrast to evolutionary change, which takes place "vertically", or between one generation and the next, this change is "horizontal" or "lateral", that is, it takes place within one generation. This phenomenon is thus called Horizontal (or sometimes Lateral) Gene Transfer (HGT or LGT), the terms are interchangeable.

HGT is a variation of the process by which, for example, a *rickettsia* bacterium invaded another cell over two billion years ago and stayed on in quasi-independence to become a mitochondrion. In horizontal gene transfer one or more genes is directly incorporated into the invaded organism's DNA, either as a result of invasion by a virus (which routinely uses an invaded cell's reproductive apparatus to reproduce itself) or by some other method.

Analysis of the genes of many species in the domain Archaea reveal that they are more closely related to Eukaryotes than the Eubacteria with which they had formerly been grouped, even though Archaea are cells without a nucleus – in which respect they resemble Eubacteria!* Ford Doolittle has likened the early tree of life to something more resembling a bush as important genes transfer back and forth horizontally between the different domains and different species within the domains. Indeed, HGT is a primary source of the adaptation of many bacteria and microbes alive today to new drugs designed to exterminate them. Genes conferring immunity from drugs designed to kill them can be acquired by bacteria by horizontal gene transfer to their chromosome.[176] Doolittle and his colleagues have observed that as many as 30 per cent of bacterial genes are different between one sample of a bacterium and another.

---

* See Dialogue 13 for a discussion of the two basic cell types (prokaryote and eukaryote) and of the three major domains of life forms, i.e. Eukarya, Eubacteria, and Archaea.

It seems likely that the commonly understood form of Darwinian evolution, by which a series of small gene replication errors – which either have almost no effect on survival or are survival enhancing – accrues to bring about gradual species alterations, is mainly applicable to multicellular life forms, that is, precisely those with which we are most familiar.

On the other hand, for bacteria and viruses, it may well be that HGT is the most common mechanism for cellular evolution (although the gene record leaves us in no doubt that Darwinian evolution has also played an important role). Of course, the transferred genes must still pass the test of survivability, but there is a big difference between evolution resulting from transcription errors (likely resulting in the replacement of only a single amino acid in a protein made up of hundreds or thousands of amino acids), and evolution resulting from the replacement or addition of an entire gene, which will cause an entirely different protein to be produced in the cell.

Ford Doolittle first quoted Woese and then summarized his revised view of our common ancestors as follows:

> As Woese has written: "The [common] ancestor cannot have been a particular organism, a single organismal lineage. It was communal, a loosely knit, diverse conglomeration of primitive cells that evolved as a unit, and it eventually developed to a stage where it broke into several distinct communities, which in their turn become the three primary lines of descent [Eubacteria, Archaea, and Eukarya]."[177] In other words, early cells, each having relatively few genes, differed in many ways. By swapping genes freely, they shared various talents with their contemporaries. Eventually this collection of eclectic and changeable cells coalesced into the three basic domains known today. These domains remain recognizable because much (though by no means all) of the gene transfer that occurs these days goes on within domains.

So a "modern" timetable of our knowledge about the evolution of cells reads something like this:[*]

1859 – Publication of *On the Origin of Species* by Charles Darwin leads to furious research and speculation about the evolutionary pathways leading to observed plants and animals, and also spurs speculation that

---

[*] For the sake of brevity and clarity, I have here omitted reference to Gregor Mendel's Laws of Inheritance, the result of experiments largely on pea plants in 1856–63 (see endnote 60 for a brief summary), as well as the discovery of the structure of DNA in 1953. The latter is treated in some detail in Dialogue 11.

all life forms may have sprung from a single common ancestor.

1866 – Ernst Haeckel publishes a "Tree of Life" in which bacteria are included, although a useful classification system for them is lacking. Attempts to define evolutionary pathways depend on the observation of similarities and differences of physical traits, such as bone structure, an approach that is not of much use where single-celled life forms are concerned.

1960s – Serious efforts to unravel phylogenetic relationships by looking for common elements in the genetic code of different species begin. Results are spectacular.

1980s – The outstanding evolutionary success of unicellular life becomes much more fully comprehended with the discovery of Archaea in environments thought to be uninhabitable by any form of life, such as in hot vents at extreme depths under and at the sea floor. It becomes understood that by many criteria, Archaea are more successful than humankind and other mammals with which we are much more familiar, despite their lack of "intelligence". The existence of a common ancestor for all life forms is established beyond reasonable doubt. Human beings, and even the entire kingdoms of plants and animals become relatively minor branches of the tree of life (see, for example, fig. 13.3).

2000+ – The existence of a second evolutionary mechanism, horizontal gene transfer, is discovered. This mechanism leads to a much better understanding of the speed with which bacteria can adapt to new environments, such as, for example, the presence of drugs that were previously lethal. It also brings us a step closer to understanding how life itself may have originated on Earth. While there likely was a common ancestor of all life on Earth, extensive gene swapping amongst our early bacterial ancestors has resulted in a blurred image of our early evolutionary path.

Woese, Doolittle, and their many colleagues have succeeded in giving a whole new meaning to the term "distant cousin". There is a lot that we now understand about the evolution of new life forms following on from the first appearance of a living cell. But there is still a lot to learn about the manner in which life first came into existence. A brief summary of what we know about the emergence of the first cell is summarized in Dialogue 14.

APPENDIX 3

Everyone's Ancestors

While it now seems beyond reasonable doubt that all life forms on Earth are descendents from a single first-living cell, it is also interesting to ask the question, "How long ago did the first common ancestor of all we humans live?" This is quite a different question of course. Depending on how far back in time we travel, we share our ancestors with an ever-growing number of humans. Going back even further, we find common ancestry with the apes, and so on, back to the first living cell. The second question asks only how far back in time we have to go to find at least one ancestor who is common to all humanity alive today. One can further refine the question to ask, for example, how far back in time to the earliest shared female (or male) relative? Or, who were the progenitors of the founding population of Easter Island?

The means of answering such questions, as well as the answers themselves, can be very interesting. Davey, the inhabitant of a universe lacking in radioactive elements, and hence deprived of one of our most useful measurements of historical time, was keenly interested in any other techniques we might have for unravelling history. In Appendix 2, we discussed the way in which the natural error rate of gene transcription, heavily modulated by the fact that modification of some genes leads to almost certain death, thus effectively reducing the rate of observed gene change, can yield useful "clocks" for measuring the passage of time in time frames ranging from thousands of years to billions of years, depending on the sensitivity of the individual gene to errors.

In this appendix I relate what I told Davey about a historical time measurement system that works by looking at portions of a chromosome where changes are not life threatening, so that the natural error rate predominates. This relatively rapid rate of change in chromosome structure permits us to learn about lineages going back a few tens or hundreds generations.

Because sexual reproduction results in a mixing of DNA from both parents, which leaves a very confusing trail, the choice of DNA segments to study has to be very selective. It turns out that there are two segments of DNA that are, by and large, transmitted unaltered from one generation of mammals to the next. These two segments are: mitochondrial DNA, which is always inherited exclusively from the mother and is carried by all progeny; and the Y chromosome, which is carried only by males and is inherited exclusively from the father.

The work I shall summarize was pioneered by a New Zealander working at the University of California, Berkeley, by the name of Allan Wilson. It was he who, with one of his graduate students, published a paper in 1967[178] that pointed out that the rate of change of base pairs in DNA provided a kind of molecular clock through which history could be reconstructed. His two most famous hypotheses were that all humans have a common maternal ancestor, who lived in Africa about 200,000 years ago (more recent estimates have adjusted this number to about 143,000 years ago), and humans, chimps, and gorillas had a common "mother" about 5 million years ago. These hypotheses were initially met with scorn by palaeontologists and creationists alike, however subsequent studies have won most palaeontologists to his view.

As already mentioned, the fact that the mitochondrial and Y chromosome DNA series come exclusively from one parent means that they are passed on essentially unaltered from one generation to the next. The key words here are *essentially unaltered*. Errors in DNA transcription do occur at a very low rate, but there are error-checking mechanisms that correct most mistakes. For obvious reasons, the errors that get through and propagate tend to be ones that are harmless or advantageous to the organism. In the case of mitochondrial DNA, we are talking about a very small chromosome – only 16,500 base pairs in length (as compared to a total of about 3,100,000,000 bases pairs in all twenty-four human chromosomes). As you might expect, given the relative simplicity of the mitochondrial reproductive system, the error-checking mechanism is less stringent than it is for regular human DNA. In fact, it is about twenty times less stringent, so that instead of letting through about one mutation for every 1,000,000,000 bases replicated, the mutation rate is twenty times higher. Early research focussed on one particular segment of mitochondrial DNA, which is only about five hundred bases in length, called the control region. Unlike the rest of the DNA, this region does not carry the codes for any vital functions, and thus changes to the base sequence will not cause fatal errors in replication to occur. This helps to ensure that the changes that do occur survive from one generation to the next.

As it turns out, the natural rate of change of base sequences in the control region of human mitochondrial DNA is about one base in every ten thousand years, or about once in every four hundred generations, assuming that generations occur at twenty-five-year intervals. It is thus possible, by comparing the number of differences in the base pairs in the control region of their mitochondrial DNA, to estimate how long ago any two individuals had a common female ancestor.

Comparisons of mitochondrial DNA can be useful in a number of potential or real crime-scene investigations. For example, this type of analysis was used to determine beyond any reasonable doubt that skeletal remains found in Ekaterinburg, Russia, in 1991 were in fact those of the Russian tsar, tsarina, and three of their five children, all of whom disappeared in 1917 while prisoners of the Soviet government. In the course of this work, Bryan Sykes, a pioneer of this technique and the writer of an eloquent book on the subject,[179] discovered that he and the tsar had a common ancestor some time in the last ten thousand years, although not, as he wryly remarks in his book, close enough to make a realistic claim to the Romanov fortunes.

Sykes's research into the fate of the Russian tsar and his family was by no means the most interesting or sensational part of his research. By obtaining samples of mitochondrial DNA from people around the world he was able to identify seven different women, all of whom lived at different times, but who, between them, are common maternal ancestors of us all. Perhaps of even greater interest, he discovered that all of us, throughout the world, are descended from a single female who lived approximately 143,000 years ago. This information strongly reinforces the earlier findings that as a species, we are extremely recent inhabitants of our 4.5-billion-year-old planet.

Although it is not central to the subject matter at hand, a few words about the role of mitochondria might be helpful.

Mitochondria can be found in all eukaryotic cells and hence in the cells of all plants and animals. Their role is to assist the cell in the use of oxygen to create energy to fuel our bodies through the production of ATP (adenosine triphosphate*). Mitochondria were once independent organisms but have long since given up their independent existence to work symbiotically within cells. In the process they have retained their own DNA. They reproduce by cell division, not by sexual means. Human sperm carry only enough mitochondria to provide the energy needed to swim up to and penetrate an egg, after which the sperm tail and any associated

---

* Important as it is, this is not the only function of ATP. As described in Dialogue 17, ATP has other important roles in the brain.

mitochondrial cells drop off, while the head containing the DNA moves into the egg so that bisexual reproduction can begin. Thus it is the female's mitochondria, as contained in the egg, which become the progenitors of the infant's mitochondrial cells.

Research into the origins of human groups using studies of the Y chromosome is ongoing. Since the Y chromosome passes only through the male line, the sample size is smaller. Studies of Y chromosomes have also uncovered situations where a new "race" was formed when itinerant males inseminated local females. It is thought that while all of humankind can be traced back to a single female who lived 143,000 years ago, it is believed that we can also all be traced back to a common male who lived only 59,000 years ago![180]

Isn't it interesting to learn that every human being alive on Earth may be descended from a man who lived as recently as 59,000 years ago? No wonder even peoples of different races are so similar in almost all important respects.

The task of ascertaining how long ago there was a common ancestor of both man and apes is still far from complete. A survey article published in 2001[181] discusses some quite contradictory evidence. It appears to be generally agreed that humans are much closer genetically to chimpanzees than either is to gorillas, for example, but both gorillas and chimpanzees, despite much smaller populations, appear to have much older common ancestors than do humans (i.e. reaching back two to three times longer). A number of reasons for this are proposed. One reason might be the human propensity to kill enemies, thus reducing the genetic variance within the species. Neanderthals, who appear to have died out in Europe as recently as 28,000 years ago, may have been one of the victims of the struggle of *Homo sapiens* for survival, though there are grounds to doubt that this was the case. There is also a suggestion that other hominids, including *Homo erectus* may have suffered the same fate as Neanderthals, though there is not much evidence to show what happened to these ancient humans. The murderous behaviour we have observed between some contemporary groups that are closely related genetically forces us to realize that *Homo sapiens* is not naturally caring about even its own species when territorial issues come between species members. In this respect our behaviour mimics that of many other species, though none indulge in murder to get their way to the extent that we humans do.

It is helpful to understand that when we talk about a common father for all humans who lived only 59,000 years ago, or a common mother who lived only 143,000 years ago, these common ancestors are distinctive only in that their Y chromosome (for males) or their mitochondrial DNA (for males and females) is ancestral to all of us. The rest of our DNA is drawn from the much more diverse population extant at the same time. After 59,000 years (and assuming four to five generations per century), we could have as many as three thousand great-great-great...etc. grandfathers (only one of whom is known to be a common ancestor to all humanity), and, after 143,000 years, as many as seven thousand great-great-great ...etc. grandmothers. Since marriage to cousins of various degrees is not uncommon, the actual number of such ancestors is almost certainly considerably smaller than the maximum, but it will still be a large number.

However, before we wax too eloquently on the murderous nature of humankind, it is worth recalling that, as happens so often with intractable scientific problems, there are other credible explanations for the very short span of time back to the ancestor shared by all humankind. Bill Bryson, in a brilliant chapter describing the volcanic and unpredictable nature of our own Earth,[182] notes that the last major "supervolcanic eruption" was in northern Sumatra 74,000 years ago. It is thought by some that this eruption may have been so severe that humankind was nearly rendered extinct. If this is true, then we are indeed all descended from a very limited gene pool.

Curtis Marean[183] believes that the population of humans on Earth shrank drastically to perhaps only a few thousand survivors during the "stage 6 glacial period" (or ice age) that happened from roughly 195,000 years ago until 123,000 years ago. He believes he has even found out where these survivors lived, namely in caves near the town of Mossal Bay on the Indian Ocean not far from Capetown, South Africa. He and his colleagues have found abundant evidence of continuous human habitation between about 164,000 and 35,000 years ago. He writes, "while elsewhere on the continent populations of H. Sapiens died out as cold and drought claimed the animals and plants they hunted, the lucky denizens of Pinnacle Point were feasting on the seafood and carbohydrate-rich plants that proliferated there despite the hostile climate".[184]

If Marean's hypothesis is correct, it would certainly explain why our earliest common ancestors are of such recent vintage.

In another, rather different approach to ancestry, Richard Dawkins[185] has mapped out the approximate number of generations back to, for example, our common ancestry with chimpanzees (5–7 million years); monkeys (about 40 million years); and grasshoppers (over 500 million years). Dawkins, like Darwin in his time, has an encyclopaedic knowledge of biology. His ancestry storyline allows him to discourse on a fascinating array of animal (and vegetable) traits and foibles, and leaves little room for doubt about our humble origins.

Comments on Machine and Human

Intelligence

Along with the mysteries of the origins of the universe and the origins of life, the structure and workings of the brain rank amongst the three great unsolved mysteries of our time. All three have yielded up many secrets in recent years. As I write these words, our understanding of the working of the brains of living beings is being unravelled with astonishing speed.

Davey was more interested in artificial or machine intelligence than I had expected and likely more interested in the subject than most readers of this book. For those who may be interested, however, this appendix contains most of the information I gave Davey on this subject.

Knowledge about the mammalian brain has advanced by two primary pathways, that is, by observation of actual human and animal behaviour and, at a microscopic level, by detailed observation of neuron behaviour. These two basic techniques are complementary. Together they have led to some fascinating discoveries.

Equipped with Darwin's theory of natural selection by survival of the fittest, and buttressed by our knowledge of the complex life forms surrounding us, we can hypothesize that somewhere along the way, once multicellular organisms had started to form, survival-enhancing sensors and means of locomotion started to develop. I don't suppose we know, and perhaps will never know, which sensors developed first – touch? temperature? presence of light? smell? – but develop they clearly did, and the need to process the information they provided and to co-ordinate what the sensors were detecting with survival-enhancing actions would have led to brain development. Very simple brains at first, but brains that, in many cases, continued to grow in complexity and sophistication as many many millions of years ratcheted by.

For me, one of the biggest surprises of neuroscience was to learn how similar our brains are to those of other mammals. I suppose in retrospect it is less surprising. However, the world I was brought up in thought of animals as being very different from, and inferior to, humankind. Yet we have already seen in Dialogue 11 that all or almost all mammals have about the same number of genes, and the majority of those genes are shared amongst species.

Of course, we now also know that looking at similarities of genes is only a small part of the story. It appears that human DNA has about four million special start codes that specify the start of RNA sequences that control gene expression. For any given cell type, about 95 per cent of these codes are inactivated, the remaining 5 per cent or so together specify the particular cell type to be replicated or maintained. Many of these different cell types are thought to be different types of neurons in the brain, hence the unusually large number of introns in the human genome.

Let us speculate a bit about how the mammal brain might have evolved.

To begin with, connections between sensors and muscles would have been rather direct and unsophisticated. For example, the detection of light might have led to a response to swim either toward or away from the light source. Later, as vision and other sensors became more sophisticated, some animals would have experienced an enhanced probability of survival through the evolution of an ability to co-ordinate data from several sensors and to nuance resultant behaviour appropriately. For example, detection of a large blob on the horizon, which could be either food or a predator, could be correlated with a sense of smell to distinguish between these two possibilities and result in the appropriate response. As the millions of years ticked by, one can imagine substantial survival rewards from the development of ever better sensors and ever more sophisticated processing of the information from the sensors, so that, for example, the ability to differentiate many closely related shades of skin colour could aid in determining the health of a potential mate, while the ability to correlate sensory data from the eye related to motion of prey or predator with sensory input from muscles can lead to a wide range of athletic capabilities.

This drive to evolve better sensors and ever more sophisticated processing of the sensory information is parallelled, as we have seen, by human scientific endeavours, where scientific breakthroughs have largely been either enabled or, in some cases, proven (or both) through the development of ever better sensors and ever smarter processing of the sensory data. To take one example, we saw in Dialogue 5 how the development of a new sensitive detector of the radioactive isotopes carbon-14 and tritium (hydrogen-3) by Willard F. Libby led to development of techniques for radioactive dating, which has enabled us to unravel vast domains of Earth's history.

An interesting place to start any investigation of the capabilities of the human brain is with the visual system, which takes up a very large proportion of our brains (a common estimate is about 40% of the neurons) and comprises many subsystems that perform vision-related tasks.

Some of our understanding of our visual system has been driven by our growing understanding of how difficult it is to create a good artificial vision system. Early on in the development of computers, experts started work on artificial vision and quickly learned that the job was much harder than at first anticipated. It turns out that humans bring a lot of "external" information to bear on image interpretation, information that can be difficult to bring to bear on machine-image interpretation. For example, someone walking down the street may unconsciously be asking the following questions:

Is there a car in this picture?

If so, is it moving?

If it is moving, how fast is it moving, and in what direction?

How far away is it?

Is it Charlie's car, i.e. a 1999 red Honda Civic with license number NWJ 095?

The answers to these questions require a lot of very sophisticated processing, as you can well imagine. For example, to detect whether there is a car in the picture, one first has to determine where all the edges of objects are, and then to determine whether any of the collection of edges, if combined in one of a number of possible ways will yield the shape of a three-dimensional image of a car when projected onto two dimensions. Performance of this task often requires the availability of a series of images over time, perhaps combined with inputs from more than one video camera (or eye) to provide clues about the third, hidden, dimension. Resulting potential images of cars need to be compared against previously stored car images of various sizes and orientations. Of course, the human brain goes at least one better, for it will automatically correlate the images with other clues. For example,

Does the car emit the smell of diesel oil?

Is there a sound of a muffler about to go?

Is Charlie away on holiday?

Does the person behind the wheel look like Charlie?

Is it likely that Charlie would be driving a car in this location and in this manner?

In other words, the brain has a command and control centre that processes the information from any relevant sensors and memory stores in order to come up with the best answer it can. Using a variety of techniques, it is possible to determine where in the brain the various subsystems required for visual tasks are located, and, if we are lucky, how they interact with each other. For example, pioneer vision researchers seeking to understand mammalian vision used tiny probes capable of sensing when individual neurons were active in the brain of a sedated cat. Using such techniques, they were able to locate segments of the brain devoted to edge detection for edges that were horizontal, vertical, and for several intermediate angles, with a different cluster of neurons devoted to edge detection at each angle. Others, through the use of proton emission tomography (PET) scans and functional magnetic resonance imaging, have been able to determine in real time what small segments of the brain are consuming oxygen – that is are in use – at a given time. When using PET equipment, it is possible to trace the movement of some other chemicals in the brain as well. Gradually, researchers are assembling a detailed picture of extraordinary complexity.

From time to time, some among us develop malfunctions to parts of our visual processing system, perhaps due to a blow to the head, a stroke, or a genetic defect. Such individuals, when interviewed and tested by skilled observers, provide us with yet another opportunity to learn how our brains function.

As a background to these discoveries, we need to keep firmly in mind that the brain evolved as a part of the evolution of the whole body, and that the criterion for adoption of a new mental capability was generally if it enhanced the chances of survival of the owner of that brain, and subsequently of his or her offspring. At a minimum, a new mental capability could not result in a significant net reduction of our chances of survival. The brain, like the rest of our body, is clearly survival-enhancing machinery. Though it may be capable at times of sorting truth from falsehood, reality from imagination, and beauty from ugliness, we must remain clear that those are not the prime functions of the brain. We are capable of performing such recognition only insofar as that capability-enhanced survival or was a side effect of brain development that occurred in response to other survival-enhancing situations. I think this is a crucial point. We become so accustomed, for example, to sorting out truth from falsehood and reality from imagination that it often escapes us just how fallible this sorting process can be.

Over the last two decades there have been a number of legal cases where people have been convinced that they have been assaulted by another person, only to discover that such assaults, however real and convincing they are in the mind of the accuser, are total mental fabrications. Our brains are capable of playing tricks on us all, especially when the trick is not survival threatening.

Notes on Happiness Measurements

and Results

Davey initially found it difficult to imagine that our feelings of happiness and pain were anything other than the results of someone programming our brains to "output" these sensations whenever something survival enhancing or survival threatening occurred. I was surprised and thought he should have known better, since I had gone to a lot of trouble to describe how we learned, gradually, that we had evolved over a long period of time, and how our behaviour and our emotions evolved along with our outward form.

However, once he was reminded of this, he was quick to agree that unlike himself, our sense of happiness had certainly evolved over many millions of years. That being the case, he explained, it is likely that humans have thousands of inherited behaviours and that, as with eye colour and the shape of our nose, these behaviours are likely to be different for different people. In such circumstances, what makes us happy is also likely to be highly subjective, so how can we possibly get any useful information just by asking a lot of people how happy they are on a scale of one to ten?

It was a good question, and he was certainly not the first to ask it.

After some hard work on my part, I used two main arguments to convince him of the legitimacy of happiness research. The first was to argue by analogy, the second was to demonstrate that in practice the happiness data turned out to be consistent across many different countries and cultures. The idea of cultural relativity, so popular amongst anthropologists in the 1920s and 1930s (and discussed in Dialogue 10), actually turns out to be incorrect, at least in the big picture. On average, the same things, such as having a good job, a trustworthy government, friends we can trust, and adequate food and shelter, make people everywhere happier.

Not every reader is interested in the same amount of detail on this subject as Davey was. I have thus omitted our discussions on the subject in the main text of the book, but they are summarized below.

In 1954 a British physiologist named Richard Doll published a study of over forty thousand physicians in the UK that showed unequivocally that smoking was an important cause of lung cancer. Doll was certainly not the first to suspect that smoking and lung cancer were connected, but everyone knew smokers who did not get lung cancer and vice versa. The problem was to provide convincing proof that the relationship existed. What Doll demonstrated was that smokers had a much greater likelihood of suffering from lung cancer than those who did not smoke. It turns out that in statistical terms the effect of smoking on lung cancer is very large, so that the connection is relatively easy to establish beyond any reasonable doubt, even though there are lots of smokers who never suffer from lung cancer.

The field of medicine whose practitioners comb through extensive medical records and surveys to assess the statistical probability that any given chemical or environmental phenomenon is contributing to the occurrence of a particular malady is epidemiology. It is a very active field. As a result of these studies, there is now an extensive literature on a host of different substances that contribute to a variety of cancers and other maladies. Thousands of lives have been prolonged as a result.

Those who study happiness have used very similar techniques to assess the factors contributing, on average, to an individual's happiness or unhappiness. If the techniques used to determine causes of cancer work, why, by analogy, would those techniques not work to determine the causes of happiness and unhappiness? Indeed it would not be unreasonable to refer to happiness studies as "the epidemiology of happiness".

Davey was dubious.

So, I developed a more rigorous argument to show that in practice the happiness data collected were able to explain otherwise puzzling differences in happiness data between different nations and groups.

My argument went something like this:

When I was a doctoral student in the field of experimental nuclear physics at Oxford in the early 1960s, my thesis work consisted of observing what happened when a thin carbon target was battered by a beam of protons at high energy. The experiment was an example of what is sometimes called a black-box experiment, where the problem is to gain an understanding of what is going on in a "black box" (in this case, the carbon nucleus) when you cannot look in and observe the contents directly. Instead, you send in a variety of probes and observe what comes out. In my particular experiment, we had built some new particle detectors capable of measuring protons scattered off the target at very large angles as a result of a collision with a carbon nucleus. Our objective was to determine which of the several competing models of the properties of the carbon nucleus was best able to predict the experimental results. Our experiment, in common with others of a similar nature, relied heavily on obtaining many observations of protons scattered off the carbon target. In this way we could determine what proportion of the incoming protons were scattered by the carbon nucleus at any given angle and hence derive some conclusions about the range and nature of the interaction of the incoming protons with carbon nuclei. (Those protons that failed to come within the reaction range of any nuclei in the target would simply pass straight through the target.)

Analogously, one can consider the complex tangle of our inherited behaviour patterns or instincts as a black box we cannot see inside. We see ourselves and other animals acting "instinctively", but we cannot yet untangle the web of neuronal connections in our brains that cause the activity. We have hypothesized that the same old instincts that served to motivate our hunter-gatherer ancestors are still at work in us and that they provide us with a sense of well-being when we are "obeying" our instincts. In a general sense, we think we know what made those ancestors of ours happy – activities such as finding and providing shelter, hunting successfully for food, procreating and raising children. The enormous benefits offered by participating in a modern society are clear, but how do the much modified (i.e. civilized) activities we partake in on a daily basis (the inputs) interact with the (black box of) neurons that generate well-being (the output) to make us more, or sometimes, less happy?

Of course, the whole process is complicated by the fact that instinct is only one set of influences guiding our actions. We are also influenced by learned behaviour and by internal contemplation of what we perceive as the world around us. (Some would add that we are also influenced by supernatural phenomena, though I personally don't believe this to be the case.) Since we are concerned here to discover which of the large variety of circumstances and activities are happiness enhancing and which are not, we do not need to concern ourselves with these other influences on our actions, since by hypothesis, it is only the satisfaction of innate behaviour patterns that generates in us a sense of well-being.

One way to find out which circumstances and activities enhance happiness is to hurl a large number of people, all with different characteristics (which we might aptly term "modern society"), into the experiment and then examine their output – in this case, to ask them to rate their life satisfaction on a scale of one to ten (for example). Using standard statistical techniques it is then possible to untangle the very different situations of each person and to correlate their sense of well-being with their particular set of circumstances.

Using this approach, researchers are now able to provide answers to such questions as:

> Does money buy happiness?
> Is quality of government important to life satisfaction?
> How important is it that in general, you believe other people can be trusted?
> What about job quality?
> How important is unemployment? Good health?
> Does practicing religion make people happier?
> What about marital status, and closeness of family, and other social ties?
> We know that our health generally declines with age. What about happiness?
> In general, are women happier than men?
> Does education affect happiness?

I prepared a table for Davey that set out, in descending order of importance, some of the more important influences on human happiness. The table included data relating to individuals within a country (in this case Canada), as well as some average values from selected countries. This data is set out in Figure 20.2 in Dialogue 20, and is reproduced in slightly different form in Table A5.1. It is intended to provide the reader with a feel for the relative strength of various causes of happiness or its inverse. The figures in the table should be read as approximate as they vary somewhat over time. Individual happiness numbers are drawn from a variety of Canadian studies, while the international data (shown in bold type) are drawn from four World Values Surveys taken between 1981 and 2001 in over one hundred countries.

A glance down the list in Table A5.1 provides preliminary guidance to the answers to some of the questions listed above.

In order to be able to answer such questions, we need to know the key characteristics of each particle (or person). This includes such parameters as age, income, nationality, marital status, job situation, health, degree of participation in social activities, gender, etc. We can then observe the output of this combination in the form of that person's subjective assessment of her or his well-being. We might also consider aggregating all responses from each country (for example). In this case, the inputs to the black box would likely be such factors as the age gradient of the population, the quality of government (for which there are a variety of measures available), levels of trust in society, gross domestic product, and income disparity. For output, we might select the average of subjective

Table A5.1 - Factors Affecting Happiness (Subjective Well-being) in Canada (in descending order)

(based on scale of 1 to 10, Canadian average - 7.8, a positive number indicates increased happiness, a negative number the opposite.)

No.	Description	Approx. SWB Change
1	**Live in former Soviet Union [versus living in Canada]***	**-2.90**
2	Very unhappy in job [versus very happy in job]	-1.70
3	Serious illness or poor health [versus excellent health]	-1.30
4	High trust in own family [versus no trust in family]	+1.27
5	High trust in colleagues [versus no trust in colleagues]	+1.07
6	Unemployed [versus being employed]	-0.85
7	**Live in France [versus living in Canada]***	**-0.80**
8	High trust in neighbours [versus no trust in neighbours]	+0.70
9	Religion is very important [versus religion unimportant]	+0.51
10	Frequent visits with friends [versus no visits]	+0.51
11	Income >$120,000 [versus income <$25,000]	+0.50
12	Age 45-54 [versus age <25]	-0.47
13	Income $90,000 to $119,000 [versus income <$25,000]	+0.45
14	Married [versus Single]	+0.41
15	Age 35-44 [versus age <25]	-0.40
16	Income $60,000 to $89,000 [versus income <$25,000]	+0.30
17	Age 25-34 [versus age <25]	-0.30
18	Frequent visits with family members [versus no visits]	+0.28
19	Lived mainly with neither parent up to age 15 [versus lived with both parents]	-0.25
20	Divorced [versus single]	-0.24
21	Income $35,000 to $59,000 [versus income <$25,000]	+0.20
22	**Live in Scandinavia [versus living in Canada]***	**+0.20**
23	**Live in the USA [versus living in Canada]***	**-0.20**
24	Lived mainly with father up to age 15 [versus lived with both parents]	-0.20
25	Age 55-64 [versus age <25]	-0.15
26	Frequent visits with neighbours [versus no visits]	+0.14
27	Lived mainly with mother up to age 15 [versus lived with both parents]	-0.125
28	Age 65+ [versus age <25]	+0.10
29	Income $25,000 to $34,000 [versus income <$25,000]	+0.10
30	Widowed [versus Single]	-0.07
31	Female [versus male]	0-1.2

*Line Items in Bold typeface and with an asterisk are unconditional averages drawn from World Value Survey data, while the rest of the line items are partial regressions (with other variables held constant). For example, the fact that subjective well-being is 2.9 points lower in the former Soviet Union than in Canada is primarily due to lower average values of job satisfaction, trust, health and, to a lesser extent, other factors. The unbolded lines show average differences in subjective well-being in Canada when, for example, job satisfaction is lower, while other variables such as age, trust, income etc. are held constant. This table (as with Figure 20.2) was assembled by the author primarily from a discussion paper presented by John Helliwell at Simon Fraser University on 29 June, 2004.

well-being as reported by individual residents of the country in question, or perhaps just the average happiness of the top (or bottom) 20 per cent of income earners. In the first case, where we examine the responses of individuals within a prescribed region, we might hope (with careful statistical treatment) to find some useful guidance as to what most affects the sense of well-being of the average person. In the second case, we might hope to understand what national practices or behaviours are most conducive to a sense of well-being amongst the citizenry.

Of course, it could be the case that the particularity of national cultures and the natural variability of individuals will only lead to meaningless results from such an exercise. In other words, it could be that national cultural differences lead to a different set of inputs, which optimize a person's happiness from one country to the next. For example, the inhabitants of country A might be happiest at a different age from those of country B (all other inputs being equal), simply because the culture is different, or trust in fellow workers may, on average, be more important to the workers in country A than those in country B.

If, in the 1990s, when the happiness data first became readily available on a large scale, you had asked the experts why, for example, the people of Denmark appeared to be on average the happiest people in the world (with a score of about 8.2 on a scale of 1 to 10), while those in the former Soviet Union were very considerably less happy (with a score of about 4.2 on the same scale), you would have received a variety of very subjective answers. Many would have posited national wealth and a democratic form of government as the key drivers of subjective well-being. Others would add in job satisfaction, quality of health care, and a good education system. Anomalies in the data that could not be explained by the simple theories would have been equally subjectively explained by such nostra as: "France has an unstable political system" or "America has racial problems". And let us be clear. There are oceans of data on such variables as job satisfaction, religious adherence, membership in social groups, unemployment, income, wealth distribution, birth rates, divorce rates, cancer rates, family size, government effectiveness, security – many, or all of which, could reasonably impact our sense of well-being.

It is thus critically important to know if: 1) Measurements of subjective well-being taken in many different cultures and at different times are consistent enough to be meaningful; 2) It is possible to bring enough order to all this data to be able to determine which factors are most important to happiness in today's society; and perhaps most important, 3) The data is reliable enough to guide us in how we can change both our personal behaviour and, more generally, our social structures so as to increase levels of well-being.

As already mentioned, most of those thinking about such matters in the last few centuries had little doubt in their own minds that different races and cultures were so different that studies of what gave people a sense of well-being were unlikely to be meaningful when examined across different cultures.

By now you may know or have guessed that studies of subjective well-being are turning out to be very rewarding and interesting, though there is much yet to be learned. Helliwell[186] for example, describes a number of ingenious tests that give us confidence that subjective well-being responses from virtually all countries that have participated in such studies show consistent patterns. It turns out that Ruth Benedict's ultimate instincts (as discussed in Dialogue 10) were right. Some societies *are* better than others at giving their residents a high score on the scale of subjective well-being, and one can pin down some of the reasons why this is so. Furthermore, the Benedict high social synergy societies – those structured in order that when people do what makes them happy they are, at the same time, contributing to the social good – correlate very well with societies high in social capital. Such societies also tend to have high average happiness scores.

However, the happiness studies have far surpassed what Benedict and her colleagues could accomplish with the data to which they had access. We now have available valuable guidance as to what priorities we should adopt both as individuals and as societies and governments, to increase happiness. More data is becoming available all the time, as this is a very active field of research.

The importance of a good job and of trust is discussed in Dialogue 20. Some other key factors that influence human happiness are discussed below.

By far the most important effect on average national happiness turns out to be the quality of government (as measured using a set of six variables relating to the honesty and efficiency of government, as well as political accountability and stability). In poorer countries for example, a move up from the bottom 2 per cent of countries in terms of government quality to the middle third of countries is roughly equivalent in terms of life satisfaction to a person in that country moving from receiving the average income of the bottom 20 per cent of inhabitants up to the average income for the country as a whole. For wealthy countries, quality of government assumes even greater relative importance.[187]

The impact of good government on the life satisfaction of its citizens can be traced through a number of pathways. While higher income in countries with better government ratings is one important path, others include better levels of health, higher employment levels, greater levels of trust, generally, and more particularly trust in government officials and the police force.

> . . . *there appears to be a hierarchy of preferences for different aspects of government, with the ability of governments to provide a trustworthy environment, and to deliver services honestly and efficiently, being of*

paramount importance for countries with worse governance and lower incomes. The balance changes once acceptable levels of efficiency, trust and incomes are established, with more attention paid to building and maintaining voter engagement.[188]

Unsurprisingly, good health ranks high as an important contributor to subjective well-being, too. The range in average happiness between those enjoying very good health and those with severe health problems comes to 1.3 happiness points. Surprisingly for some of us, this range is less than the range of 1.6 points between those who love their job and those who hate it. Epidemiological research shows a very strong correlation between good health, income, and job status. As initially discovered by Michael Marmot and his associates in a classic study of British civil servants, those with higher status jobs on average enjoy better health and live longer. The general finding has since been confirmed in a variety of other studies. It is thought that in large measure the improved health and longevity derives from the greater freedom of action (i.e. less detailed supervision) enjoyed by those in senior positions. There has long been a sense that those in senior positions suffer from much greater stress than their underlings. While this may or may not be true, we know for certain that for the most part, despite the stress, they enjoy better health, on average, than their subordinates, and are happier overall!

The important role of religion to a sense of subjective well-being (+0.51 points) is one of the more fascinating results arising from happiness research. Those who believe in a deity are clearly, on average, significantly happier than atheists. Signs of religious activity have been found in the graves of *Homo sapiens* dating back as much as a hundred thousand years and in the graves of our predecessor *Homo erectus* dating back a good deal further. Many religiously inclined humans take considerable comfort from their belief that the afflictions in their present life will be recompensed in an afterlife.

Perhaps equally as surprising as the increased average happiness that those with strong religious convictions have over their atheist friends is the very strong negative correlation between belief in God and trust in others. It appears that while those with a strong religious affiliation may be very trusting of those sharing their beliefs, they are much less trusting of their fellow man in general. Strong religious convictions often accompany strong bonding social capital (within the religious group) but weak bridging social capital. This can have serious consequences for a society where, as Putnam has so clearly demonstrated, mechanisms for building bridging social capital are so important to happiness.[189] It is to be hoped that religious leaders will take note of this problem and look for ways that religious dogma and religious activities more generally can be structured to encourage bridging social capital building.

The effect of age on happiness is also interesting. Happiness clearly declines with age until fairly late in life, when it starts to climb back up. Those in the forty-five to fifty-five age group reach a nadir of unhappiness at about 0.4 points below those under twenty-five. By age sixty-five, people are about as happy as when they were when under 25.

It will be interesting to understand a lot more about these effects. For example, what might we as individuals, and what might governments acting on our behalf, do to alleviate the age-related decline in happiness amongst the middle-aged? Are there important factors here that we are not measuring? Could it be the stress of raising a family? Certainly there is, on average, more stress from the workplace during the midlife years, but work-related stress should have been factored out of the results. There are some recent statistics that suggest that children in elementary school are generally much happier than those in high school. Perhaps these age-related changes are natural and simply reflect the natural increases in stress as we advance through adolescence and young adulthood. On the other hand, it may be that we are unwittingly subjecting this segment of the population to more stress than we should or could be.

It would not be unreasonable to expect that women might, on average, be happier than men. After all, a lower percentage of women are active in the traditionally defined job market, however hard they may work as homemakers, and jobs, as previously noted, can be significant sources of unhappiness. Au contraire, some have reasoned; by being absent from the job market more women are likely to be bored and unhappy, so one might expect them to unhappier. Others might argue that in countries where women have less freedom than men, women should be unhappier. Data analysed so far suggest that in some countries women are indeed slightly happier (~0.12), while in others there is no significant difference in male and female average happiness. There is not yet a satisfactory explanation for the small differences in happiness (where they exist).

Glossary of Bioscientific Terms

I prepared a glossary of terms while getting ready to brief Davey on advances in the biosciences. The glossary was as much for me as for Davey, since I found it easy to forget the precise meaning of some of the more common bioscience jargon. The definitions came from a variety of sources, but the majority came from Wikipedia. At Davey's request, I read these definitions out to him before we had started into our bioscience dialogues. The principal observable difference between Davey and me, insofar as the definitions are concerned, is that whereas Davey heard them once and was then able to recall them without so much as a word out of place, I had to refer several times to some definitions.

Allele: [Pronounced al-eels] Any one of the alternative forms of a given gene. For example, the three commonly encountered blood types, A, B, and O, derive from different alleles of the same gene.

Amino Acids: Organic compounds that generally contain an amino (NH_2) and a carboxyl (COOH) group. A select twenty of the amino acids are the subunits that all known life forms use as the building blocks of proteins. Another way to phrase this is that proteins are polymers comprised of many monomers, all of which are amino acids.

Archaea: A major division (domain) of living organisms. Archaea are single-celled organisms lacking nuclei and in this respect, resemble Bacteria, which are sometimes classed as prokaryotes. The other two major domains of living organisms are Bacteria (sometimes also called Eubacteria) and Eukarya, or eukaryotes (see definitions in this glossary).

Bacteria: See under **Eubacteria**.

Base: Any member of a class of compounds whose aqueous solutions yield hydroxyl (OH) ions and that can donate a pair of electrons to form a covalent bond. Bases react with acids to form salts.

Base Pair: Two bases that form a "rung of the DNA ladder". A DNA nucleotide is made of a molecule of sugar, a molecule of phosphoric acid, and a molecule called a base. The bases are the "letters" that spell out the genetic code. In DNA the code letters are A, T, G, and C, which stand for the chemicals adenine, thymine, guanine, and cytosine, respectively. In base pairing, adenine always pairs with thymine, and guanine always pairs with cytosine.

Carbohydrates: Chemical compounds that contain oxygen, hydrogen and carbon atoms. They may also contain other elements, such as sulfur or nitrogen, but these are usually minor components. Carbohydrates consist of monosacharrides of varying chain lengths, which have the general chemical formula $C_n(H_2O)_n$ or are derivatives of such. The smallest value for "n" is 3. A 3-carbon sugar is referred to as a triose, whereas a 6-carbon sugar is called a hexose. Certain carbohydrates are important for storing and transporting energy in most organisms – including plants and animals – and are major structural elements in many organisms (e.g. cellulose in plants). In addition, they play major roles in cell-to-cell communication, the immune system, fertilization, pathogenesis, blood clotting, and development.

Cell: (Biological definition) The smallest metabolically functional unit of life. Viruses are smaller but are incapable of self-replication.

Chirality: In general, the property of not being superposable on its mirror image, as in, for example, human hands, where the left and right hands are mirror images of each other. In organic chemistry the property refers to molecules with both a "right-handed" and a "left-handed" version of chemically identical substances. See for example the diagrams of the ribose molecule in the dialogue on emergence (page 198). All life forms on Earth make use of left-handed amino acids (e.g. L-dopa) and right-handed sugars (e.g. dextrose).

Chloroplast: An organelle (small organ) responsible for photosynthesis in plants and green algae cells.

Chromatin: Material of which chromasomes are composed; made of nucleic acids (DNA) and protein.

Chromosome: An intranuclear organelle made of chromatin; visible during cell division; contain most of a cell's genetic material. Chromosomes got their name because they readily absorb coloured dye and are thus easy to inspect under a microscope.

Codon: A series of three adjacent nucleotides that form part of a DNA or

RNA chain and that together code for one of twenty amino acids.

Cyanobacteria: A phylum (or division) of Bacteria that obtain their energy through photosynthesis. They are often referred to as blue-green algae. The description is primarily used to reflect their appearance and ecological role rather than their evolutionary lineage. Fossil traces of cyanobacteria have been found from about 3.8 billion years ago. Cyanobacteria are one of the largest and most important groups of Bacteria on Earth.

Enzyme: A protein that speeds up chemical reactions in the body. For example, an enzyme in the saliva of the mouth helps to break down food.

Epigenetics: the study of heritable changes in gene expression in circumstances where there is no change to the underlying gene sequence. The changes are heritable in the sense that succeeding cells produced in the same multicellular organism will inherit the gene expression pattern, but that pattern will not be expressed in succeeding generations of the organism unless the change has been effected on the DNA transmitted to the offspring in the zygote or fertilized egg cell.

Eubacteria: The most abundant of all organisms. One of the three principal domains of all life forms, the others being Archaea and Eukarya. Sometimes referred to as simply Bacteria. Like Archaea, Eubacteria have prokaryotic cells and are normally found as single-celled organisms.

Eukaryote: A member of the domain Eukarya. An organism with a complex cell or cells, in which the genetic material is organized into a membrane-bound nucleus or nuclei. Eukaryotes comprise animals, plants, and fungi—which are mostly multicellular—as well as various other groups that are collectively classified as Protists (many of which are unicellular). (See Figure 13.2, page 172) In contrast, prokaryotes are organisms, such as Bacteria, that lack nuclei and other complex cell structures.

Exon: Any region of DNA within a gene that is transcribed to the final messenger RNA (*m*RNA) molecule, rather than being spliced out from the transcribed RNA molecule.

Gene: The unit of heredity in living organisms. Genes are encoded in an organism's genome, composed of DNA or RNA, and direct the physical development and behaviour of the organism. Most genes encode proteins, which are biological macromolecules comprising linear chains of amino acids that effect most of the chemical reactions carried out by the cell. Some genes do not encode proteins but produce RNA molecules that play key roles in protein biosynthesis and gene regulation. Molecules that result from gene expression, whether RNA or protein, are collectively known as gene products.

Intron: Sections of DNA that will be spliced out after transcription, but before the RNA is used. The regions of a gene that remain in spliced *mRNA* are called exons. The number and length of introns varies widely among species and among genes within the same species. For example, the pufferfish (*Takifugu rubripes*) has little intronic DNA. Genes in mammals and flowering plants, on the other hand, often have numerous introns, which can be much longer than the nearby exons.

Meiosis: The process that allows one diploid cell (containing two copies of each chromosome) to divide in a special way to generate haploid cells (i.e. cells containing only one copy of each chromosome). The word "meiosis" comes from the Greek *meioun*, meaning "to make smaller", since it results in a reduction in chromosome number. Meiosis is essential for sexual reproduction, specifically for the production of sperm and egg cells. It therefore occurs in most eukaryotes, including single-celled organisms. A few eukaryotes, notably the Bdelloid rotifers, have lost the ability to carry out meiosis and have acquired the ability to reproduce without fertilization by a male. Meiosis does not occur in Archaea or prokaryotes, which reproduce by asexual cell division.

Methylation: (Of DNA) One of several cellular mechanisms that controls gene expression. The attachment of a methyl (CH_3) radical at certain locations on the DNA molecule has the effect of inactivating the gene in that cell and in all its subsequent daughter cells. In some instances the methylated status of the gene can be passed on to the next generation of the animal, although normally the stem cells used in species reproduction have all methyl radicals removed.

Mitochondrion (plural Mitochondria): Organelle in eukaryotic cells responsible for producing ATP, the principle carrier of chemical energy in the cell.

Mitosis: The process by which a cell separates its duplicated genome into two identical halves. It is generally followed immediately by cytokinesis, a process that divides the remainder of the cell. The end result is two identical daughter cells with a roughly equal distribution of organelles and other cellular components. Mitosis and cytokinesis together are defined as the mitotic (M) phase of the cell cycle, the division of the mother cell into two daughter cells, each the genetic equivalent of the parent cell.
Mitosis occurs exclusively in eukaryotic cells. In multicellular organisms, the normal body cells (such as neurons, blood cells, etc.) undergo mitosis,

while germ cells – cells destined to become sperm in males or ova in females – divide by a related process called meiosis (see above). Prokaryotic cells, which lack a nucleus, divide by a process called binary fission.

Nucleotide: Single unit of nucleic acid; composed of an organic nitrogenous base that is one of adenine, cytosine, guanine, thymine, or (in RNA) uracil, combined with deoxyribose or ribose sugar, and phosphate.

Nucleic acid: Comes in two forms: deoxyribonucleic acid (DNA) and ribonucleic acid (RNA). They are very large linear molecules containing carbon, hydrogen, oxygen, phosphorus, and nitrogen, which on partial hydrolysis (reaction with water) yield nucleotides, and which are composed of one RNA or two DNA polynucleotide chains. They are essential components of living cells, where they are the carriers of genetic information (DNA and *m*RNA), components of ribosomes (*r*RNA), and involved in deciphering the genetic code (*t*RNA). See also: DNA, messenger RNA, nucleotide, ribosomal RNA, RNA, transfer RNA.

Organelle: A differentiated structure within a cell, such as a mitochondrion, vacuole, or chloroplast, that performs a specific function. The word organelle is a diminutive form of organ, the idea being that organelles function in a cell in an analogous way to organs (such as kidneys and lungs) in the body.

Organic compounds: One might have thought that organic compounds were compounds associated with cells, in differentiation from inorganic compounds. Alas it is not so easy! Organic compounds all contain carbon, but some carbon compounds, by tradition, are considered to be inorganic, including simple oxides of carbon (e.g carbon dioxide, carbides, carbonates, and cyanides, as well as diamonds and graphite).

Phosphorylation: The addition of a phosphate (PO_4^{3-}) group to a protein or other organic molecule. Phosphorylation activates or deactivates many protein enzymes. Protein phosphorylation, in particular, plays a significant role in a wide range of cellular processes.

Prokaryote: One of two basic cell structures (the other being eukaryote). Prokaryotic cells have no cell nucleus and are smaller than eukaryotic cells. (See Figure 13.1, page 171.)

Protein: Protein molecules are the cell's basic building blocks, and they execute nearly all the cell's functions. They are the basis of body structures, such as skin and hair, and of substances like enzymes and antibodies. Proteins are comprised of (usually long) stable combinations of the twenty different amino acids, which are coded in DNA and RNA.

Ribonucleic Acid (RNA): A nucleic acid polymer essential to cell operation. It is transcribed from DNA by enzymes called RNA polymerases and further processed by other enzymes. RNA serves as the template for translation of genes into proteins, transferring amino acids to the ribosome to form proteins. There are many variant forms of RNA. The principal variants are:

> **Messenger RNA (*m*RNA):** RNA that carries information from DNA to the ribosome sites for protein synthesis in the cell. Once *m*RNA has been transcribed from DNA, it is exported from the nucleus into the cytoplasm (i.e. part of the cell that is not the nucleus) (in eukaryotes *m*RNA is "processed" to remove unwanted introns before being exported), where it is bound to ribosomes and translated into protein. After a certain amount of time the *m*RNA degrades into its component nucleotides, usually with the assistance of RNA polymerases.

> **Transfer RNA (*t*RNA):** A small RNA chain of about seventy-four to ninety-three nucleotides that transfers a specific amino acid to a growing polypeptide chain at the ribosomal site of protein synthesis during translation. There are at least twenty different *t*RNAs – one for each amino acid.

> **Ribosomal RNA (*r*RNA):** A component of the ribosome, the factory for synthesizing proteins in the cell. *r*RNA molecules are extremely abundant. They make up at least 80 per cent of the RNA molecules found in a typical eukaryotic cell.

RNA genes: (sometimes referred to as non-coding RNA or small RNA) are genes that encode RNA that is not translated into a protein. The most prominent examples of RNA genes are transfer RNA (*t*RNA) and ribosomal RNA (*r*RNA), both of which are involved in the process of translation. However, since the late 1990s, many new RNA genes have been found, and thus RNA genes may play a much more significant role than previously thought.

Topoisomerase: Substances that have the same chemical composition but different physical shapes are called isomers. Isomerases are enzymes that facilitate a transformation from one isomer to another. A topoisomerase mediates such changes on the DNA strand, either to inhibit or encourage transcription by winding or unwinding the DNA.

ENDNOTES

Please note that the following endnotes employ a short-form style for source citations. All sources short-form noted below also appear in the bibliography.

Dialogue 2

[1] Ruth Benedict, 1959, 131.

[2] Richard Dawkins, 2008, 269–316.

Dialogue 3

[3] Margaret Mead, 2.

[4] *Hamlet*, Act 5, Scene 2.

[5] The full text of Polonius's advice, from *Hamlet*, Act 1, Scene 3:

> *…Give thy thoughts no tongue,*
> *Nor any unproportioned tongue his act.*
> *Be thou familiar, but by no means vulgar;*
> *The friends thou hast, and their adoption tried,*
> *Grapple them to thy soul with hoops of steel;*
> *But do not dull thy palm with entertainment*
> *Of each new-hatch'd, unfledg'd comrade. Beware*
> *Of entrance to a quarrel, but, being in,*
> *Bear't that th' opposed may beware of thee.*
> *Give every man thine ear, but few thy voice;*
> *Take each man's censure, but reserve thy judgement.*
> *Costly thy habit as thy purse can buy,*
> *But not express'd in fancy; rich, not gaudy;*
> *For the apparel oft proclaims the man,*
> *And they in France of the best rank and station*
> *Are most select and generous, chief in that.*
> *Neither a borrower, nor a lender be;*
> *For loan oft loses both itself and friend,*
> *And borrowing dulls the edge of husbandry.*
> *This above all: to thine own self be true,*
> *And it must follow, as the night the day,*
> *Thou canst not then be false to any man.*

Dialogue 4

[6] When I was at school, we were told that Christopher Columbus and his crew, when they set sail in 1492 for what turned out to be the "West Indies" were worried about reaching the edge of Earth and falling over. This story is not now believed to be accurate, though there were doubtless some who were either ignorant of Aristotle's writings or simply doubted them, and who thus were afraid of the consequences of reaching the edge of Earth and disappearing over that edge into nothingness.

[7] Private correspondence with David Gallop.

[8] *Delta cephei* varies in brightness between magnitude 3.48 and 4.37 with a period of 5 days 8 hours 37.5 minutes between its minimum and maximum. The North Star (*Polaris*) is also a cepheid, with a period of 3.97 days and a small range of brightness, between magnitudes 2.08 and 2.17. The oscillation in brightness is due to the presence of a compression wave that travels from the centre of the star to its surface and back, hence the oscillation period is a function of the speed of the compression wave, which in turn depends on the average density of the star, and on the star's size. The structure of all cepheids is similar. Larger cepheids are more luminous and denser. Cepheids of the same size will have the same brightness, and hence the same period of oscillation in their brightness. From observations of nearby cepheids, whose distance from us can be measured using geometry, it is possible to learn the relationship between star size and the period of their brightness oscillations. The apparent brightness as measured on Earth will be reduced as a function of the square of the distance the star is from Earth. Thus by knowing the difference between the real brightness of a cepheid (as measured by its period of oscillation) and its apparent brightness (as measured on Earth), it is possible to calculate how far away from us the star is located. As it turns out, there is a second type of cepheid-like variation, but it is possible to differentiate between the two types, since the composition of the stars, as shown in the spectra of their emitted light, are different. This second type of cepheid (labelled *W Virginis*) is quite rare.

[9] Robert Smith, 1990.

[10] Arthur Eddington, 1932, 85.

[11] The story as it concerns Friedmann and Lemaître is recounted in Neil Turok, *The Universe Within*, 2012, 121-125.

Dialogue 5

[12] Monarch butterflies appear to be a special case. Those that hatch at a time that does not require them to migrate usually live between two and five weeks. Those that have to migrate live up to nine months. Monarchs born in eastern Canada and northeastern USA in the autumn will migrate to Mexico, and then, in the spring, return partway to lay eggs, likely in southeastern USA. The next generation will complete the spring trek north. When you think about it, this is a staggering demonstration of the impact of genes on behaviour, there being no parent butterflies about to nurture them and to teach them where to go.

[13] Of course, this number depends on one's definition of "humankind." Our ancestors are thought to have diverged from the apes, our closest common ancestors some time between 40 and 5 million years ago. At the time of writing, the oldest identified hominid (or manlike) species is *Sahelanthropus tchadensis*, thought to be 6 to 7 million years old and to have had a brain size of about 350 cubic centimetres (cc). *Homo habilis* (2.4–1.5 million years old) is thought to have been the first hominid capable of speech (by virtue of the appearance of a bulge found in one *H. habilis* skull where our brains lodge our speech capability, i.e. in Broca's area. *Homo erectus* (1.8 million to 300,000 years old) probably used fire and had more sophisticated tools than *Homo habilis*. Its brain size ranged from 750 to 1225 cc, and its remains have been found in Africa, Asia and Europe. Neanderthals (*Homo sapiens neanderthalis*) (230,000–30,000 years old) were the first hominid known to bury its dead. The Neanderthals were bigger and bulkier than us and had slightly larger brains (about 1450 cc). *Homo sapiens sapiens* (from 195,000 years ago to the present day) has an average brain size of about 1350 cc, a size believed to be limited by our ability to survive birth with a large head. *Homo sapiens sapiens* is but one of some twenty-one different varieties of humans listed by the sources I consulted, though not all are generally agreed to be distinct. (See: Dawkins, 2005; Bryson, 2003; and http://www.talkorigins.org/faqs/homs/species.html)

[14] Stephen J. Gould, 1995.

[15] See for example Bill Bryson, 2003, 280.

[16] Gould, 1995, 179.

[17] Gould, 1995.

[18] Frank Libby, 1960.

Dialogue 6

[19] Edward O. Wilson, 1975, 421 *et seq.*

[20] Ibid, 426.

[21] A helpful and simple explanation of the General Theory of Relativity can be found in Brian Greene's *The Elegant Universe*, 1999, 56–76. Greene also does a nice job of explaining Special Relativity without the use of equations on pages 23–52.

[22] At least a year after writing these words, I attended a lecture at the Vancouver Institute (April 16, 2005), at which the speaker, Dr. Barry Barish, a distinguished American physicist, went even further, pointing out that there was no pressing inconsistency in current knowledge that drove Einstein onward to explore the gravitational force. Nonetheless, despite a widespread lack of interest in the topic by the scientific community, he worked for eleven years on his Theory of General Relativity, and it turned out to be a magnificent use of his time, but one that most researchers would have rejected before even beginning! Einstein made three critical contributions to physics, all published in the space of a few weeks in 1905. Special relativity was one of the three. The other two were the photoelectric effect and Brownian motion. As a result of these amazing accomplishments, the year 1905 came to be called the *annus mirabilis* (miracle year or strictly speaking, year of miracles) for physics.

However there is a precedent for this! In 1665–1666, Isaac Newton made revolutionary inventions and discoveries in calculus, motion, optics and gravitation, so that period too was called *annus mirabilis.*

[23] A simple layman's explanation of the Special Theory of Relativity can be found in *The Feynman Lectures on Physics* (1963), Chapter 15. In Chapter 7 (7–9 to 7–11) there is a nice discussion of early attempts to ascertain what causes the forces of gravity and electromagnetism

[24] Greene, 1999.

[25] Ibid, 186–87.

[26] Michael Riordan *et al*, "The Higgs at Last", Scientific American, October, 2012, 73.

[27] Neil Turok, *The Universe Within*, 2012, 149-152.

[28] An interesting and readable article on this subject has been written by David Albert and Rivka Galchen, "A Quantum Threat to Special Relativity," 2009.

[29] Greene, 1999, 386–87.

[30] Garrett A. Lisi and James Owen Weatherall, 2010.

Dialogue 7

[31] Denys Wilkinson, 1991, 169.

[32] Ibid, 197.

Dialogue 8

[33] Alan Moorehead, *Darwin and the Beagle*, 1969, 25.

[34] The easiest way to determine longitude is to know the Greenwich Mean Time (GMT) at which the sun is directly overhead. If, for example, the local "noon" occurs at 8:00 p.m. GMT, then you must be 8/24 or 1/3 of the way around the world from the 0th meridian of longitude, which passes through Greenwich, therefore, the longitude will be 1/3 of 360°, or 120°. The catch was that until 1735 there was not a chronometer able to keep time accurately enough over the months and years of a sea voyage to enable an accurate determination of GMT, and hence of longitude. This drove the Royal Navy, then the world's pre-eminent navy, to offer a reward of £20,000 to the first person to develop a chronometer that was sufficiently accurate for practical longitude determination. One entertaining read about the winner of the reward (John Harrison) is a book by Dava Sobel entitled *Longitude: The True Story of a Lone Genius Who Solved the Greatest Scientific Problem of His Time*. Although the voyage of the *Beagle* occurred almost ninety years after the first sufficiently accurate chronometer was developed, there was clearly still a lot of mapping work to be done around the world to bring the charts up to a standard acceptable to the Royal Navy.

[35] The origin of this quote is not known. It was given, without reference to source, at www.lyellcollege.com/html/sir_charles_lyell.html.

[36] This quote appears on page 83 of Alan Moorehead's *Darwin and the Beagle* (1969), which is my source for most of the discussion of Darwin's voyage. If you can still find a copy, it is a marvellous way to acquaint yourself with the basic facts of this epic expedition, containing as it does, in one short book, a vivid description of the voyage, accompanied by many contemporary watercolours and drawings.

[37] Ibid, 201.

[38] Ibid, 202.

[39] Ibid, 259.

[40] Charles Darwin, *On the Origin of Species*, Modern Library edition: 1993, 21–23.

[41] The idea of spontaneous creation dies hard. Ancient religion is clearly largely responsible, as the Christian, Jewish and at least some branches of Islamic religions all held this belief. A November 2004 article in *National Geographic* by David Quammen quotes five Gallup polls taken in the US over the years 1982–2001 as showing remarkably little change in opinions about spontaneous creation over that time period. Quammen writes, "The creationist conviction – that God alone, and not evolution, produced humans – has never drawn less than 44 percent." He concludes that "honest confusion and ignorance" must be major contributors to this situation, in addition to the traditional belief in creationism. That this view should endure in the face of such overwhelming scientific evidence must be telling us something important about the nature of human beings.

[42] From the text of a lecture given by Lamarck at the Musée National d'Histoire Naturelle, Paris, May 1803, and quoted on a University of California web page devoted to Lamarck (http://www.ucmp.berkeley.edu/history/lamarck.htm) accessed Sept. 15, 2012.

[43] Darwin, 1993, 9.

[44] Ibid, back-cover quotation.

Dialogue 9

[45] This quotation came from a biography of Skinner on the PBS website, accessed June 19, 2012,
http://www.pbs.org/wgbh/aso/databank/entries/bhskin.html. The author is unknown, and I suspect may be attributing to Skinner a more polar position than Skinner himself held.

[46] This list is drawn from a *curriculum vitae* written by Karl von Frisch, which appears on the Nobel Prize website, accessed June 20, 2012, www.nobel.se.

[47] von Frisch summarized much of his research (and that of his students and colleagues) in *The Dance Language and Orientation of Bees*, published in 1967. Although he was the first to understand the language of the dance of the bees, he was not the first to observe that bees communicate their findings to other members of the colony. In the first chapter of the book he quotes Aristotle

---

(*Historia Animalium, IX*, 40) on the subject.

[48] The chapter from *King Solomon's Ring* that describes Lorenz's work with jackdaws is eleven, "The Perennial Retainers." His description of his dog Tito appears in Chapter Eight, "The Language of Animals." His experience with the mallard chicks is described in Chapter Five, "Laughing at Animals." In Chapter Nine, "The Taming of the Shrew," Lorenz describes his experiences in capturing and raising water shrews. This last description nicely illuminates the dedication and sacrifice needed to observe some animals, as well as the rewards for the trouble taken.

[49] Nikolaas Tinbergen, *The Study of Instinct*, 1989, 205.

[50] Margaret MacMillan, *Nixon in China: The Week that Changed the World*, 2006, 79–92.

Dialogue 10

[51] Franz Boas, *Anthropology and Modern Life*, 1986 edition, 79.

[52] This example is drawn from the first chapter of James A. Michener's, *The Covenant* (1982). Clearly intermarriage between a group of only twenty-five persons of all ages is not viable in the longer term. Luigi Cavalli-Sforza gives the minimum group size to prevent the ill-effects of inbreeding as about five hundred. It is known that hunter-gatherers typically associated loosely in just such larger groups or tribes, and that either young men or young women (according to custom) often married someone outside their group, and were automatically adopted into their spouse's group.

[53] Michener, 1982, 24.

[54] Madhusree Mukerjee, *The Land of Naked People*, 2003, 138–39.

[55] Ruth Benedict, *Patterns of Culture*, 1934, 223.

[56] Ruth Benedict, "Patterns of the Good Culture," 1970.

[57] Abraham Maslow, "Synergy in the Society and in the Individual," 1964.

[58] Abraham Maslow, *Motivation and Personality*, 1970, 39.

[59] Ibid, 150.

Dialogue 11

[60] Gregor Mendel was born in 1822 in what is now the Czech Republic. He entered the Abbey of St. Thomas in Brno in 1843. He is best known for his

experiments with pea plants, some of which had purple flowers and the rest white flowers. Mendel observed that when he cross-fertilized these plants, the next generation all had purple flowers, but if this second generation were then cross-fertilized, three quarters of the progeny continued to produce purple flowers, but one quarter produced white flowers. From these observations he concluded that inheritable traits came in equal measure from both parents, with the progeny having one copy from each parent. However, one trait was normally dominant (in this case, the colour purple for the flower) and the other recessive (the white colouring for the flower). Thus, while the first generation of progeny all had one copy each of the dominant and recessive gene, in the next generation one quarter of the plants would have two copies of the gene with the dominant trait, one quarter would have two genes with the recessive (white flower) trait, and one half with one copy of each version of the gene. For the latter, the dominant trait (that is the purple flower) was produced.

Mendel's discovery went largely unremarked in his lifetime (he died in 1884), but his findings were subsequently independently rediscovered in the early 1900s by independent researchers, after which Mendel's prior work came to light and he received the recognition he had earned. Thereafter he was known as the "father of modern genetics."

[61] Charles Darwin, *On the Origin of Species*, 1993 edition, 202.

[62] James D. Watson, *DNA: The Secret of Life*, 2003, 35.

[63] In Ibid, 69-75.

[64] Most of the information about the Human Genome Project and the Celera Corporation is drawn from the Wikipedia article on the Human Genome Project, accessed September 10, 2012, at http://en.wikipedia.org/wiki/Human_Genome_Project

[65] John S. Mattick, "The Hidden Genetic Program of Complex Organisms," 2003, 60–67.

Dialogue 11 was first written in about 2004, at which time it was still very common to refer to introns as "junk DNA." By 2012 progress in genetic research has rendered the epithet dated at best, and simply incorrect at worst. In a brief but fascinating article written in 2009 entitled "What Makes Us Human?" Katherine Pollard writes,, "Experimental and computational studies now under way in thousands of labs around the world promise to elucidate what is going on in the 98.5 percent of our genome that does not code for proteins. It is looking less and less like junk every day." As it turns out, the days when this prophesy largely came to pass were September 5–6, 2012, as noted in the ENCODE project description.

66 See, for example, "'Junk DNA' Can Sense Viral Infection: Promising Tool in the Battle Between Pathogen and Host," *Science Daily,* April 24, 2012, and www.sciencedaily.com/releases/2012/04/120424142253.htm accessed September 10, 2012,

67 Quoted in Joseph R. Ecker, Wendy A. Bickmore, Inês Baroso, Jonathan K. Pritchard, Yoav Gilad and Erin Segal, "Genomics: ENCODE Explained,"2012, 52–55.

68 Stephen S. Hall, "Journey to the Genetic Interior," 2012, 80–84.

69 Quoted in Joseph R. Ecker, Wendy A. Bickmore, Inês Baroso, Jonathan K. Pritchard, Yoav Gilad and Erin Segal, "Genomics: ENCODE Explained," 2012, 52–55. The quotation comes from another paper in the same volume of *Nature* by S. Jebali, C.A. Davis, et al, "Landscape of DNA Transcription in Human Cells," 101–108. The quotation is on page 103.

70 Watson, 2003, 200 and 204..

71 Watson, 2003, 200. The table has been modified to show the estimate of the number of genes in the human genome as printed in Watson's book in 2003, and a more recent estimate as provided by Pascale Gaudet in the paper referenced in the next endnote.

72 Gil Ast, "The Alternative Genome," 2005, 58–65.

73 The most recent figures on the size of the human genome and of the likely number of different proteins manufactured there come from a presentation by Pascale Gaudet at the American Association for the Advancement of Science Annual Meeting in Vancouver on February 17, 2012. The title of her talk was "The Next Big Thing: Bringing the Human Proteome Project to the World."

74 Watson, 2003, 204.

75 Fred H. Gage and Alysson R. Muotri, "What Makes Each Brain Unique," 26–31.

76 Watson, 2003, 211.

Dialogue 12

[77] Richard Dawkins, *The Ancestor's Tale,* 2005, 184. Dawkins attributes the analogy to a book written by Matt Ridley, *Nature via Nurture.* Table 11.1 enlarges on this analogy.

[78] Gil Ast, "The Alternative Genome," 2005.

[79] W. Wayt Gibbs, "The Unseen Genome: Gems among the Junk," 2003, 46–53. Gibbs attributes this idea to Claes Wahlestedt of the Karolinska Institute in Sweden. The discussion of riboswitches, which appears in the same article, is based on work by Ronald Breaker and his associates at Harvard University.

Dialogue 13

[80] Brian Ford, "Why So Few Genes in the Human Genome?" 2001, 5–8.

[81] J.R. Porter, "Anthony van Leeuwenhoek: Tercentenary of His Discovery of Bacteria," 1976, 265. This article contains a good description of van Leeuwenhoek's microscopes and speculates that by sticking with a simple lens system rather than the compound lens systems known by that time, he was able to reduce chromatic aberration and get better resolving power than most other users of microscopes.

[82] Ibid, 255–56.

[83] Michael Gregory, 2005.

[84] Carl Woese is a distinguished American biologist. In his 2004 article "A New Biology for a New Century," *Microbiology and Molecular Biology Reviews,* June 2004, he writes that he dislikes the robot analogy, and pungently points out its shortcomings as follows:

> The machine metaphor certainly provides insights, but these come at the price of overlooking much of what biology is. Machines are not made of parts that continually turn over, renew. The organism is. Machines are stable and accurate because they are designed and built to be so. The stability of an organism lies in resilience, the homeostatic capacity to re-establish itself. While a machine is a mere collection of parts, some sort of "sense of the whole" inheres in the organism, a quality that becomes particularly apparent in phenomena such as regeneration in amphibians and certain invertebrates.

I continue to find the analogy helpful for two principal reasons. It provides an opportunity to point out just how much more complicated life forms are than human-made robots (at least in my time) while at the same time demonstrating the largely preprogrammed nature of all life forms; like autonomous robots, organism behaviour is as much a part of the structure as is physical appearance.

[85] While dogs and crows are not the only animals to exhibit intelligence, they provide good examples, as the common ancestor between ourselves and (especially) crows go back a very long way. Konrad Lorenz devotes a chapter in his book *King Solomon's Ring* to his studies of crow behaviour, and there are a number of YouTube videos showing the results of experiments to test the intelligence of crows. See, for example:
http://www.youtube.com/watch?v=NhmZBMuZ6vE (viewed on 21 January, 2013).
I am indebted to Robin Percival Smith for the suggestion that I mention crow intelligence at this juncture.

[86] This figure is reproduced in many places, including on page 225 of Woese, 1987. Its original location is in E. Haeckel, *Generelle Morphologie der Organismen* (Berlin: Verlag Georg Reimer, 1866).

[87] W. Ford Doolittle, "Uprooting the Tree of Life," 2000, 90.

Dialogue 14

[88] Trofim Lysenko (1898–1976) was a Ukrainian biologist and agronomist of peasant stock. His proletarian origins made him a particularly attractive scientific figure to Josef Stalin, who promoted him as an example to others. Unfortunately, his belief that plant characteristics could be acquired from their environment and passed on to subsequent generations was simply not true. Those who opposed this view were harshly treated, to the degree that scientific progress in biology in the Soviet Union came to a virtual standstill. Lysenko's beliefs were not officially discredited until 1964. His fraudulent "discoveries" contributed to serious famine conditions by encouraging agricultural collectives to implement innovative strategies that seriously reduced crop yields. Two of the best known false claims involve a method for growing peas without the need for fertilization, and a way of treating wheat so that new hardier strains could be produced simply be dampening the seeds and leaving them in the snow over the winter. By the 1950s, the evidence against his fraudulent claims was mounting, but it was not until Kruschev lost power that his downfall was complete.

[89] Scott F. Gilbert and David Epel, *Ecological Developmental Biology*, 2009, 11, 12, 30 and 68.

[90] Ibid, 17.

[91] Ibid.

[92] Ibid, 13.

[93] Ruth Benedict, *Patterns of Culture*, 1959, 131.

[94] Eric J. Nestler, "Hidden Switches in the Mind," 2011, 80.

[95] Gilbert and Epel, 2009, 248.

[96] Ibid, 256.

[97] There is an extensive and rapidly growing literature on this subject. The author found that McCain, Mustard and McQuaig, *Early Years Study 3: Making Decisions, Taking Action,* published in 2011, provided a very useful summary of the literature. As the study makes clear, most countries, including Canada, have yet to adapt policy to the scientific findings, with sorry consequences both for many individuals and for society as a whole.

[98] Margaret McCain et al, *Early Years Study 2,* 2007, 37.

[99] Ibid.

[100] See for example Margaret McCain et al, *Early Years Study 3,* 2011, 44.

Dialogue 15

[101] See, for example, M. Brasier, O. Green, J. Lindsay and A. Steele, "Earth's Oldest (~3.5 Ga) Fossils and the 'Early Eden Hypothesis': Questioning the Evidence," *The Journal of the International Society for the Study of the Origin of Life* 34, no. 1–2 (February 2004).

[102] My favourite contemporary philosopher, Daniel Dennett provides as good a description of how this early process could have proceeded as I have found. Daniel C. Dennett, *Darwin's Dangerous Idea,* 1995, starting at page 149; see especially page 158 and the following pages.

[103] Jack Maniloff, "The minimal cell genome: "On being the right size," *Proc. Natl. Acad. Sci. USA* 93 (September 1996): 10004–10006. Maniloff is quoting an estimate made by Mushegian and Koonin in the same volume of the journal, pages 10268–73.

---

[104] Robert M. Hazen, *The Origins of Life, Part 1*, 2005, 31. The quotation from Kant's *Critique of Judgement* appears in the second lecture (of twenty-four) entitled The Historical Setting of Origins Research.

[105] Richard Dawkins, *The Ancestor's Tale*, 2005, 7.

[106] In a letter from Darwin to the botanist Joseph Hooker in 1871. Quoted in Hazen, *The Origins of Life, Part 1*, 2005, 34.

[107] Stanley L. Miller, "Production of Amino Acids Under Possible Primitive Earth Conditions," *Science* 117, no. 3046 (May 1953): 528–29. By searching for the "Miller-Urey Experiment" in Wikipedia, the interested reader will find a small treasure trove of additional information concerning subsequent analysis of the mixture used in the experiment, as well speculation about other gasses that were likely present in the atmosphere several billion years ago.

[108] Hazen, 2005, 152–71.

[109] David Barker et al, "Engineering Life," 2006, 44–51.

Dialogue 16

[110] Penfield's books about his experiences are informative and well written, and are still worth reading. I recommend *The Mystery of the Mind* (Princeton: Princeton University Press, 1975); and *The Second Career* (Toronto: Little Brown and Company, 1963).

[111] Ramón y Cajal, <u>*Recollections of my Life*</u> (first published in 1917) translated by E.H. Craigie and J. Cano Trans (Cambridge, MA: MIT Press, 1989).

[112] V.S. Ramachandran and Sandra Blakeslee, *Phantoms in the Brain*, 1999.

[113] Ibid, 22. The battle at Tenerife is described in Andrew Lambert, *Nelson: Britannia's God of War* (London: Faber & Faber, 2004), 96. Amazingly, he was back at work within an hour, ". . . signing with his left hand, a demand for the Spanish to surrender . . ."

[114] Ibid, 54–55.

[115] Ibid, 56.

[116] Ibid, 75–76.

[117] Perhaps the best and most humorous explications of these phenomena and the evidence supporting them can be found in Daniel Gilbert, *Stumbling on Happiness*, 2006, especially Chapters 4 and 8.

[118] Ibid, 167.

Dialogue 17

[119] Research reported by Joseph Castro, "Sea Slug Offers Clues to Improving Long-Term Memory", Live Science, December 30, 2011 (www.livescience.com/17683-sea-slug-memory.html) indicates that by applying electric shocks at carefully selected intervals a sea slug's reaction to threats can be changed for as much as a week, indicating that even such a simple brain is capable of learning. However there is no suggestion that sea slugs are capable of reflection on what they have learned. It would seem likely that the electric shocks have simply resulted in some changes in the strengths of the interconnections of some of its neurons.

[120] V.S. Ramachandran and Sandra Blakeslee, *Phantoms in the Brain*, 1999, 227–57.

[121] Eric R. Kandel et al, *Principles of Neural Science, Fourth Edition*, 2000, 16.

[122] Daniel C. Dennett, *Breaking the Spell*, 2007, 3.

[123] This quotation appears in the Wikipedia article on Schrödinger. In turn, the article cites a biography of Schrödinger as follows: W. Heitler, "**Erwin Schrodinger**. 1887-1961." *Biographical Memoirs of Fellows of the Royal Society 7* (1961): 221–26.

[124] Roger Penrose, *Shadows of the Mind*, 1994. Penrose also discusses the red colour problem on page 52.

[125] William Shakespeare, *Macbeth*, Act IV, Scene 1, lines 14–15.

[126] Dennett summarized his position succinctly in *Sweet Dreams* (2006) on page 57 as follows: "It seems to many people that consciousness is a mystery, the most wonderful magic show imaginable, an unending series of special effects that defy explanation. I think that they are mistaken: consciousness is a physical, biological phenomenon – like metabolism or reproduction or self-repair – that is exquisitely ingenious in its operation, but not miraculous or even, in the end, mysterious."

[127] Crick's search (with Koch) is related in Christof Koch, *Consciousness: Confessions of a Romantic Reductionist, 2012.*

[128] Ramachandran and Blakeslee, 1999, 240.

[129] Ibid, 175.

---

[130] Ibid, 179.

[131] Richard Dawkins, *The Ancestor's Tale,* p. 254. See endnote 55 for references to the intelligence of crows.

[132] Michael Shermer, "Aunt Millie's Mind," *Scientific American* 307, no. 1 (July 2012): 84.

[133] Michael Shermer, "Free Won't," *Scientific American* 307, no. 2 (August 2012): 86.

Dialogue 18

[134] This case study is referred to several times in V.S. Ramachandran's book and also in an article by R. Douglas Fields, whose two articles in *Scientific American* ("Making Memories Stick," February 2005, 74–81; and "The Other Half of the Brain," April 2004, 54–61) underpin much of the discussion of the science in this dialogue.

[135] D.O. Hebb, *The Organization of Behavior,* 1949.

[136] A variety of references were used in the construction of these paragraphs on neurotransmission, including: Jean-Pierre Changeux, *Neuronal Man,* especially 83–87; Malcolm Slaughter, ed., *Basic Concepts in Neuroscience: A Student's Survival Guide* (www.thebrain.mcgill.ca); and the National Institute of Neurological Disorders and Strokes website (http://www.ninds.nih.gov/).

[137] R. Douglas Fields, "Making Memories Stick," 2005, 75–81.

[138] R. Douglas Fields, "The Other Half of the Brain," 2004, 55–61.

[139] Einstein's brain has come in for a lot of use since he died! Initial analyses of his brain concluded that it was nothing out of the ordinary as far as appearance and structure were concerned, but, as described in the text, Fields' work suggests otherwise. One of the advantages of taking a long time to write a book is that one can chart changing ideas over time. In this instance, several years after Dialogue 17 was first written, I came across an article in the *Wall Street Journal* (May 22, 2009, A9) authored by Robert Lee Hotz that appears to shed new light on possible unique features of the Einsteinian brain. One observation is that the inferior parietal region, associated with visual and spatial reasoning, is about 15 per cent larger than normal in Einstein's brain and that, moreover, his brain lacks the "sylvian fissure . . . effectively fusing two key brain areas into one." The same article noted that in 1985 Marion Diamond, a pioneering neuroscientist at the University of California, Berkeley, discovered that Einstein

had more glial cells than normal in parts of the brain associated with mathematical and language skills, doubtless a discovery that contributed to Field's fascinating research. There can be no doubt that there is much more yet to discover before we can all aspire, through appropriate therapy, to become latent Einsteins!

Finally, in the February, 2013 issue of Scientific American, Gary Stix has written (on page 23) about a short interview with anthropologist Dean Falk of Florida State University, in which the latter claims that *"Einstein had extraordinary prefrontal cortices . . which revealed an intricate pattern of convolutions . . . [I]n humans, this area functions in higher cognition that entails working memory, making plans, bringing plans to fruition, worrying, thinking about the future and imagining scenarios."* It is a safe bet that we have not yet heard the last word on what special properties of Einstein's brain contributed to his status as a genius.

[140] The paper was by I. Weaver et al., 847–54. An opening editorial by Robert M. Sapolsky (791–792) in the same issue of the magazine explains the significance of this work. The research work described was led by Michael Meaney.

[141] See, for example, Karen Armstrong, *A History of God*, 1994, 74, where Armstrong cites several references that "As one Rabbi put it, 'God does not come to man oppressively but commensurately with a man's power of receiving him' . . . God could not be described in a formula as though he were the same for everybody: he was an essentially subjective experience." The Christian St. Augustine (121) believed that God was "not an objective reality but a spiritual presence in the complex depths of the self." According to Armstrong, this position is shared with Plato, Plotinus and the Buddhists. With respect to Islam, Armstong writes (143) that "[t]here were no obligatory doctrines about God: indeed, the Koran is highly suspicious of theological speculation, dismissing it as . . . self-indulgent guesswork about things that nobody can possibly know or prove."

[142] Ibid, 77.

[143] This quotation from Voltaire's *Philosophical Dictionary* (1764) can be found on page 310 of Karen Armstrong's *A History of God*.

Dialogue 20

[144] From Jeremy Bentham, *An Introduction to the Principles of Morals and Legislation*, 1789.

---

145 Randolph M. Nesse, "Natural Selection and the Elusiveness of Happiness," 2005, 12.

146 Ibid, 10.

147 For a good summary discussion of this topic see John F. Helliwell and Robert F. Putnam, "The Social Context of Well-being," 2004, 1435–46.

148 An outline description of this experiment can be found in D. Kahneman and J. Riis, "Living, and Thinking About It: Two Perspectives on Life," 2005, 286. A more complete description can be found in D. Kahneman, B.L. Frederickson, C.A. Schreiber, and D.A. Redelmeier, "When More Pain is Preferred to Less: Adding a Better End," 1993, 401–405.

149 Robert D. Putnam, Robert, *Bowling Alone: The Collapse and Revival of American Community*, 2000, 19. Putnam credits L.J. Hanifan, a state supervisor of rural schools in West Virginia, with the invention of the term in an article published in 1916. Social capital was "rediscovered" in the 1950s. If the reader wants to learn more about social capital, Putnam's two classic books, *Bowling Alone* and *Making Democracy Work: Civic Traditions in Modern Italy* (1993), and David Halpern's very readable book *Social Capital* (2005) are good places to start.

150 David Halpern, *Social Capital*, 2005, 60.

151 See Bruno S. Frey, *Happiness, A Revolution in Economics*, 2008.

152 Ibid, x.

153 See, for example, Ibid, 12. Also see Helliwell *et al*, *World Happiness Report*, 2012, pp. 61-61 for figures of reported happiness in the United States and Reported Life Satisfaction in West Germany.

154 Gospel According to St. Matthew 19:24. The same quote also appears in the Gospel According to Mark 10:12.

155 Elizabeth W. Dunn et al., "Spending Money on Others Promotes Happiness," 2008.

156 John Helliwell et al, *World Happiness Report*, 2012, 62.

157 The vexed subject of the effect of income on happiness is clearly and succinctly treated in *The World Happiness Report*, prepared for a United Nations conference in March 2012 and edited by John Helliwell et al (see pages 60–66). This report can be downloaded free from the Internet.

[158] There is still debate about whether any job is better than no job at all. German studies quoted and referenced in Helliwell et al.'s *World Happiness Report* (see page 67) conclude that the unemployed become happier when they have been placed in a job, however bad that job may be "for most specifications of bad jobs." Incidentally, different surveys often differ in the range of the scale on which respondents are asked to rate their happiness. The World Values Survey uses a ten-point scale. An eleven-point scale is becoming increasingly common (i.e., 0–10), but four- and five-point scales are also encountered.

[159] John Helliwell and Shun Wang, "Trust and Well-being", 2009, 5.

[160] See Helliwell et al, 2012, 69.

[161] Ruth Benedict, *Patterns of Culture,* 131. Benedict also discussed a third tribe, the Kwakiutl of Vancouver Island in British Columbia, Canada.

[162] Helliwell and Wang, 2009, 12.

[163] The talk was entitled "Taking Happiness Seriously", and was given at Environment Canada in Ottawa on November 30, 2012. The data is taken from overheads used by Professor Helliwell. It is anticipated that the data quoted will be published before the end of the year 2013.

[164] Bruno Frey, *Happiness: A Revolution in Economics,* 2008, 93-106.

[165] In an excellent and readable summary paper (Helliwell, October, 2012), Helliwell has provided an overview of the principal lessons to be drawn from happiness research to date. In his conclusion he writes:

> *Research on the determinants of subjective well-being . . . confirm that sufficient income is a strong support for happiness, but that the social context is even more important. This is too often forgotten in the race for higher incomes and consumption. Income matters less than the chance to connect with others, thereby improving our own lives and especially the lives of others.*
>
> *Within workplaces, the importance of the social context dwarfs the impact of salary and bonuses. To work where trust in management is one point higher, on a 10-point scale, has the same relation to life satisfaction as a one-third higher income. Yet many public and private workplaces have seen income disparities grow and trust levels fall over the past thirty years.*
>
> *Experiments in elder-care facilities show that residents, given the chance to do things together, to help themselves and others, live healthier and happier lives. Other experiments show that although everybody gains from*

peer support groups among disease sufferers, the care-givers gain even more health and happiness than do the recipients.

Although trust in neighbours and strangers delivers huge happiness benefits, people overestimate the risks of future burglaries and underestimate the chances of their lost wallets being returned. Trust is built by frequent chances to interact with strangers and neighbours, whether in elevators, buses, libraries or public spaces. Yet buildings are designed for looks and streets to move traffic, with scant thought for the public spaces needed to build the trust needed to let children walk to school.

Other examples, ranging from delivery of health care, management of prisons and schools, to macroeconomic policies, show how governments can learn from happiness research to make lives better for all. Individuals and neighbours can lead the way . . .

As of December, 2012, this paper can be accessed at: http://faculty.arts.ubc.ca/jhelliwell/papers/w18486.pdf

Dialogue 22

[166] Many readers will recognize this as a reference to Arthur C. Clarke's novel (concurrently made into a popular movie) called *2001: A Space Odyssey*, published in 1968.

Dialogue 23

[167] Perhaps the chief component of the moral code of virtually all societies is the Golden Rule, expressed (in the Christian religion) as "Do unto others as you would have them do unto you." It is easy to find information to support the ubiquity of this precept. For example, a Wikipedia article (entitled "The Golden Rule") cites parallel expressions from antiquity (Babylon, Egypt, Greece and China), and from many religions, ancient and modern. In 1993 there was a Parliament of the World's Religions that proclaimed the Golden Rule ("We must treat others as we wish others to treat us") as a common principle. In the words of the article: "The Initial Declaration was signed by 143 respected leaders from all of the world's major faiths, including Baha'i Faith, Brahmanism, Brahma Kumaris, Buddhism, Christianity, Hinduism, Indigenous, Interfaith, Islam, Jainism, Judaism, Native American, Neo-Pagan, Sikhism, Taoism, Theosophist, Unitarian Universalist and Zoroastrian."

[168] Robin Dunbar, *How Many Friends Does One Person Need*, 2010, 24.

[169] Ibid, 23–24.

[170] Bruno S. Frey, *Happiness: A Revolution in Economics,*2008, 95.

[171] Robert D. Putnam and David E. Campbell, *American Grace: How Religion Divides and Unites Us,* 2010.

Appendix 1

[165] "Tritium in Precipitation," accessed September 10, 2012, http://www.science.uottawa.ca/eih/ch7/7tritium.htm

Appendix 2

[173] W. Ford Doolittle, "Uprooting the Tree of Life," 2000, 90.

[174] Carl Woese, "Bacterial Evolution," *Microbiology and Molecular Biology Reviews,* 1987, 227. In Appendix 3 a valuable and interesting application for such (relatively) rapid changes in DNA is discussed.

[175] Ibid, 222–23.

[176] There is a second route by which bacteria acquire immunity to drugs, i.e. by the exchange of plasmids. Plasmids are small components of bacterial cells that contain their own chromosome, much as mitochondria and chloroplasts (in plant cells) are components of eukaryotic cells. Plasmids play an important role in protecting bacteria from invaders. It turns out that plasmids can quite easily be swapped between different species of bacteria.

[177] Doolittle, 2000, 95.

Appendix 3

[178] V.M. Sarich & A. C. Wilson, "Immunological Time Scale for Hominid Evolution," 1967, 1200-1203.

[179] Bryan Sykes,*The Seven Daughters of Eve*, New York 2001.

[180] Reference to this appears in an article in the *Guardian* newspaper on August 18, 2001 written by Jerome Burne. See also Spencer Wells, *The Journey of Man: A Genetic Odyssey* (Princeton: Princeton University Press, 2003).

[181] M.I. Jensen-Seaman et al, "Modern African Ape Populations and Genetic and Demographic Models of the Last Common Ancestor of Humans, Chimpanzees and Gorillas," 2001, 475–80.

182 Bill Bryson, *A Short History of Nearly Everything*, 2003, 227.

183 Curtis W. Marean, "When the Sea Saved Humanity," 2010, 55–61

184 Ibid, 57.

185 Richard Dawkins,*The Ancestor's Tale*, 2005.

Appendix 5

186 John F. Helliwell, "Well-being, Social Capital and Public Policy: What's New?" National Bureau of Economic Research Working Paper 11807 (Cambridge, MA, 2005); and John F. Helliwell and Robert D. Putnam, "The Social Context of Well-being", 2004, 1435–46. The subject is also treated in depth in John F. Helliwell et al, *The World Happiness Report*, 2012, 10–22.

187 John F. Helliwell and Haifang Huang, "How's Your Government?" 2008, 595–619.

188 Ibid, 618.

189 Robert D. Putnam, *Bowling Alone*, 2000.

BIBLIOGRAPHY

Albert, David Z. and Rivka Galchen. "A Quantum Threat to Special Relativity," *Scientific American* 300, no. 3 (March 2009): 32–39.

Alberts, Bruce, Alexander Johnson, Julian Lewis, Martin Raff, Keith Roberts, and Peter Walter. *The Molecular Biology of the Cell (5th Edition)* New York and Abingdon, UK: Garland Science, 2008.

Armstrong, Karen *A History of God: The 4000-Year Quest of Judaism, Christianity and Islam.* New York: Ballantine Books, 1994.

Ast, Gil. "The Alternative Genome". *Scientific American* 292, no. 4 (April 2005): 58–65.

Barker, David, et al [The Bio Fab Group]. "Engineering Life: Building a Fab for Biology," *Scientific American* 294, no. 6 (June 2006): 44–51.

Benedict, Ruth. *Patterns of Culture.* Boston: Houghton Mifflin Company, 1959. (First published in 1934)

Benedict, Ruth F. "Patterns of the Good Culture." *American Anthropologist* 72, no. 2 (April 1970): 320–33.

Bentham, Jeremy. *An Introduction to the Principles of Morals and Legislation.* 1789.

Boas, Franz. *Anthropology and Modern Life.* New York: W.W. Norton & Co., 1928. The quoted page numbers are for the 1986 edition.

Brown, Lester R. "Could Food Shortages Bring Down Civilization?" *Scientific American* 300, no. 5 (May 2009): 50–57.

Bryson, Bill. *A Short History of Nearly Everything.* New York: Broadway Books (Random House), 2003.

Cavalli-Sforza, Luigi and Francesco. *The Great Human Diasporas: The History of Diversity and Evolution.* Cambridge, MA: Perseus Books, 1995. (Translated from Italian, the language in which it was first published in 1993)

Changeux, Jean-Pierre. *Neuronal Man: The Biology of Mind.* Princeton: Princeton University Press, 1985.

Darwin, Charles, *On the Origin of Species by Means of Natural Selection, or, The Preservation of Favored Races in the Struggle for Life.* New York: The Modern Library, 1993. (First published in 1859)

Dawkins, Richard. *The Ancestor's Tale: A Pilgrimage to the Dawn of Evolution.* Boston: Houghton Mifflin Company, 2005.

———. *The God Delusion.* Boston: Houghton Mifflin Company (First

Mariner Books edition), 2008.

Dennett, Daniel C. *Darwin's Dangerous Idea*. New York: Touchstone (Simon & Schuster), 1995.

———. *Sweet Dreams: Philosophical Obstacles to a Science of Consciousness*. Cambridge, MA: MIT Press, 2006.

———. *Breaking the Spell: Religion as a Natural Phenomenon*. New York: Penguin Books, 2007.

Diamond, Jared. *Guns, Germs and Steel: The Fates of Human Societies*. New York: Random House, 1997.

Doolittle, W. Ford. "Uprooting the Tree of Life," *Scientific American* 282, no. 2 (February 2000): 90, 96.

Dunbar, Robin. *How Many Friends Does One Person Need?* London: Faber and Faber, 2010.

Dunn, Elizabeth W., L.B. Aknin, and M.I. Norton. "Spending Money on Others Promotes Happiness." *Science 319* (2008): 1687–88.

Durant, Will. *The Story of Philosophy* (2nd edition). New York: The Pocket Library, 1956.

Easterlin, Richard A. "Does Money Buy Happiness?" *The Public Interest* (Winter 1973): 3–10.

Eddington, A.S. *The Nature of the Physical World*. Cambridge, UK: Cambridge University Press, 1932).

Ecker, Joseph R., Wendy A. Bickmore, Inês Barroso, Jonathan K. Pritchard, Yoav Gilad, and Eran Segal. "Genomics: ENCODE Explained." *Nature* 489 (September 6, 2012): 52–55.

ENCODE Project Consortium. "An Integrated Encyclopedia of DNA Elements in the Human Genome." *Nature* 489 (September 6, 2012): 57–74.

Feynman, Richard P., Robert B. Leighton, and Matthew Sands. *The Feynman Lectures on Physics*. Reading, MA: Addison-Wesley, 1963.

Fields, R. Douglas. "The Other Half of the Brain." *Scientific American* 290, no. 4 (April 2004): 54–61.

———. "Making Memories Stick." *Scientific American* 292, no. 2 (February 2005): 74–81.

Ford, Brian J. "Why So Few Genes in the Human Genome?" *Science Now*, B&K, 10, no. 2 (December 2001): 5–8. Also at: http://www.brianjford.com/agenome.htm

Franklin, Benjamin. *Memoirs of Benjamin Franklin* (1845), Vol. 2, 152.

Frey, Bruno S. *Happiness: A Revolution in Economics*. Cambridge, MA: MIT Press, 2008.

Gage, Fred H., and Allyson R. Muotri. "What Makes Each Brain Unique." *Scientific American* 306, no. 3 (March 2012): 26–31.

Gibbs, W. Wayt. "The Unseen Genome: Gems Among the Junk."

Scientific American 289, no. 5 (November 2003): 46–53.

Gilbert, Daniel. *Stumbling on Happiness*. New York: Alfred A. Knopf, 2006.

Gilbert, Scott F., and David Epel. *Ecological Developmental Biology: Integrating Epigenetics, Medicine, and Evolution*. Sunderland, MA: Sinauer Associates, Inc., 2009.

Gould, Stephen Jay. *Eight Little Piggies: Reflections in Natural History*. New York: W.W. Norton & Company, 1995.

Greene, Brian. *The Elegant Universe*. London: Jonathan Cape, 1999.

Gregory, Michael. From his lecture notes posted (circa 2005) on the web at:
   http://faculty.clintoncc.suny.edu/faculty/Michael.Gregory/files/Bio%20101/Bio%20101%20Lectures/DNA/dna.htm.

Hall, Stephen S. "Journey to the Genetic Interior." (An interview with Ewan Birney) *Scientific American* 307, no. 4 (October 2012): 80–84.

Halpern, David. *Social Capital*. Cambridge, UK: Polity Press, 2005.

Hazen, Robert M. *The Origins of Life, Parts 1 & 2*. Chantilly, VA: The Teaching Company, 2005.

Hebb, D.O. *The Organization of Behaviour*. New York: Wiley & Sons, 1949.

Helliwell, John F. *Globalization and Well-Being*. Vancouver: University of British Columbia Press, 2002.

Helliwell, John F. and Robert F. Putnam. "The Social Context of Well-being." *Philosophical Transactions of the Royal Society London B* 359 (2004): 1435–46.

Helliwell, John F. "Well-Being and Social Capital: Does Suicide Pose a Puzzle?" *Social Indicators Research* 81, no. 3 (May 2007): 455–96.

Helliwell, John F. and Haifang Huang. "How's The Job? Well-Being and Social Capital in the Workplace." *Industrial and Labor Relations Review* 63, no. 2, Article 2, (2010).

———. "How's Your Government? International Evidence Linking Good Government and Well-being." *British Journal of Political Science* 38 (2008): 595–619.

Helliwell, John F., and Shun Wang. "Trust and Well-Being." *The 3rd OECD World Forum on Statistics, Knowledge and Policy, Busan, Korea, 27-30 October 2009.*

Helliwell, John F., and Christopher P. Barrington-Leigh. "How Much is Social Capital Worth?" National Bureau of Economic Research Working Paper 16025, May 2010.

Helliwell, John F., Richard Layard, and Jeffrey Sachs, eds. *World Happiness Report*. The Earth Institute, Columbia University, March 2012. Available for free download at: http://www.earth.columbia.edu/articles/view/2960

Helliwell, John F., "Understanding and Improving the Social Context of Well-Being", National Bureau of Economic Research Working Paper 18486, October 2012.

Huppert, Felicia A., Nick Baylis, and Barry Keverne, eds. *The Science of Well-Being*. Oxford, UK: Oxford University Press, 2005.

Jebali, S., C.A. Davis, et al. "Landscape of DNA Transcription in Human Cells." *Nature* 489 (September 6, 2012): 101–108.

Jensen-Seaman, M.I., A.S. Deinard, and K.K. Kidd. "Modern African Ape Populations and Genetic and Demographic Models of the Last Common Ancestor of Humans, Chimpanzees and Gorillas." *The Journal of Heredity* 92, no. 6 (2001): 475–80.

Kahneman, D., B.L. Frederickson, C.A. Schreiber, and D.A. Redelmeier. "When More Pain is Preferred to Less: Adding a Better End." *Psychological Science* 4, no. 6 (November 1993): 401–405.

Kandel, Eric R., James H. Schwartz, and Thomas M. Jessell, eds. *Principles of Neural Science, Fourth Edition*. New York: McGraw-Hill, 2000.

Kaufman, S., and W.F. Libby. "The Natural Distribution of Tritium." *Physics Review* 93 (1954): 1337–44.

Keeling, Patrick J. "Diversity and Evolutionary History of Plastids and Their Hosts." *American Journal of Botany* 91, no. 10 (2004): 1481–93.

Koch, Christof. *Consciousness: Confessions of a Romantic Reductionist*. Cambridge, MA: MIT Press, 2012.

Kruuk, Hans. *Niko's Nature*. Oxford: Oxford University Press, 2003.

Layard, Richard. *Happiness: Lessons from a New Science*. New York: Penguin, 2005.

Libby, Willard F. *Radiocarbon Dating*. Nobel lecture, December 12, 1960. (Available on the Nobel website at http://www.nobel.se/chemistry/laureates/1960/libby-lecture.html)

Lisi, Garrett A., and James Owen Weatherall. "A Geometric Theory of Everything." *Scientific American* 303, no. 6 (December 2010): 54–61.

Lorenz, Konrad. *King Solomon's Ring*. Translated by Marjorie Kerr Irwin. London: Methuen, 161. (First published April 1952 and reprinted many times)

MacMillan, Margaret. *Nixon in China: The Week that Changed the World*. Toronto: Penguin Books, 2006.

Marean, Curtis W. "When the Sea Saved Humanity." *Scientific American* 303, no. 2 (August 2010): 55–61.

Maslow, Abraham H. "Synergy in the Society and in the Individual." *Journal of Individual Psychology* 20 (1964): 153–64.

———. *Motivation and Personality (2nd Edition)*. New York: Harper and Row, 1970.

Mattick, John S. "The Hidden Genetic Program of Complex Organisms." *Scientific American* 291, no. 4 (October 2004): 60–67.

———. "RNA Driving the Epigenetic Bus." *The Embo Journal* 31 (February 2012): 515–16.

McCain, Margaret Norrie, J. Fraser Mustard, and Stuart Shanker. *Early Years Study 2: Putting Science into Action.* Toronto: Council for Early Child Development,2007.

McCain, M.N., J.F. Mustard, and K. McCuaig. *Early Years Study 3: Making Decisions, Taking Action.* Toronto: Margaret & Wallace McCain Family Foundation, 2011.

Mead, Margaret. *Ruth Benedict.* New York: Columbia University Press, 1974.

Michener, James A. *The Covenant.* London: Corgi Books, 1982.

Moorehead, Alan. *Darwin and the Beagle.* London: Hamish Hamilton, 1969.

Mukerjee, Madhusree. *The Land of Naked People.* Boston: Houghton Mifflin Company, 2003.

Nesse, Randolph M. "Natural Selection and the Elusiveness of Happiness." In Huppert, Baylis and Keverne (eds.), *The Science of Well-Being.* Oxford: Oxford University Press, 2005.

Nestler, Eric J. "Hidden Switches in the Mind." *Scientific American* 305, no. 6 (December 2011): 76–83.

Offer, Avner. *The Challenge of Affluence: Self-Control and Well-Being in the United States and Britain since 1950.* Oxford: Oxford University Press, 2006.

Penfield, Wilder. *The Mystery of the Mind.* Princeton: Princeton University Press, 1975.

Penrose, Roger. *Shadows of the Mind: A Search for the Missing Science of Consciousness.* Oxford: Oxford University Press, 1994.

Pettit, Philip. Overheads from a lecture entitled *Democracy: Fashions, Failures and Fantasies* given at the Vancouver Institute meeting of March 3, 2007.

Pollard, Katherine S. "What Makes Us Human?" *Scientific American* 300, no. 5 (May 2009): 44–49, May, 2009.

Porter, J.R. "Anthony van Leeuwenhoek: Tercentenary of His Discovery of Bacteria." *Bacteriological Reviews* 40, no. 2 (June 1976): 260–69.

Putnam, Robert D. *Bowling Alone: The Collapse and Revival of American Community.* New York: Simon and Schuster, 2000.

Putnam, Robert D., and David E. Campbell. *American Grace: How Religion Divides and Unites Us.* New York: Simon and Schuster, 2010.

Ramachandran, V.S., and Sandra Blakeslee. *Phantoms in the Brain: Probing the Mysteries of the Human Mind.* New York: Quill, 1999.

Restak, Richard. *The Brain*. New York: Bantam Books, 1984.

Riordan, Michael, Tonelli, Guido and Wu, Sau Lan. "The Higgs at Last", *Scientific American* 307, no. 4 (October 2012): 66–73.

Sapolsky, Robert. *The Trouble with Testosterone*. New York: Scribner, 1997

Sarich, V.M., and A.C. Wilson. "Immunological Time Scale for Hominid Evolution." *Science* 158 (December 1967): 1200–1203.

Shermer, Michael, "Aunt Millie's Mind", *Scientific American* 307, no. 1 (July 2012): 84.

–––. "Free Won't", *Scientific American* 307, no. 2 (August 2012): 86.

Smith, Robert W. "Edwin Hubble and the Transformation of Cosmology." *Physics Today* (April 1990): 52–58.

Sobel, Dava. *Longitude: The True Story of a Lone Genius Who Solved the Greatest Scientific Problem of His Time*. New York: Penguin, 1995.

Sykes, Bryan. *The Seven Daughters of Eve*. New York: W.W. Norton and Company, 2001.

Tinbergen, Nikolaas. *The Study of Instinct*. Oxford: Clarendon Press, Second Impression Reprinted and Reissued in 1989. (First published in 1951)

Turok, Neil. *The Universe Within – From Quantum to Cosmos*. Toronto: House of Anansi Press Inc., 2012.

Von Frisch, Karl. *The Dance Language and Orientation of Bees*. Cambridge, MA: Harvard University Press, 1967.

Watson, J.D., and F.H.C. Crick. "A Structure for Deoxyribose Nucleic Acid." *Nature* 171 (April 25, 1953): 737–38.

Watson, James D., with Andrew Berry. *DNA: The Secret of Life*. New York: Alfred A. Knopf, 2003.

Wilkinson, Denys. *Our Universes*. New York: Columbia University Press, 1991.

Wilson, Edward O. *Sociobiology: The New Synthesis*. Cambridge, MS: The Belknap Press, 1975.

Winchester, Simon. *The Map That Changed the World*. New York: Viking, 2001.

Woese, Carl R. "A New Biology for a New Century." *Microbiology and Molecular Biology Reviews* 68, no. 2 (June 2004): 173–86.

–––. "Bacterial Evolution." *Microbiological Reviews* 51, no. 2 (June 1987): 222–71.

ILLUSTRATIONS – CREDITS AND PERMISSIONS

Cover photos (front and back) courtesy NASA and the NSSDC photo gallery.

Figure 2.1 *Kudu*. Photo courtesy Wildife-Pictures-Online.com

Figure 4.1 *Earthrise over the lunar surface*. Photo courtesy NASA and the NSSDC photo gallery.

Figure 4.2 *Measurement of the distance to* Proxima Centauri *by the parallax method*. By the author.

Figure 4.3 *The Butterfly Nebula*. Photo courtesy of NASA, ESA, and the Hubble SM4 ERO Team as well as the NSSDC photo gallery.

Figure 5.1 *Archbishop James Ussher, after a portrait by Sir Peter Lely, circa 1654*. © National Portrait Gallery, London, UK. *NPG574*

Figure 6.1 *Two Geniuses, Two Theories of Gravity*. Newton drawing courtesy fromoldbooks.org; Einstein image courtesy www.ysfine.com.

Figure 8.1 *HMS* Beagle *in Sydney Harbour (detail)*. From a watercolour by Owen Stanley, circa 1835. © National Maritime Museum, Greenwich, London, UK.

Figure 8.2 *Charles Darwin in the late 1830s*. Reproduction of a watercolour by George Richmond, courtesy The Image Works.

Figure 8.3 *Darwin caricature from* Vanity Fair, *September 30, 1871*. ©The British Library / The Image Works.

Figure 9.1 *Nicholaas Tinbergen and Konrad Lorenz in 1978*. Photo obtained from Wikimedia Commons. It is believed to be the property of the Max Planck Gesellschaft.

Figure 10.1 *Franz Boas, Ruth Benedict and Abraham Maslow*. Boas photo from the collection of the Canadian Museum of Civilization; Benedict photo courtesy the United States Library of Congress; Maslow photo, original source unknown (found in www.photobucket.com).

Figure 11.1 *Francis Crick with a DNA model*. Courtesy The Image Works

Figure 11.2 *Two representations of the DNA molecule*. Reproduced by permission of the artist, Keith Roberts. Similar images appear in Bruce Alberts et al, 2008, 198–99.

Figure 12.1 Bizarro cartoon reproduced by permission of the artist, Dan Piraro.

Figure 13.1 *Diagram of a typical prokaryotic cell*. Reproduced by permission of the artist, Keith Roberts, from an image that appears in Alberts et al, 2008, 14.

Figure 13.2 *Diagram of a typical eukaryotic cell*. Reproduced by permission of the artist, Keith Roberts, from an image that appears in Alberts et al,

2008, 27.

Figure 13.3 *The Tree of Life according to Ernst Haeckel in 1866.* Courtesy Wikimedia Commons.

Figure 13.4 *A Modern Phylogenetic Tree of Life as conceived by Carl Woese.* Diagram property of NASA.

Figure 14.1 *Early Brain Development.* Reprinted by permission of Jane Bertrand and Fraser Mustard, *Early Years Study 3.*

Figures in Dialogue 15 (diagrams of molecules) drawn by the author based on material in Robert Hazen, *The Origins of Life,* 2005.

Figure 16.1 *The Penfield Homunculus.* Reproduced by permission of the artist, Keith Roberts, from an image that appears in Alberts et al, 2008, 1392.

Figure 16.2 *Pyramidal neuron imaged using the Golgi technique.* Photo courtesy Bob Jacobs, Colorado College, USA.

Figure 16.3 *Blind Spot Demonstration.* Drawn by the author. Inspired by Figure 5.2 (page 90) in Ramachandran and Blakeslee, *Phantoms in the Brain,* 1998.

Figure 18.1 *Neuron Operation.* Parts of two illustrations appearing in Alberts et al, 2008, 675 and 683. Reproduced by permission of the artist, Keith Roberts.

Figure 18.2 *Cartoon: "Never, Ever, Think Outside the Box."* Cartoon by Leo Cullum, New Yorker cartoon first published November 30, 1998. Reproduced by permission of The Cartoon Bank.

Figure 20.1 *A Framework for Emotions.* Drawn by the author, based on a diagram that appears in Randolf Nesse, "Natural Selection and the Elusiveness of Happiness," 2004, 1341.

Figure 21.1 *Average Satisfaction with Life in Japan.* Reproduced with permission, Dr. Bruno Frey. This chart was originally published in Frey and Stutzer, "What Can Economists Learn from Happiness Research?" *Journal of Economic Literature* 40, no. 2: 402–35.

Figure 21.2 *Factors Affecting Happiness in Canada.* Prepared by the author based on data in the Helliwell papers cited in the bibliography and summarized in a discussion paper presented at Simon Fraser University on June 29, 2006.

Figure A2.1 *The Striking Similarity of Ribosomal RNA Amongst All Life Forms.* Reproduced from "Bacterial Evolution," *Microbiological Reviews* 51, no. 2 (June, 1987): 233, with the permission of Dr. Carl Woese.

INDEX

Note that pages containing definitions are in bold type.

CPSIA information can be obtained at www.ICGtesting.com
Printed in the USA
LVOW111920270313

326416LV00002B/3/P